Steroid Hormones and the T-Cell Cytokine Profile

Springer
London
Berlin
Heidelberg
New York
Barcelona
Budapest
Hong Kong
Milan
Paris
Santa Clara
Singapore
Tokyo

G.A.W. Rook and S. Lightman (Eds)

Steroid Hormones and the T-Cell Cytokine Profile

With 51 Figures

Springer

Professor G.A.W. Rook
Department of Bacteriology, UCL Medical School,
Windeyer Building, 46 Cleveland Street, London W1P 6DB, UK

Professor S. Lightman
Department of Medicine, Dorothy Hodgkin Crowfoot Laboratories,
Bristol Royal Infirmary, Marlborough Street, Bristol BS2 8HW, UK

ISBN-13:978-1-4471-1238-9

British Library Cataloguing in Publication Data
Steroid hormones and the T-cell cytokine profile
1.Immune response – Regulation 2.Steroid hormones – Therapeutic use 3.T-cells
I.Rook, Graham A. W. II.Lightman, Stafford L. (Stafford Louis)
571.9'646
ISBN-13:978-1-4471-1238-9

Library of Congress Cataloging-in-Publication Data
Steroid hormones and the T-cell cytokine profile / G.A.W Rook and S. Lightman, eds.
p. cm.
Includes bibliographical references and index.
ISBN-13:978-1-4471-1238-9 e-ISBN-13:978-1-4471-0931-0
DOI:10.1007/978-1-4471-0931-0

1. Glucocorticoids – Immunology. 2. T cells. 3. Cytokines. 4. Immune response – Regulation. I. Rook, G. A. W. (Graham A. W.), 1946– . II. Lightman, Stafford L.
[DNLM: 1. Immunity, Cellular – drug effects. 2. Steroids – immunology. 3. Steroids – pharmacology. 4. Cytokines – drug effects. 5. T-Lymphocytes – drug effects. 6. Anti-Inflammatory Agents, Steroidal – pharmacology. QW 568 S838 1997]
QP572.G54S74 1997
616.07'9 – DC21
DNLM/DLC
for Library of Congress 97–15904

Softcover reprint of the hardcover 1st edition 1997

Typeset by EXPO Holdings, Malaysia

28/3830-543210 Printed on acid-free paper

Preface

The fact that certain adrenal steroid hormones are immunosuppressive and anti-inflammatory has been known for many years, and is routinely exploited by physicians. These effects are attributable to glucocorticoid steroids such as cortisol. Pharmacological doses of glucocorticoids inhibit most or all T cell types. However we now know that the effect of exposure to raised physiological levels is mainly to drive developing lymphocyte responses towards a Th2 cytokine profile (interleukin-4 secreting) while suppressing the development of Th1 (gamma interferon-secreting) lymphocytes.

Only recently have two further regulatory mechanisms become apparent. First, these effects of cortisol are balanced by pro-inflammatory and Th1-enhancing effects of another adrenal steroid, dehydroepiandrosterone sulfate (DHEAS). Second, the activity of cortisol is directly modulated by enzymes in the target organs and lymphoid tissues that convert it into inactive cortisone.

This new information leads to the possibility that immunoregulatory steroids could be used by physicians in novel ways. We can envisage steroid combinations that exploit the anti-inflammatory effects of cortisol, while the Th1-suppressing and Th2-promoting properties of these hormones are opposed by derivatives of DHEAS. Such therapies are already proving effective in animal models of infection, and could revolutionize treatment of Th2-mediated diseases such as asthma, where the anti-inflammatory effects of cortisol are desirable but the effects on T lymphocyte differentiation are not. Alternatively, steroid analogues may be used to inhibit the enzymes that inactivate cortisol, or those that generate the DHEAS derivatives that oppose cortisol's functions, and in this way the effects of endogenous cortisol may be enhanced without administration of cortisol itself.

This book provides an overview of these new concepts, starting with accounts of how the immune response triggers adrenal steroid production, and finishing with *in vivo* models of infection and autoimmunity that demonstrate the clinical importance of these new insights into the physiological control of glucocorticoid function. The information contained is mostly very new, some of it unpublished, and will open up new horizons for immunologists, endocrinologists, physiologists, cell biologists and clinicians.

Contents

Contributors

K. Arriaga
Department of Pathology, Instituto Nacional de la nutricion, Salvador Zubiran, Calle Vasco de Quiroga 15, Delegacion Tlalpan, 14000 Mexico DF

R. Baker
Academic Infectious Disease Unit, Department of Medicine, University College Medical School, Windeyer Building, 46 Cleveland Street, London W1P 6DB, UK

R.-M. Bluthe
Neurobiologie integrative, INSERM U394, Rue Camille Saint-Saens, 33077 Bordeaux Cedex, France

D.H. Brown
Life Sciences Test Facility, U.S. Army Dugway Proving Ground, Dugway, Utah, USA

R. Dantzer
Neurobiologie integrative, INSERM U394, Rue Camille Saint-Saens, 33077 Bordeaux Cedex, France

R.A. Daynes
Geriatric Research, Education and Clinical Center, Veterans Affairs Medical Center, Salt Lake City, Utah 84112, USA

D.L. Felten
Center for Psychoneuroimmunology Research and Department of Neurobiology and Anatomy, University of Rochester School of Medicine, 601 Elmwood Avenue, Rochester, NY 14642, USA

R. Foulkes
Celltech Therapeutics, 216 Bath Street, Slough SL1 4EN, UK

E. Goujon
Centre G.I.P. Cyceron, Boulevard Henri Becquerel, Caen, France

M.S. Harbuz
Department of Medicine, Bristol Royal Informary Bristol BS2 8HW, UK

J.D. Hennebold
Department of Pathology, University of Utah School of Medicine, Salt Lake City, Utah 84122, USA

R. Hernandez-Pando
Department of Pathology, Instituto Nacional de la nutricion, Salvador Zubiran, Calle Vasco de Quiroga 15, Delegacion Tlalpan, 14000 Mexico DF

K.W. Kelley
Laboratory of Immunophysiology, University of Illinois, Urbana, IL 61801, USA

J.P. Konsman
Neurobiologie integrative, INSERM U394, Rue Camille Saint-Saens, 33077 Bordeaux Cedex, France

B. Kruzewska
Center for Psychoneuroimmunology Research and Department of Neurobiology and Anatomy, University of Rochester School of Medicine, 601 Elmwood Avenue, Rochester, NY 14642, USA

S. Laye
Neurobiologie integrative, INSERM U394, Rue Camille Saint-Saens, 33077 Bordeaux Cedex, France

S.L. Lightman
Department of Medicine, Bristol Royal Infirmary, Bristol BS2 8HW, UK

D. Mason
MRC Cellular Immunology Unit, Sir William Dunn School of Pathology, University of Oxford, Oxford OX1 3RE, UK

J.A. Moynihan
Department of Psychiatry, University of Rochester, 300 Crittenden Boulevard, Box Psych, Rochester, New York 14642, USA

H. Orozco
Department of Pathology, Instituto Nacional de la nutricion, Salvador Zubiran, Calle Vasco de Quiroga 15, Delegacion Tlalpan, 14000 Mexico DF

P. Parnet
Neurobiologie integrative, INSERM U394, Rue Camille Saint-Saens, 33077 Bordeaux Cedex, France

L. Pavon
Department of Pathology, Instituto Nacional de la nutricion, Salvador Zubiran, Calle Vasco de Quiroga 15, Delegacion Tlalpan, 14000 Mexico DF

F. Ramirez
MRC Cellular Immunology Unit, Sir William Dunn School of Pathology, University of Oxford, Oxford OX1 3RE, UK

G.A.W. Rook
Department of Bacteriology, University College Medical School, Windeyer Building, 46 Cleveland Street, London W1P 6DB, UK

S. Shaw
Celltech Therapeutics, 216 Bath Street, Slough SL1 4EN, UK

M. Streber
Department of Pathology, Instituto Nacional de la nutricion, Salvador Zubiran, Calle Vasco de Quiroga 15, Delegacion Tlalpan, 14000 Mexico DF

A. Suitters
Celltech Therapeutics, 216 Bath Street, Slough SL1 4EN, UK

B.R. Walker
University Department of Medicine, Western General Hospital, Edinburgh EH4 2XU, UK

B. Zwilling
Ohio State University, Department of Microbiology, 484 West 12th Avenue, Columbus, OH43210-1292, USA

Chapter 1

Mechanisms of action of cytokines on the central nervous system. Interaction with glucocorticoids

Robert Dantzer, Sophie Layé, Emmanuelle Goujan, Rose-Marie Bluthé,
Jan Pieter Konsman, Patricia Parnet and Keith W. Kelley

Introduction

Interleukin-1 (IL-1), IL-6 and tumour necrosis factor-alpha (TNFα) are multifunctional cytokines which are released by activated monocytes and macrophages and play a pivotal role in inflammation and infection [1]. Although their biological activities widely overlap, each cytokine has its own characteristic properties. IL-1 exists in two molecular forms, IL-1α and IL-1β, which are encoded by different genes. Another member of the IL-1 family is the interleukin-1 receptor antagonist (IL-1ra). This cytokine behaves as a pure endogenous antagonist of IL-1 receptors and blocks most biological effects of IL-1α and IL-1β in vivo and in vitro. IL-6 is mainly responsible for the synthesis of acute phase proteins by hepatocytes. TNFα plays a major role in the pathogenesis of the acute shock syndrome. Although all of these cytokines are produced mainly by activated monocytes and macrophages, they are also synthesized by a wide variety of immune and non-immune cells, including glia and neurons [2]. IL-1, IL-6 and TNFα have potent effects in the central nervous system, resulting in fever, induction of sickness behaviour, and activation of the hypothalamic-pituitary-adrenal (HPA) axis [3].

Fever corresponds to an increased set point for thermoregulation [4]. Feverish organisms feel cold in previously thermoneutral environments and, as a result, they increase their heat production and decrease heat losses. Fever is an important response for the survival of the host, since many pathogenic micro-organisms stop proliferating at temperatures higher than 36 °C. Fever is accompanied by profound behavioural alterations which help sparing energy and decreasing heat losses [5]. Sick individuals show piloerection and are lethargic and anorexic. They adopt a curled posture, and sleep more. Activation of the HPA axis during the host response to infection was first recognized by Hans Selye [6] as another example of the non-specificity of the stress response. It plays an important role in limiting the magnitude of the acute phase reaction, as evidenced by the higher sensitivity of adrenalectomized animals to endotoxin-induced septic shock [7].

Our views on the mechanisms by which peripherally released cytokines act on the brain to induce fever, sickness behaviour and activation of the HPA axis have been profoundly modified by the demonstration that peripherally released cytokines induce the expression of cytokines in the brain and that cytokine receptors are expressed in the brain. The objective of this chapter is to summarize the current knowledge on the mechanisms of action of cytokines on the brain and the way the cytokine compartment in the brain is regulated. Emphasis will be put on cytokines of the IL-1 family since they are fully representative of the class of pro-inflammatory cytokines and have profound effects on the brain.

Cytokines and cytokine receptors are expressed in the central nervous system

IL-1α and IL-1β are expressed in the brain in two pools: a constitutive form, which appears to be mainly present in neurons, and an inducible pool which is expressed in brain macrophages and microglial cells. Peripheral administration of the active fragment of endotoxin, lipopolysaccharide (LPS), induces the synthesis and release of IL-1 and other pro-inflammatory cytokines by activated monocytes and macrophages. These peripheral cytokines, in turn, induce the expression of IL-1, TNFα, IL-6 and IL-1ra at the mRNA and protein levels, in various structures of the brain (Fig. 1.1) [8,9]. In the hypothalamus for example, TNFα and IL-1β transcripts appear within 1 h and decrease after 2 h whereas IL-6 and IL-1ra take longer to appear and remain stable for several hours [9].

IL-1β is synthesized as an inactive precursor form which requires proteolytic cleavage to produce the mature biologically active protein. The enzyme responsible for this cleavage is called IL-1β converting enzyme (ICE). It is also present in the brain [10] although the necessity of its intervention for the processing of brain pro-IL-1β into IL-1β has not yet been demonstrated.

Two subtypes of IL-1 receptors have been identified, an 80-kDa IL-1 receptor, now known as the type I IL-1 receptor (IL-1RI), and a 68-kDa IL-1 receptor, known as the type II IL-1 receptor (IL-1RII) [11]. The two IL-1 receptors are similar in overall organization and derive from a common ancestor. Both contain a ligand-binding domain composed of three immunoglobulin-like domains. They differ most strikingly in the cytoplasmic segment. The intracellular portion of the human IL-1RI consists of 213 amino acids and is essential for cellular responses to IL-1 in vitro while the type II receptor has a cytoplasmic domain of only 29 residues and does not appear to transduce a signal. Since it binds IL-1 and thus prevents it from reaching the type I (signalling) receptor, while leading to no response itself, the type II IL-1 receptor is believed to serve as a negative regulator of IL-1 action. The type II receptor can serve its regulatory role either as a cell-bound molecule or as a soluble receptor that is generated by shedding of the ligand-binding portion from the cell surface.

The binding characteristics of the two IL-1 receptors have been extensively studied in immunocytes. IL-1RI binds IL-1α, IL-1β and IL-1ra with high affinity. The dissociation rate of IL-1ra from IL-1RI is very slow, making the binding essentially irreversible. IL-1RII has a high affinity only for IL-1β. It binds IL-1α poorly and has a very low affinity for IL-1ra.

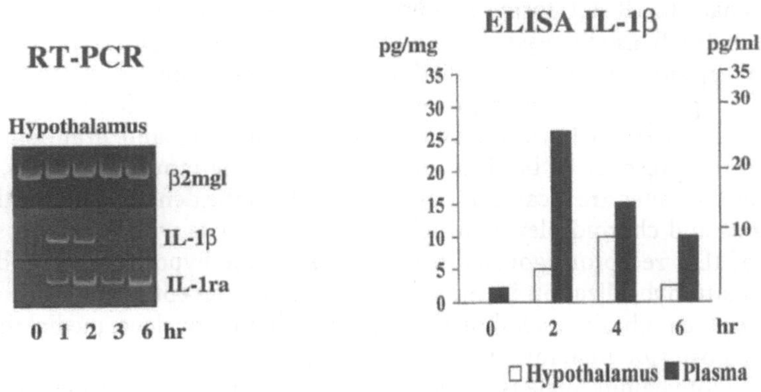

Fig. 1.1. Systemic administration of LPS induces IL-1β at the mRNA and protein levels in the brain of mice. Mice were injected intraperitoneally with LPS and killed just before or at different times after the injection. Total RNA was extracted from the hypothalamus and submitted to RT-PCR to determine transcripts for IL-1β and IL-1ra, using β2-microglobulin (β2mgl) as an internal standard. Levels of IL-1β were measured in the hypothalamus (pg/mg protein) and in the plasma (pg/ml) using a validated ELISA.

Early attempts to determine the number of IL-1R per cell revealed that most cell types expressed <500. Further investigations involving receptor occupancy studies and transfection of receptor negative cells with the IL-1RI gene demonstrated that less than ten receptors per cell were sufficient to cause cellular activation. These data suggest that a major amplification in signal occurs following IL-1 binding. However, the exact mechanisms of IL-1 signalling are still largely unknown. Although addition of IL-1 to cells leads to increased phosphorylation of a number of different proteins on serine and threonine residues, the receptor itself is not a kinase. IL-1 does activate MAP kinase, and MAP kinase activity is required for some, but not all, responses to IL-1, indicating that multiple signalling pathways exist [12]. Possible second messengers include cyclic adenosine monophosphate (cAMP), diacylglycerol, ceramide, PGE2 and nitric oxide.

More recently, another level of complexity in IL-1 signalling has been evidenced with the discovery of a whole family of IL-1R-like proteins [13,14]. The IL-1RI signalling domain shares significant homology with the cytoplasmic region of the *Drosophila melanogaster* transmembrane protein Toll. Toll is involved in the establishment of dorsal/ventral polarity in the *Drosophila* embryo. Site directed mutagenesis and deletion analysis have demonstrated that the cytoplasmic domain of IL-1RI, and in particular, residues conserved between IL-1RI and Toll, are essential for transduction of intracellular IL-1 stimulated signals. In addition to Toll, five other characterized cDNAs have been recognized as homologous to the cytoplasmic domain of IL-1RI. These include in particular T1/ST2, the IL-1 receptor accessory protein (IL-1R AcP) [15], and the IL-1R related protein [14]. T1/ST2 is a transmembrane protein that has been characterized as a novel primary response gene expressed in BALB/c-3T3 cells. The IL-1R AcP is also a transmembrane protein and was identified through a monoclonal antibody that blocked the binding of Il-1β to IL-1R1, but recognized a protein distinct from the receptor itself. Both T1/ST2 and IL-1R AcP share homology to IL-1RI intracellularly and extracellularly.

Recombinant IL-1R AcP forms a complex with IL-1RI and either IL-1α or IL-1β, but not IL-1ra. It also increases the binding affinity of IL-1RI for IL-1β.

IL-1 receptors have been identified in brain tissues, using radioligand-binding studies, immunocytochemistry, or expression of receptor mRNA. Equilibrium binding of ^{125}I-IL-1α by brain membrane homogenates and brain slices has revealed the presence of binding sites for IL-1 in the mouse brain [16,17,18]. These binding sites are located almost exclusively in the dentate gyrus of the hippocampus and choroid plexus, which does not correlate with proposed sites of action of IL-1 receptor agonists in such areas as the hypothalamus and brain stem. In situ hybridization identified the type I IL-1 receptor mRNA in several regions of the rat brain, including the anterior olfactory nucleus, medial thalamic nucleus, posterior thalamic nucleus, basolateral amygdaloid nucleus, ventromedial hypothalamus nucleus, arcuate nucleus, median eminence, mesencephalic trigeminal nucleus, motor trigeminal nucleus, facial nucleus and Purkinje cells of the cerebellum [19]. Despite their importance for the understanding of the way IL-1 affects brain functions, the mechanisms of cell-to-cell communication in the nervous system by IL-1 have not yet been elucidated. IL-1RI is present on neurons, astrocytes, epithelial cells of the choroid plexus and ventricles and endothelial cells. Using a rat anti-mouse type I IL-1 receptor monoclonal antibody, neurons and oligodendrocytes have been found to express type I IL-1 receptor protein [18]. In addition, transcripts for the type I but not the type II IL-1 receptors are expressed in murine neuronal cell lines whereas transcripts for both types of receptors are present in mouse brain. Northern blot analysis indicates that IL-1R AcP is constitutively expressed at high levels in mouse brain [15] but its cellular localization and its coexpression with IL-1RI are still unknown. Little information is available on the factors that regulate IL-1 receptors in the nervous system. IL-1 receptor expression on the cell surface is negatively regulated by IL-1 in monocytes and fibroblasts. Indirect evidence in favour of a similar mechanism in the brain comes from experiments showing a decrease in the number of IL-1 binding sites in the hippocampus of mice administered an intraperitoneal injection of LPS [16]. The same phenomenon occurs in astrocytes and could be due to internalization of the ligand-receptor complex.

Currently, relatively little information is available as to whether IL-1 receptors in the nervous system are the same as those identified in other tissues. Using polymerase chain reaction (PCR) Bristulf et al. [20] cloned a type II IL-1 receptor of cDNA from an excitable rat insulinoma cell line. This receptor has structural features that are not found in the previously identified type II IL-1 receptor, suggesting that different subtypes of type II IL-1 receptors could be expressed in the central nervous system (CNS).

Brain IL-1 mediates the neural components of the host response to infection

Interest and research in the neurobiology of IL-1 has experienced an exponential increase over the last few years. It is now clear that IL-1 represents a major molecular signal between the immune system and CNS in the induction of metabolic, neuroendocrine and behavioural components of the host response to infection. Using a number of approaches, it has been amply demonstrated that IL-1 activates the HPA axis and induces fever, malaise, lethargy, decreased appetite, loss of interest in usual activities and learning and memory disorders. For example, intraperitoneal injection of IL-1β to mice and rats increases plasma levels of adrenocorticotrophin (ACTH) and corticosterone [21], induces increases in body temperature and oxygen consumption [4], and decreases social interactions and feeding behaviour [22]. More importantly, neutralization of endogenous IL-1 by passive immunization, administration of IL-1ra, or injection of neutralizing antibody to IL-1 receptors, attenuates the neural effects of cytokine inducers such as LPS, the active fragment of endotoxin [23].

When injected directly into the brain, IL-1α and IL-1β induce the same biological effects as when injected at the periphery, but are effective at much lower (100–1000 fold less) doses, indicating that these effects are centrally mediated. Furthermore, blockade of IL-1 receptors in the brain by local administration of IL-1ra, which binds both to both types of IL-1 receptor, abrogates the effects of peripherally injected IL-1β on behaviour [24] and fever [25]. Blockade of the brain type II IL-1 receptor by a monoclonal antibody abrogates the fever response, suggesting that this receptor might be functional in the CNS [26]. The results showing the possibility of blocking the effect of peripherally injected IL-1 by central administration of IL-1 antagonists are important because they imply that fever, sickness behaviour and activation of the HPA axis are mediated by IL-1 produced in the brain in response to a peripheral immune stimulus but may depend on different receptor subtypes. However, the exact nature and localization of the receptors that mediate these effects of IL-1 are not known.

Neural afferent pathways transmit the immune message from the periphery to the brain

Since cytokines cannot cross the blood–brain barrier, it has been proposed that these molecules act at the level of the brain where this barrier is deficient. These privileged sites for communication with the internal milieu are known as circumventricular organs. The organum vasculosum of the lamina terminalis (OVLT) has long been considered as the main site of action of peripherally released cytokines, based on results of experiments using fever as an end point in animals in which the OVLT has been lesioned [27]. Circulating cytokines would pass from the general circulation into the brain side of the fenestrated blood–brain barrier and by interacting with IL-1 receptors located on parenchymal astrocytes, generate the synthesis of prostaglandins of the E2 family, which would then freely diffuse to nearby structures such as the anterior preoptic area of the hypothalamus.

This hypothesis was elaborated at a time at which there was no knowledge available on the presence of cytokines and cytokine receptors in the brain. The existence of a central cytokine compartment which is inducible by peripheral cytokines requires a communication pathway between these two cytokine compartments. As proposed by Kent et al. [22], neural afferents are able to play this role since inflammation involves two important sensory components, *calor* and *dolor* (heat and pain). In the case the peripheral immune stimulus originates from the abdominal cavity, this communication takes place via the vagus nerve, as demonstrated by c-*fos* mapping experiments using animals in which this nerve has been sectioned right under the diaphragm. The immediate early gene c-*fos* is differentially expressed in many regions of the CNS following different physiological challenges. Intraperitoneal administration of LPS to rats induces the expression of c-*fos* protein in the primary and secondary projections areas of the vagus nerve (Fig. 1.2). This effect is completely abrogated by subdiaphragmatic vagotomy [28].

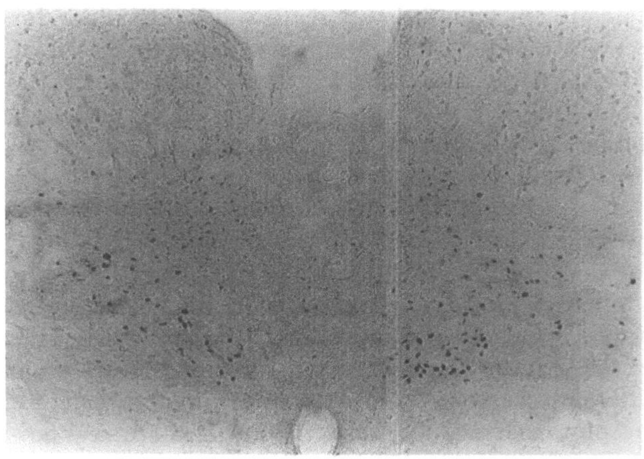

(a)

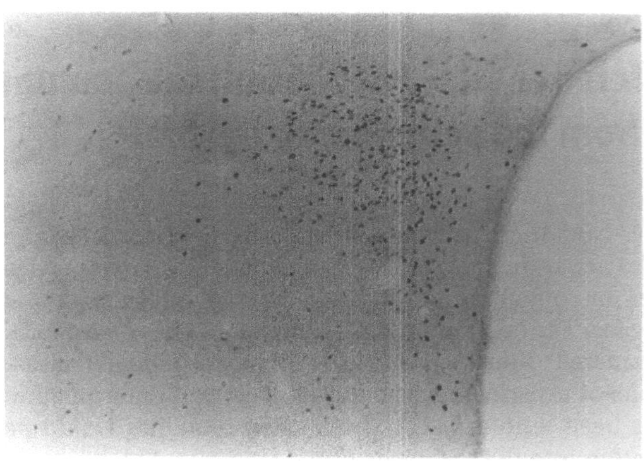

(b)

Fig. 1.2. Systemic administration of LPS induces the expression of c-*fos* protein in the projection areas of the vagus nerve. Rats were injected intraperitoneally with LPS and killed 2 h later. Brains were removed and studied for Fos immunocytochemistry. Illustrated in the top figure (a) is the presence of numerous Fos-positive cells in the nucleus tractus solitarius, which is the primary projection area of the vagus nerve. The lower picture (b) shows Fos positive cells in the paraventricular nucleus of the hypothalamus, which is one of the secondary projection areas of the vagus nerve.

Vagotomy also abrogates the depressing effects of LPS and IL-1 on behaviour in rats and mice (Fig. 1.3) [29,30]. This protecting effect is not due to an impaired peripheral immune response since the increase in the levels of IL-1β that is induced by intraperitoneal injection of LPS in the peritoneal macrophages and plasma of mice and rats was not impaired by vagotomy [29,31]. The involvement of the vagus nerve in the behavioural effects of cytokines is specific to cytokines sensed by nerve terminals in the abdominal cavity since vagotomy did not block sickness behaviour induced by subcutaneous and intravenous injections of IL-1β [32] and had no effect on sickness behaviour induced by central injection of IL-1β [33].

According to the hypothesis of a neural transmission pathway of immune signals from the periphery to the brain, section of vagal afferents should abrogate the induction of expression of IL-1β mRNA in the brain in response to intraperitoneal administration of LPS. In accordance with this prediction, vagotomy was demonstrated to abrogate LPS-induced increases in hypothalamic and hippocampal but not pituitary IL-1β mRNA (Fig. 1.3) [31].

The brain actions of cytokines that are mediated by neural afferents also include fever and activation of the HPA axis since vagotomy attenuates LPS-induced fever and increase in plasma ACTH levels [34].

All these data converge to support the hypothesis that vagal afferents transmit the immune message from the abdominal cavity to the brain. Whether other neural afferents are involved in the transmission of the immune message originating from other bodily sites to the brain and whether the circumventricular organs can also be involved in the relatively rare case cytokine levels dramatically increase in the general circulation remain to be determined.

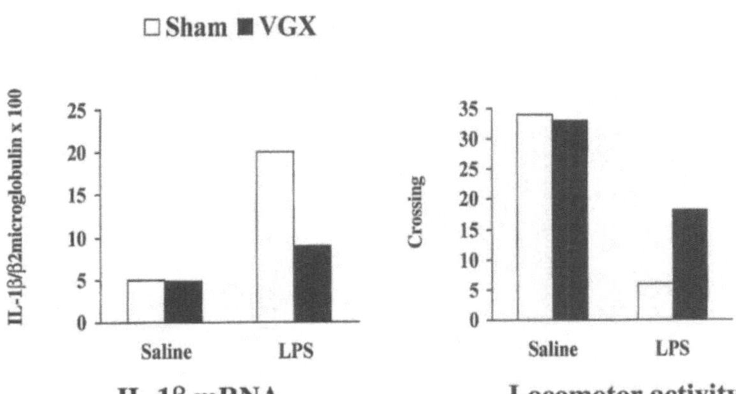

Fig. 1.3. Section of the vagus nerve abrogates the induction of IL-1β mRNA in the hypothalamus in response to intraperitoneal injection of LPS and attenuates the behavioural effects of this treatment. Mice which had been submitted to subdiaphragmatic vagotomy (vgx) or sham surgery 3 weeks before the experiment were injected intraperitoneally with LPS and killed 2 h later, after their locomotor activity had been measured in their home cage. The left figure represents hypothalamic IL-1β mRNA measured by comparative RT-PCR and expressed as percentage of β2-microglobulin mRNA. The right figure represents the mean locomotion scores measured over a 4-min session in the different experimental groups (adapted from [31]).

Glucocorticoids regulate cytokine expression and actions in the brain

Pro-inflammatory cytokines have potent biological effects which play a key role in the development of the host response to infection but can also lead to cell death by necrosis or apoptosis. It is therefore important for the activity of pro-inflammatory cytokines to be tightly regulated. In immune cells, pro-inflammatory cytokine gene expression is mainly regulated by glucocorticoids. Glucocorticoids downregulate the synthesis and secretion of IL-1, IL-6 and TNFα by activated monocytes and macrophages [35]. LPS-induced increases in plasma levels of TNFα and IL-6 are much higher in adrenalectomized (ADX) than in sham-operated animals (Fig. 1.4) [36]. The enhanced release of these cytokines is inhibited by administration of glucocorticoids. At the molecular level, glucocorticoids inhibit transcriptional and post-transcriptional expression of the IL-1β gene and decrease stability of IL-1β mRNA. Regulation of the synthesis and release of pro-inflammatory cytokines by glucocorticoids has marked functional consequences since adrenalectomy sensitizes experimental animals to the septic shock syndrome whereas administration of glucocorticoids has the opposite effect [7,37].

To assess the possible influence of endogenous glucocorticoids on cytokine expression in the brain, ADX mice and sham-operated mice were injected with saline or LPS and the levels of transcripts for IL-1α, IL-1β, IL-1ra, IL-6 and TNFα were subsequently determined in the spleen, pituitary, hypothalamus, hippo-

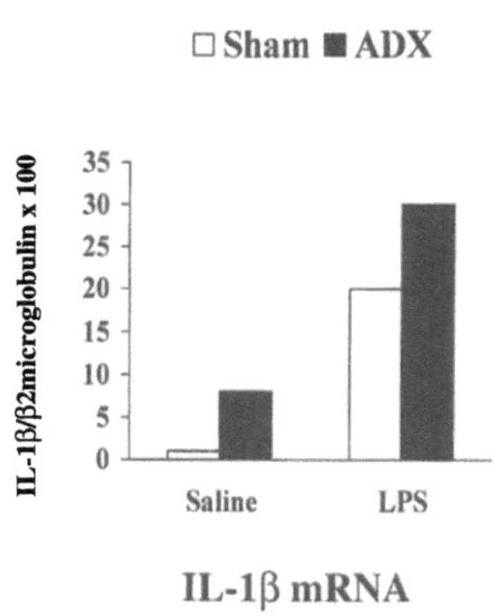

Fig. 1.4. Adrenalectomy increases expression of IL-1β mRNA in mice. Adrenalectomized (ADX) and sham-operated mice were injected subcutaneously with saline or LPS and killed 2 h later. Total RNA was extracted from the hypothalamus and IL-1β mRNA (expressed as percentage of β2-microglobulin mRNA) was measured by comparative RT-PCR. The figure represents the mean results of three different experiments. Note that LPS increases hypothalamic IL-1β mRNA and that adrenalectomy increases the expression of this cytokine in saline- and LPS-treated mice (adapted from [37]).

campus and striatum, using comparative reverse transcription polymerase chain reaction (RT-PCR) [38]. Levels of IL-1β were also measured by a specific ELISA in the plasma and tissues of experimental animals. In accordance with the results of previous investigations, LPS induced the expression of pro-inflammatory cytokines at the mRNA level in most tissues under investigation. This effect was potentiated by adrenalectomy in the plasma and different tissues, including brain structures, under investigation, although the magnitude of this enhancement differed according to the tissue. LPS increased plasma and tissue levels of IL-1β, as determined by ELISA, and this effect was potentiated by adrenalectomy in plasma and tissues other than the spleen. Since IL-1β and other pro-inflammatory cytokines are mainly synthesized in the brain by microglial cells and meningeal and perivascular macrophages, and since glucocorticoids have been shown to inhibit in vitro the synthesis and secretion of IL-1, IL-6 and TNFα by activated monocytes and macrophages, it is tempting to propose that the enhancing effects of adrenalectomy on cytokine gene expression in peripheral and central tissues is the result of the lack of glucocorticoids.

This interpretation is strengthened by the observation that the effects of stress are contrary to those of adrenalectomy on the LPS-induced expression of cytokines in the brain [39]. In this experiment, mice injected with LPS or saline, were submitted to a 15-min restraint stress which resulted in a significant increase in plasma corticosterone levels. They were subsequently killed to assess the effects of this stressor on the induction of pro-inflammatory cytokines in the spleen, pituitary, hypothalamus, hippocampus and striatum. LPS-induced cytokine gene expression, as determined by comparative RT-PCR, was lower in restrained than in non-restrained mice. LPS increased plasma and tissue IL-1β levels, as determined by ELISA. This effect of LPS was also less marked in restrained than in non-restrained mice. It is important to note that stress has also been reported to induce an increase in IL-1β mRNA levels in the hypothalamus of rats [40] and an enhancement of plasma levels of IL-6 [41,42]. The reason for these contradictory results is not known.

As mentioned in the first section of this chapter, IL-1β is synthesized as an inactive precursor which needs to be cleaved to be secreted in an active form. The enzyme which is responsible for this cleavage is a protease, called interleukin-1β converting enzyme (ICE). To determine whether those factors which regulate the expression of IL-1 in immune and non-immune tissues are also able to regulate the expression of ICE, mice were injected with LPS and the levels of ICE mRNA in the spleen, pituitary and brain of experimental animals were measured by comparative RT-PCR. ICE mRNAs were more abundant in the spleen and hippocampus than in the pituitary and hypothalamus, but they were not significantly altered by LPS treatment. In another experiment, mice were submitted to adrenalectomy or a 15-min restraint stress and injected with saline or LPS. ADX mice had significantly higher ICE mRNA levels whereas stressed mice had significantly lower ICE mRNA levels than their respective controls. These results can be interpreted to suggest that endogenous glucocorticoids regulate the expression of ICE in peripheral and brain tissues, in a manner which is commensurate with their effects on IL-1β gene expression [10].

In addition to their action on the synthesis of pro-inflammatory cytokines in the periphery and the brain, glucocorticoids are also able to modulate the expression of IL-1 receptors. Although glucocorticoids induce the expression of IL-1 receptors in various immune and non-immune tissues, mixed results have been reported in the brain [43,44]. The problem with the radioligand-binding techniques which have been used for these studies is that they do not allow an accurate description of what is going on at the receptor level, since the same effect, e.g. decreased binding, can be seen as a result of very different processes, e.g. down-regulation of receptors, or activation of receptors with internalization of receptor-ligand complexes. Studies at the mRNA level revealed that adrenalectomy was accompanied by a decrease in the levels of IL-1RI and IL-1RII transcripts in the mouse pituitary and brain, whereas stress had the opposite effects (E. Goujon et al., personal communication). When ADX mice were implanted with a corticosterone pellet to induce plasma levels of corticosterone intermediate between baseline and stress levels, IL-1RI and IL-1RII mRNAs were increased above baseline levels. In all cases, the effects of corticosterone supplementation and stress were more marked for IL-1RII than for IL-1RI mRNA, which would fit with the down-regulatory effects of glucocorticoids on IL-1 biological activity.

The down-regulatory effects of glucocorticoids on the pro-inflammatory cytokine network at the periphery and in the brain should lead to an increased sensitivity of ADX animals to the brain actions of these cytokines. This appears to be the case since adrenalectomy increases the febrile response to LPS whereas administration of glucocorticoids has the opposite effect [45,46]. In the same manner, adrenalectomy enhances the depression of social exploration induced by peripheral injection of IL-1β or LPS [47]. This effect is mimicked by administration of the glucocorticoid type II receptor antagonist, RU 38486. Chronic replacement with a 15-mg corticosterone pellet which yields plasma corticosterone levels intermediate between baseline and stress levels, abrogates the enhanced susceptibility of ADX mice to the lower dose of IL-1β but has only partial protective effects on the response to a higher dose of IL-1β and to LPS.

Taken together, these findings can be interpreted to suggest that the phasic response of the pituitary-adrenal axis to cytokines has an important regulatory role on the neural effects of cytokines. However, they do not reveal the level (peripheral or central) at which glucocorticoids are acting. Central sites of action appear to be involved in the regulatory effects of glucocorticoids on the neural effects of cytokines, since central injection of RU 38486 potentiates fevers induced by a peripheral injection of LPS to rats [48]. In the same manner, adrenalectomy sensitizes mice to the depressing effects of intracerebroventricular administration of IL-1 on social exploration and this effect is attenuated by corticosterone compensation [49]. Central administration of RU 38486 mimics the effect of adrenalectomy.

Conclusion

Although the scope of this review has been restricted mainly to members of the interleukin-1 family and sickness behaviour, sufficient evidence is now available to indicate that the pro-inflammatory cytokines which are released at the periphery by activated monocytes and macrophages activate peripheral afferent nerves to induce the synthesis and release of pro-inflammatory cytokines in the brain. These centrally released cytokines act probably by volume transmission on those neural structures that are involved in the control of thermoregulation, metabolism and behaviour, resulting in the development of the neural components of the host response to infection. Glucocorticoids that are released by the adrenal cortex in response to the hypothalamic effects of pro-inflammatory cytokines regulate the expression and actions of cytokines not only in the periphery but also in the brain.

Acknowledgements

Supported by INSERM, INRA, Université de Bordeaux II, Pôle Médicament Aquitaine, Ministère de l'Environement, DRET and NIH (MH51569-02 and DK49311)

References

1. Nicola NA (ed) (1994) Guidebook to cytokines and their receptors. Oxford University Press, Oxford
2. Hopkins SJ, Rothwell NJ (1995) Cytokines and the nervous system. I. Expression and recognition. Trends Neurosci 18:83–87
3. Rothwell NJ, Hopkins SJ (1987) Cytokines and the nervous system. II. Action and mechanisms of action. Trends Neurosci 18:130–136
4. Kluger MJ (1991) Fever: role of pyrogens and cryogens. Physiol Rev 71:93–127
5. Hart BL (1988) Biological basis of the behaviour of sick animals. Neurosci Biobehav Rev 12:123–137
6. Selye H (1956) The stress of life. McGraw Hill, New York
7. Bertini R, Bianchi M, Ghezzi P (1988) Adrenalectomy sensitizes mice to lethal effects of interleukin-1 and tumour necrosis factor. J Exp Med 167:1708–1712
8. Gatti S, Bartfai T (1993) Induction of tumour necrosis factor-A mRNA in the brain after peripheral endotoxin treatment: comparison with interleukin-1 family and interleukin-6. Brain Res 624:291–295
9. Layé S, Parnet P, Goujon E, Dantzer R (1994) Peripheral administration of LPS induces the expression of cytokine transcripts in the brain and pituitary of mice. Mol Brain Res 27:157–162
10. Layé S, Goujon E, Combe C, VanHoy R, Kelley KW, Parnet P, Dantzer R (1996) Effects of lipopolysaccharide and glucocorticoids on expression of interleukin-1β converting enzyme in the pituitary and brain of mice. J Neuroimmunol 68:61–66
11. Dinarello CA (1996) Biologic basis for interleukin-1 in disease. Blood 87:2095–2147
12. Saklatvala J (1995) Intracellular mechanisms of interleukin-1 and tumour necrosis factor: possible targets for therapy. BMJ 51:402–418
13. Mitcham J, Parnet P, Bonnert T, Garka K, Gerhart M, Slack J, Gayle M, Dower S, Sims J (1996) T1/ST2 signalling establishes it as a member of an expanding IL-1 receptor family. J Biol Chem 271:5777–5783

14. Parnet P, Garka K, Bonnert T, Dower S, Sims J (1996) IL-1Rrp is a novel receptor-like molecule similar to the type I IL-1 receptor and its homologues T1/ST2 and IL-1R AcP. J Biol Chem 271:3967–3970
15. Greenfeder SA, Nunes P, Kwee L, Labow M, Chizzonite RA, Ju G (1995) Molecular cloning and characterization of a second subunit of the interleukin-1 receptor complex. J Biol Chem 270:13757–13765
16. Haour FG, Ban EM, Milon GM, Baran D, Fillion GM (1990) Brain interleukin-1 receptors. Characterization and modulation after lipopolysaccharide injection. Prog NeuroEndocrinoImmunol 3:196–204
17. Takao T, Tracey DE, Mitchell WM, De Souza EB (1991) Interleukin-1 receptors in mouse brain: characterization and neuronal localization. Endocrinology 127:3070–3078
18. Parnet P, Amindari S, Wu C, Brunke-Reese D, Goujon E, Weyhenmeyer JA, Dantzer R, Kelley KW (1994) Expression of type I and type II interleukin-1 receptors in mouse brain. Mol Brain Res 27:63–70
19. Ericsson A, Liu C, Hart RP, Sawchenko PE (1995) Type 1 interleukin-1 receptor in the rat brain: distribution, regulation, and relationship to sites of IL-1 induced cellular activation. J Comp Neurol 361:681–698
20. Bristulf J, Gatti S, Malinowsky D, Bjork L, Sundgren AK, Bartfai T (1994) Interleukin-1 stimulates the expression of type I and type II interleukin-1 receptors in the rat insulinoma cell line Rinm5F: sequencing a rat type II interleukin-1 receptor cDNA. Eur Cytokine Network 5:319–330
21. Besedovsky HO, del Rey A, Sorkin E, Dinarello CA (1986) Immunoregulatory feedback between interleukin-1 and glucocorticoid hormones. Science 233:652–654
22. Kent S, Bluthé RM, Kelley KW, Dantzer R (1992) Sickness behaviour as a new target for drug development. Trends Pharmacol Sci 13:24–28
23. Bluthé RM, Dantzer R, Kelley KW (1992) Effects of interleukin-1 receptor antagonist on the behavioural effects of lipopolysaccharide in rat. Brain Res 573:318–320
24. Kent S, Bluthé RM, Dantzer R, Hardwick AJ, Kelley KW, Rothwell NJ, Vannice JL (1992) Different receptor mechanisms mediate the pyrogenic and behavioural effects of interleukin-1. PNAS USA 89:9117–9120
25. Luheshi GN, Miller AJ, Brouwer S, Dascombe MJ, Rothwell NJ, Hopkins SJ (1995) Interleukin-1 receptor antagonist inhibits endotoxin fever and systemic interleukin-6 induction in the rat. Am J Physiol 270:E91–E95
26. Luheshi GN, Hopkins SJ, LeFeuvre RA, Dascombe MJ, Ghiara P, Rothwell NJ (1993) Importance of brain IL-1 receptors in fever and thermogenesis in the rat. Am J Physiol 265:E585–E591
27. Blatteis CM (1990) Neuromodulative actions of cytokines. Yale J Biol Med 63:133–142
28. Wan W, Wetmore W, Sorensen CM, Greenberg AH, Nance DM (1994) Neural and biochemical mediators of endotoxin and stress-induced c-fos expression in the rat brain. Brain Res Bull 34:7–14
29. Bluthé RM, Walter V, Parnet P, Layé S, Lestage J, Verrier D, Poole S, Stenning BE, Kelley KW, Dantzer R (1994) Lipopolysaccharide induces sickness behaviour in rats by a vagal mediated mechanism. CR Acad Sci Paris, Sciences de la Vie 317:499–503
30. Bret-Dibat JL, Bluthé RM, Kent S, Kelley KW, Dantzer R (1995) Lipopolysaccharide and interleukin-1 depress food-motivated behaviour in mice by a vagal-mediated mechanism. Brain Behav Immun 9:242–46
31. Layé S, Bluthé RM, Kent S, Combe C, Médina C, Parnet P, Kelley KW, Dantzer R (1995) Subdiaphragmatic vagotomy blocks the induction of interleukin-1β mRNA in the brain of mice in response to peripherally administered lipopolysaccharide. Am J Physiol 268:R1327–R1331

32. Bluthé RM, Michaud B, Kelley KW, Dantzer R (1996) Vagotomy blocks behavioural effects of interleukin-1 injected via the intraperitoneal route but not via other systemic routes. NeuroReport 7:2823–2827
33. Bluthé RM, Michaud B, Kelley KW, Dantzer R (1996) Vagotomy attenuates behavioural effects of interleukin-1 injected peripherally but not centrally. Neuroreport 7:1485–1488
34. Watkins LR, Maier SF, Goehler LE (1995) Cytokine-to-brain communication: a review & analysis of alternative mechanisms. Life Sci 57:1011–1026
35. Lee SW, Tsou AP, Chan H, Thomas J, Petrie K, Eugui EM, Allison CA (1988) Glucocorticoids selectively inhibit the transcription of the interleukin-1β gene and decrease the stability of interleukin-1β mRNA. PNAS USA 85:1204–1208
36. Zuckerman SH, Shellhaas J, Butler LD (1989) Differential regulation of lipopolysaccharide-induced interleukin-1 and tumour necrosis factor synthesis: effects of endogenous glucocorticoids and the role of pituitary-adrenal axis. Eur J Immunol 19:301–305
37. Rachamandra RN, Sehon AH, Berczi I (1992) Neuro-hormonal host defence in endotoxin shock. Brain Behav Immun 6:157–169
38. Goujon E, Parnet P, Layé S, Combe C, Dantzer R (1996) Adrenalectomy enhances proinflammatory cytokine gene expression in the spleen, pituitary and brain of mice in response to lipopolysaccharide. Mol Brain Res 36:53–62
39. Goujon E, Parnet P, Layé S, Combe C, Kelley KW, Dantzer R (1995) Stress downregulates lipopolysaccharide-induced expression of pro-inflammatory cytokines in the spleen, pituitary and brain of mice. Brain Behav Immun 9:292–303
40. Minami M, Kuraishi Y, Yamaguchi T, Nakai S, Hirai Y, Satoh M (1991) Immobilization stress induces interleukin-1 beta mRNA in the hypothalamus. Neurosci Lett 123:254–256
41. Le May LG, Vander AJ, Kluger MJ (1990) The effect of psychological stress on plasma interleukin-6 activity in rats. Physiol Behav 47:957–961
42. Zhou D, Kusnecov AW, Shurin MR, DePaoli M, Rabin BS (1993) Exposure to physical and psychological stressors elevates plasma interleukin-6: relationship to the activation of hypothalamic-pituitary-adrenal axis. Endocrinology 133:2523–2530
43. Ban E, Marquette C, Sarrieau A, Fitzpatrick F, Fillion G, Millon G, Rostène W, Haour F (1993) Regulation of interleukin-1 receptor expression in mouse brain and pituitary by lipopolysaccharide and glucocorticoids. Neuroendocrinology 58:581–587
44. Betancur C, Lledo A, Borrell J, Guaza C (1994) Corticosteroid regulation of IL-1 receptors in the mouse hippocampus: effects of glucocorticoid treatment, stress and adrenalectomy. Neuroendocrinology 59:120–128
45. Coehlo MM, Souza GEP, Pela IR (1982) Endotoxin-induced fever is modulated by endogenous glucocorticoids in rats. Am J Physiol 263:R423–R427
46. Morrow LE, McClellan JL, Conn CA, Kluger MJ (1993) Glucocorticoids alter fever and IL-6 responses to psychological stress and to lipopolysaccharide. Am J Physiol 264:R1010–R1016
47. Goujon E, Parnet P, Aubert A, Goodall G, Dantzer R (1995) Corticosterone regulates behavioural effects of lipopolysaccharide and interleukin-1β in mice. Am J Physiol 269:R154–R159
48. McClellan JL, Klir JJ, Morrow LE, Kluger MJ (1994) Central effects of glucocorticoid receptor antagonist RU-38486 on lipopolysaccharide and stress-induced fever. Am J Physiol 267:R705–R711
49. Goujon E, Parnet P, Cremona S, Dantzer R (1995) Endogenous glucocorticoids downregulate central effects of interleukin-1β on body temperature and behaviour in mice. Brain Res 702:173–180

Chapter 2

Signals from the hypothalamus to the pituitary during chronic immune responses

Michael S. Harbuz and Stafford L. Lightman

The question of whether stress is good or bad is a pertinent one in relation to pathology. A major component of the host response to stress is the activation of the hypothalamo-pituitary-adrenal (HPA) axis. The end point of this activation is the release of glucocorticoid steroid hormones from the adrenal cortex (predominantly corticosterone in the rat and cortisol in man). The importance of this system to the integrity of the individual can be demonstrated by investigating the extreme situation where the adrenal is removed or no longer functioning. This situation can be created surgically in experimental animals and occasionally in humans, although more frequently it occurs as a result of defective function of the adrenals due to autoimmune disease or tuberculosis leading to adrenal insufficiency (Addison's disease). In patients with Addison's disease minor challenges which might only produce mild symptoms in a normal individual can be life threatening. An interesting parallel to this can be seen following injection of the cytokine interleukin-1 (IL-1) or the immune stimulant lipopolysaccharide (LPS) an agent which produces flu-like symptoms for a few hours when given to humans. Doses which only result in "sickness behaviour" in adrenal-intact animals prove fatal in the absence of adrenal glands. In a number of animal models of immune-mediated diseases such as experimental allergic encephalomyelitis (EAE) and adjuvant-induced arthritis (AA) it has been demonstrated that in adrenalectomized animals following induction of the disease there is an earlier onset, an increased severity and, if left untreated, a fatal outcome. In all these cases treatment with steroids diminishes the severity of disease and prevents the fatal outcome. This chapter will draw together our current understanding of the mechanisms, particularly at the central level, underlying the activation of the HPA axis following immune activation. Much of our understanding of these mechanisms comes from studies in experimental animal models and this chapter will concentrate principally but not exclusively on the evidence from these.

Normal regulation of the HPA axis

Glucocorticoids released from the adrenal cortex are able to regulate their own synthesis and release by completing a negative feedback loop at the levels of the anterior pituitary, hypothalamic paraventricular nucleus (PVN) and higher centres. In addition, and pertinent to this discussion, glucocorticoids exert a wide range of actions affecting many aspects of bodily function including metabolism, inflammation and immunity. In this chapter we will concentrate principally on the activation of the HPA axis in response to acute and chronic (disease) immune activation.

Geoffrey Harris [1] first postulate a functional link between the CNS and the adenohypophysial endocrine system. He proposed that the release of hormones from the anterior pituitary gland was under the control of factors of hypothalamic origin released into the hypothalamic portal vasculature – despite previous opinions that portal blood flow was in the other direction. It was not until the isolation of corticotrophin-releasing factor (CRF) in the 1950s [2,3] and the subsequent characterization of CRF1-41 some 25 years later [4] that detailed studies on the release and the interactions of the many factors which have now been isolated in hypophysial portal blood (HPB) and which are known to influence adrenocorticotrophin (ACTH) release could be undertaken.

It is now generally accepted that CRF synthesized in the parvocellular cells of the PVN (pPVN) is the major corticotrophin-releasing factor responsible for the release of ACTH from the anterior pituitary in response to acute stress (Fig. 2.1). Axons from these CRF neurons terminate in the external zone of the median eminence. CRF1-41 is released into the HPB to evoke the release of ACTH and induce transcription of proopiomelanocortin (POMC) mRNA, the ACTH precursor. CRF1-41 is currently the only hypothalamic factor which has been demonstrated to induce POMC gene expression [5]. Arginine vasopressin (AVP), another ACTH secretagogue, is also synthesized in the pPVN. These parvocellular cells are distinct from the classical magnocellular cells of the PVN which synthesize both AVP and oxytocin and project through the internal zone of the median eminence to terminate in the posterior pituitary. In the parvocellular cells AVP coexists with CRF in approximately 50% of CRF-positive axons and terminals in resting normal animals [6,7] and in humans [8]. CRF and AVP have been shown to act synergistically at the corticotroph to evoke the release of ACTH in vitro [9–11], and also in vivo [12].

→

Fig. 2.1. Corticotrophin-releasing factor (CRF) is synthesized in the parvocellular cells of the paraventricular nucleus (PVN) located within the hypothalamus. Axons from these neurons terminate in the external zone of the median eminence where CRF, arginine vasopressin (AVP) and enkephalin have all been localized in the same secretory vesicles. Proenkephalin A (PEA) is the enkephalin precursor. CRF and AVP are released into the hypophysial portal blood and carried to the corticotrophs of the anterior pituitary to synergistically evoke release of adrenocorticotrophin (ACTH). Pro-opiomelanocortin (POMC) is the ACTH precursor. ACTH evokes the synthesis and release of corticosterone from the adrenal cortex which has an action at the hypothalamic and pituitary levels to negatively control its release. The PVN receives inputs from a number of brain regions believed to mediate the response to stress (see text). Immune activation results in the release of interleukin-1 (IL-1) which initiates cytokine cascades to exert immune effects. IL-1 is able to stimulate the HPA axis through an action on CRF neurons. The evidence for the action of other cytokines on HPA axis activation is less well established. Corticosterone is able to inhibit synthesis and release of IL-1.

In addition to CRF and AVP a number of other factors have also been detected in HPB and have also been shown to affect ACTH release. However, the evidence that these might have a role in the physiological regulation of ACTH secretion is less well-established. Candidates which may affect corticotroph function include oxytocin [13–15], adrenaline [16], opiates [17,18], substance P [19] and atrial natriuretic peptide [20].

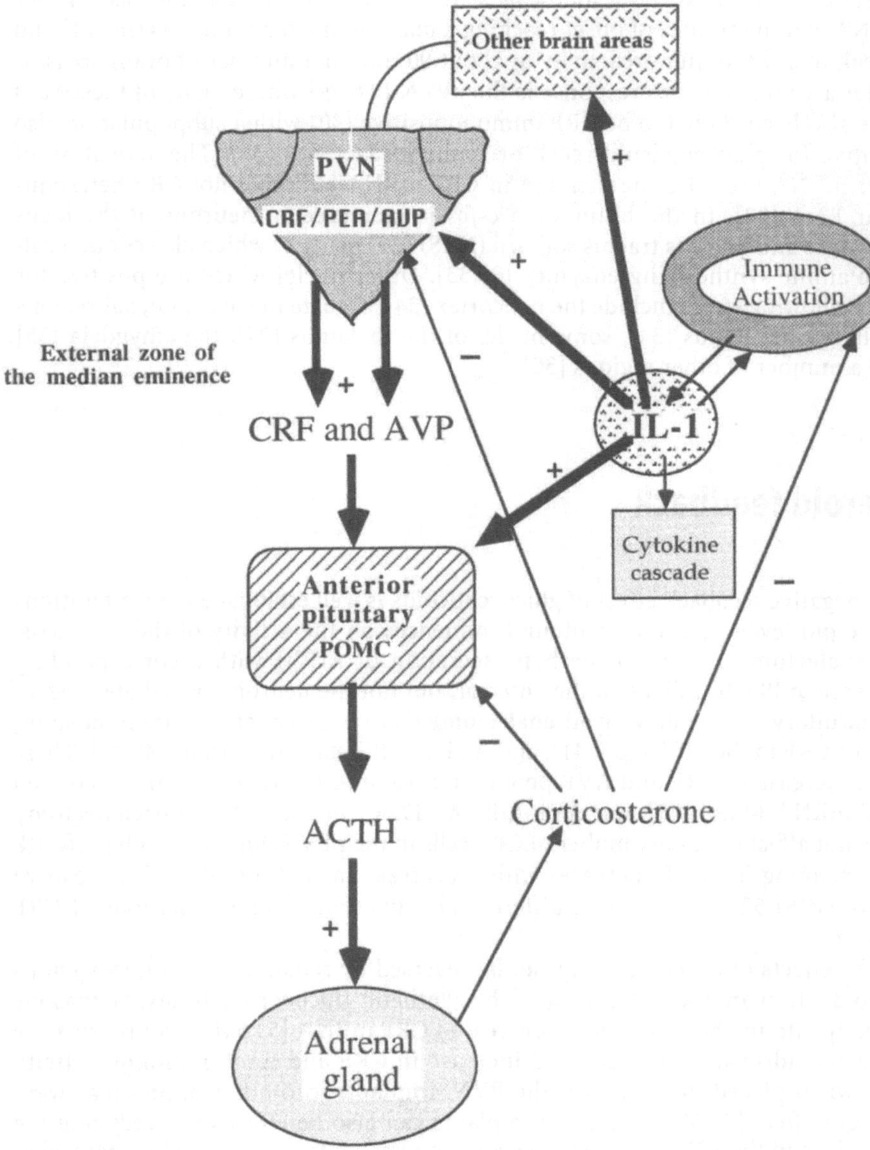

The PVN receives inputs from a number of brain regions believed to mediate the response to stress. These include catecholaminergic projections from the locus coeruleus, A1, A2, C1 and C2 regions of the brain stem [21], serotoninergic inputs from the raphe [22,23], substance P from the arcuate nucleus [19] together with inputs from the hippocampus, amygdala and bed nucleus of the stria terminalis.

The proto-oncogene c-*fos* has proved a useful tool as a marker of neuronal activation to identify central nuclei involved in mediating stress responses. C-*fos* immunoreactivity and c-*fos* mRNA are generally below the limits of detection in most brain nuclei. However following acute stress there is a rapid increase in c-*fos* mRNA within minutes of onset reaching a maximum after about 30 min [24] and a peak in c-*fos* immunoreactivity at about 90 min in a number of brain areas. A major area for the c-*fos* response is the pPVN [24–29] where many of these cells have also been shown to be CRF immunopositive [30] with a subpopulation also positive for glucocorticoid receptor immunoreactivity [31]. The activation of c-*fos* mRNA precedes the increase in CRF mRNA although not CRF heteronuclear RNA [32]. In the brain stem c-*fos* is increased in neurons of the locus coeruleus and nucleus tractus solitarii (NTS) [29] many of which also contain catecholamine synthesizing enzymes [30,33]. Other nuclei which are positive for c-*fos* following stress include the neocortex [34], arcuate nucleus [30], subregions of the hippocampus [34], some nuclei of the thalamus [24], the amygdala [35] and a number of other regions [30].

Steroid feedback

The negative feedback effect of glucocorticoids is well established and alterations in steroid levels can have profound influences on the activity of the HPA axis. Adrenalectomy results in the hypersecretion of ACTH with a corresponding increase in POMC mRNA in the anterior, but not the neurointermediate lobe, of the pituitary [36]. Following adrenalectomy there is also enhanced immunostaining for CRF in the pPVN [37–41], and AVP in CRF-stained neurons [42,43]. There is an increase in CRF and AVP peptide release into the HPB [44] and increased CRF mRNA [36,45–48] and AVP mRNA [49,50] in the pPVN. Adrenalectomy does not affect the total number of CRF cells in the pPVN, but the number of CRF cells staining for AVP increases with a decrease in AVP-negative/CRF-positive neurons [51,52], suggesting a differential activation of subpopulations of CRF neurons.

The effects of adrenalectomy can be reversed by replacement with exogenous steroids. Implantation of pellets of the synthetic glucocorticoid dexamethasone directly into the brain results in decreased CRF mRNA [53,54], and prevents the expected adrenalectomy-induced increase in CRF and AVP immunoreactivity [55] when placed directly over the PVN. Implants into other brain areas were without effect [53,55,56]. Steroid implants can also be effective in reducing the activation in the HPA axis when placed in higher centres known to be involved in mediating signals concerned with the activation of the HPA axis such as the dorsal hippocampus [56], which contains both type-I (mineralocorticoid) and type-II (glucocorticoid) receptors. There is evidence for both type-II and type-I corticosteroid receptors in cultured fetal hypothalamic cells [57], which can inhibit the release of CRF into the culture medium. These data suggest that corticosterone and dexamethasone have different feedback potencies and act at different sites within the brain to affect the HPA axis. In addition to the evidence for the PVN as a site for glucocorticoid feedback evidence suggests that both total and anterior hypothalamic deafferentation increase CRF and AVP mRNAs in the pPVN, while posterior deafferentation has no significant effect. These data support a role for feedback control via an extrahypothalamic neural input coming from anterior structures [58].

Stress

Hans Selye [59] coined the term stress when he first described a syndrome produced by "diverse noxious agents" such as exposure to cold, surgical injury, spinal shock, excessive or sublethal doses of a variety of drugs. The acute phase, occurring over minutes and hours, he termed the "general alarm reaction". The chronic phase which occurred in response to repeated acute challenges or a stimulus present for a number of days, he regarded as a generalized effort of the organism to adapt itself to new conditions and this he termed the "general adaptation syndrome".

Many acute stressors activate the HPA axis including cold, ether, footshock, subcutaneous (s.c.) formalin, intraperitoneal (i.p.) histamine, i.p. hypertonic saline, insulin-induced hypoglycaemia, restraint, swimming and laparotomy or surgical stress. All of these result in an increase in the activity of the HPA axis with increases in circulating ACTH and corticosterone. At the hypothalamic level there is a rapid transient increase in bioassayable or immunoreactive CRF in the median eminence, lasting 2–4 min after which median eminence CRF levels decline, presumably in response to increased release of CRF1-41 into the HPB [review: 60,61]. Immunoreactive CRF increases 6 h after hypertonic saline stress which correlates well with the increase in CRF mRNA in the pPVN occurring from 2–4 h after onset of stress [62,63]. AVP mRNA in the pPVN is also increased in response to hypertonic saline stress [62].

There is a differential activation of hypothalamic-releasing factors in response to stress. Predominantly psychological stressors such as restraint and swim stress activate CRF mRNA, while physical stressors like footshock, i.p. hypertonic saline and naloxone-induced morphine withdrawal increase proenkephalin A (PEA) mRNA in addition to CRF mRNA [62–66]. Ether stress while having no effect on CRF mRNA increases PEA mRNA [67]. These differential responses in mRNA levels presumably related to activation of different patterns of hypothalamic afferents may also be reflected in changes in the release of ACTH secretagogues into the HPB although the precise role of PEA mRNA in this response remains to be determined. The differential release and hence the response of the anterior pituitary could be regulated to reflect the type and severity of the challenge. Our understanding of the role of these factors is limited by the technology available to measure these changes in vivo and the interpretation put on these. A recent review presented the various methods used in neuroendocrine studies and discussed their limitations [61].

Physiologically relevant release of releasing factors into the HPB has been demonstrated in a number of studies. Antiserum to CRF substantially reduces the ACTH response to s.c. formalin or restraint stress. Antiserum to AVP is not as effective. However, combining both antisera results in a greater reduction than either alone [68] reflecting the synergistic activation on the corticotrophs by these releasing factors. In the rat, HPB sampling can only be undertaken in the anaesthetized animal which has also undergone extensive cranial surgery [review: 61]. This limitation prevents investigation of mild or psychological stressors. CRF and PEA mRNAs can be increased in anaesthetized rats over and above the increases induced by anaesthesia suggesting that a conscious appreciation of the stressor is not essential for activation to occur [66]. In the adrenalectomized animal and in animals treated chronically with high-dose steroids, acute stress is able to evoke an increase in CRF mRNA whatever the level determined by the steroid milieu [47,69] suggesting that stress-responsive transsynaptic activation of hypothalamic neurons can overcome steroid feedback at the hypothalamic level. These data support the idea of mechanisms being in place to maintain stress-responsiveness in the face of elevated endogenous corticosterone secretion [70].

Taking together the available evidence it seems that in response to acute stress there is an immediate (1–4 min) increase in CRF in the median eminence which is released from neurosecretory vesicles into the HPB together with AVP, enkephalins and possibly oxytocin presumably dependent on the nature and severity of the stressor. These peptides are carried in the HPB to the anterior pituitary where they act on corticotrophs to evoke the release of ACTH and induce POMC gene expression. CRF may act in either a passive or dynamic fashion, depending on the nature of the stressor and possibly reflecting the composition of the secretagogue cocktail present in the HPB. Following release of ACTH, there is a subsequent increase in circulating corticosterone levels.

While the above studies have concentrated on the immediate effects of acute stress on the activity of the HPA axis it should also be noted that a single exposure to an acute stimulus such as insulin-induced hypoglycaemia, LPS, IL-1 i.p., brain surgery with or without central injection of IL-1 and footshock can all produce significant long-lasting changes in the intracellular stores of AVP and CRF in the external zone of the median eminence [71]. In all instances, with the exception of insulin, AVP stores were increased in the external zone of the median eminence 7 and 11 days after single exposure to these stimuli. CRF stores were only enhanced in response to footshock. These data confirm an important role for AVP and also provide further evidence of the independent control of AVP and CRF stores in the median eminence. Following a single exposure to IL-1β 11 days previously a further injection of IL-1β provokes a twofold greater increase in ACTH and corticosterone than is seen in naive animals suggesting a hyperresponsiveness of the HPA axis associated with the increased production and storage of AVP in these animals [72].

Chronic stress

The study of chronic stress has been hampered by the lack of a suitable model. Most studies concerned with chronic stress utilize an acute stress repeated for a number of days. This raises a further problem as repeated stress may produce habituation or adaptation of the HPA axis resulting in attenuated responses. For example, with repeated footshock [73], restraint [74] or ethanol stress [75] plasma corticosterone may be elevated for up to 1 week, but levels subsequently return towards control levels with repeated exposures. The habituation appears to be stressor specific. Although there is no longer a response to the same stressful stimulus in chronically stressed animals, alternative acute stressors may be able to elicit a normal stress response [73,75], or even an exaggerated response in terms of increased peak release and/or greater duration of release [74,76–79]. This is presumably due to stimulation of alternative afferent pathways. In spite of the return to normal levels of circulating hormone this parameter may not be the best indicator of the neuroendocrine status of the animal, since POMC mRNA and ACTH content of the anterior pituitary, together with adrenal weight, are all increased.

An important role for AVP in chronic or repeated stress is supported by a number of studies. Exogenous CRF has been reported to not affect circulating ACTH levels in chronically restrained rats even though these animals are able to respond to an acute stress [74]. Exogenous AVP (but not oxytocin), however, does evoke an increase in hormone release. It would appear that endogenous AVP is necessary for sustaining the stress response in circumstances when the pituitary is refractory to CRF stimulation [80]. Repeated restraint stress results in increased AVP stores and increased co-localization within CRF positive neurons of the median eminence [81]. Chronic restraint stress also results in a loss of anterior pituitary CRF receptors, but this loss does not prevent a significant potentiation of ACTH release with an alternative acute stress suggesting that the adaptation to repeated stress is stressor specific [82]. These data provide compelling evidence that AVP may be an important regulator of pituitary responsiveness in the face of repeated or chronic stress.

Effects of stress and steroids on immune function

Steroids and immune function

The bi-directional communication between immune and neuroendocrine systems has attracted considerable interest. The modulatory effects of a variety of stressors on the functioning of the immune system is well established [83]. Glucocorticoids reduce the number of circulating macrophages and monocytes, lyse immature T-cells, inhibit IL-1 and IL-2 production and block phospholipase

A2 activity. The crucial role of glucocorticoids has been demonstrated in a number of studies where following adrenalectomy challenge with IL-1 or tumour necrosis factor at doses which would be well tolerated in adrenal-intact animals prove fatal [84]. The lethal effects can be prevented by steroid treatment. The ability to respond to an acute immune challenge is critical but the mechanisms underlying this are not yet clear. It has been proposed that the function of the increase in circulating steroids is not to protect against the source of the challenge itself but rather is necessary to protect against an excessive reaction resulting from the activation of the organisms own defence mechanisms. In the absence of the steroid response to suppress these defence mechanisms and hence prevent their overreaction the result might prove life threatening. These glucocorticoid-mediated immunosuppressive effects of stress are well characterized [85] but, in addition, the finding that adrenalectomized rats also show immunosuppression in response to stress, strongly suggests the involvement of other undefined factors [86,87].

Dehydroepiandrosterone sulfate

The most abundant adrenal steroid found in the circulation is dehydroepiandrosterone sulfate (DHEAS) which is converted to the active form DHEA by the enzyme DHEAS sulfatase. Specific receptors for the free hormone are found in T-cells where it is considered to be involved in IL-2 production [88]. DHEA is thus able to enhance the Th1 T-cell response and it is believed that DHEAS sulfatase activity regulates the Th1/Th2 cytokine balance. DHEA has also been reported to antagonize the suppressive effects of dexamethasone on lymphocyte proliferation [89]. DHEA levels fall with increasing age and DHEA replacement is able to reverse many of the immunological deficits of aging in mice [90], and also in humans [91,92].

Immunoneuropeptides

One possible mechanism may operate through the modulatory effects of peptides synthesized within the immune tissues. CRF, AVP, and the POMC products ACTH, β-endorphin and α-melanocyte stimulating hormone (α-MSH) are synthesized in peripheral blood lymphocytes, splenocytes and thymocytes [93]. Specific receptors for these hormones are also found on immune tissues suggesting the possibility of a functional interaction between immune and neuroendocrine systems in these tissues. The levels of these immunoneuropeptides are very low. However, stimulation of these cells with a mitogenic activator such concanavalin A results in increased production and release [94]. In humans insulin-induced hypoglycaemia can produce an increase in leukocyte ACTH content demonstrating an increase in POMC gene expression [95] although no clear role for the products of this transcript in leukocytes has been defined.

CRF as an immunomodulator

CRF is an important factor in stress-induced immunomodulation. Infusion of CRF centrally for 1 week reduced both concanavalin A-induced lymphocyte proliferation and LPS-induced B lymphocyte mitogenesis [94]. These CRF-treated animals showed a marked increase in expression of IL-1β mRNA in splenocytes. Stress in both normal and adrenalectomized rats results in decreased T-cell proliferation and natural killer cytotoxicity [96]. These effects are blunted by the administration of CRF antibodies or antagonists which are effective whether given intravenously (i.v.) or intracerebroventricularly (i.c.v.) suggesting a central mechanism. Administration of CRF produces immunosuppression similar to that caused by stress. Exogenous CRF enhances the splenic antibody response to sheep red blood cells in the rat, an effect which can be blocked by prior administration of CRF antibodies, acting via an as yet undetermined pathway through which CRF is able to control the humoral response to stress. CRF has also been shown to increase natural killer cell activity, stimulate B lymphocyte proliferation and increase cytokine production suggesting a pro-inflammatory role for CRF. There is also evidence for a direct anti-inflammatory role for CRF. CRF is able to inhibit adrenaline-induced pulmonary oedema and decrease vascular leakage in the paws of anaesthetized rats exposed to heat or cold. This effect is independent of corticosterone or hypotensive effects and could be antagonized by α-helical CRF(9–41) [97]. There is a good correlation between the neuroendocrine effects of CRF analogues and their anti-oedema effects [98]. CRF has also been shown to inhibit natural killer cell activity in vitro. These contrasting data suggest a complex role for CRF in immunomodulation.

Footshock stress blunts the response of both plasma ACTH and IL-1 to an injection of endotoxin, providing evidence for stress-mediated modulation of the capacity of immune cells to produce and/or secrete immune regulatory cytokines [99]. IL-1 administered i.c.v. decreases natural killer cell activity and suppresses both the immune response to mitogen and the IL-2-induced production of splenic and blood lymphocytes. These effects could be completely prevented by prior administration of CRF antibody. Blocking sympathetic ganglia partially blocked the immunosuppressive effects of IL-1 suggesting that IL-1 in the brain acts to suppress immune response by activation of both the HPA axis and the sympathetic nervous system. These effects appear to be mediated via hypothalamic CRF [100].

In addition to central effects, local effects of CRF may also be important in inflammation as i.p. administration of CRF antiserum reduces the inflammatory exudate volume and cell concentration in chemically induced aseptic inflammation. CRF is present in the inflamed area but not present in the general circulation suggesting that CRF may have autocrine or paracrine inflammatory actions directly at the site of inflammation [101]. The presence of increased CRF receptors in inflamed rat paw supports this contention. The origin of this CRF has not been established. However, CRF-like mRNA and a peptide which reacted with CRF antiserum have been reported in unstimulated normal leukocytes [102]. The recent identification of a variant form of CRF (urocortin) in the rat suggests other forms may also play a role [103]. The ability of urocortin to interfere with

CRF-binding protein may also implicate urocortin in this response [104]. Together these data strongly suggest a role for CRF or a CRF-like compound in immunomodulation and suggest that different stressors may influence immune function by altering the capacity of immune cells to produce and/or secrete immunoregulatory cytokines and/or other immune modulators. The presence of CRF receptors in mouse spleen also suggests the possibility of direct effects [105].

HPA axis response to cytokines

In addition to the effects of stress and stress hormones on the immune system the cytokines are able to influence the HPA axis (Fig. 2.2). IL-1 increases plasma corticosterone and/or ACTH whether administered by an i.v., i.p. or i.c.v. route. CRF in the PVN is essential for this activation as deafferentation of the medial basal hypothalamus [106], or immunoneutralization of endogenous CRF [107–109] blocks this effect. In addition, CRF release into the HPB [108] and CRF turnover [110,111] are both increased by IL-1 in vivo, and in vitro [112–114]. The presence of circulating glucocorticoids also appears to be important as following adrenalectomy IL-1β is no longer able to evoke an ACTH [106; 115] or CRF mRNA [115] response to this cytokine. This response returns after steroid replacement to physiological levels.

Within the CNS a similar pattern of activation of c-*fos* to that induced by acute stress has been seen following acute injection of either LPS or IL-1β [116–120] although in addition to areas activated in response to stress there is also activation in the meninges, and circumventricular organs such as the organum vasculosum of the lamina terminalis (OVLT) and subfornical organ [119; 120]. The route of administration of these immune mediators is important as there is a differential activation comparing i.p. with i.c.v. routes of administration [117; 120].

The route by which cytokines injected peripherally are able to activate central mechanisms has been the subject of much debate. A number of mechanisms have been proposed including activation of IL-1 receptors on endothelial cells on the circumventricular organs (CVOs) and stimulation of brain pathways [121,122], leakage through extracellular pathways at the CVOs [123], active transport across the blood–brain barrier (BBB) via saturable transport mechanisms [123,124], loss of BBB integrity allowing monocytes to cross and release cytokines [125,126] and via activation of vagal afferents [127–129]. These mechanisms are discussed more fully by Dantzer and colleagues in Chapter 1. Readers are also directed to the review of Hori and co-workers [130], concerning the role of the autonomic nervous system in immune nervous system communication and the effects of stress and central injection of cytokines and CRF on immune responses.

An adaptation of the HPA axis after repeated endotoxin or IL-1 administration has been demonstrated [131–134]. Following repeated endotoxin injection the ability to respond to further endotoxin declines as does the response to a novel neuroendocrine challenge such as insulin-induced hypoglycaemia [135].

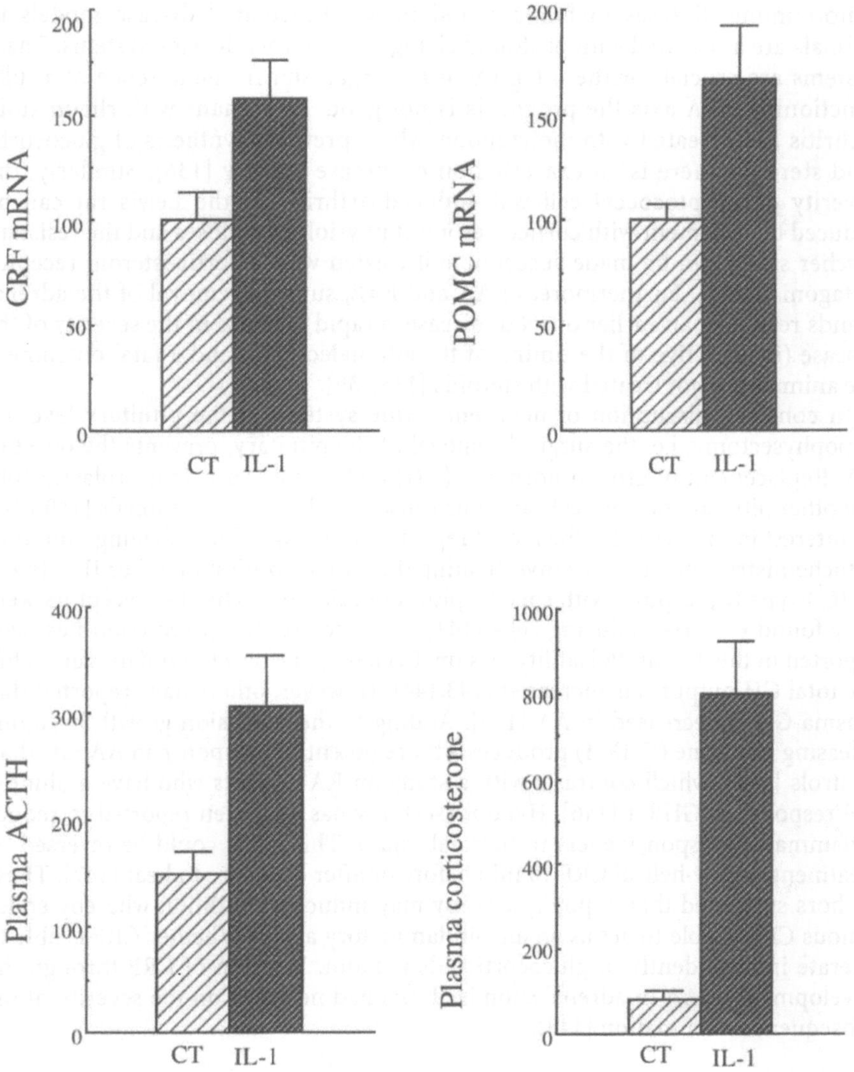

Fig. 2.2. The effects of a single i.p. injection (2 μg) of interleukin-1β (IL-1) or vehicle (CT) on the activation of the HPA axis. CRF mRNA in the paraventricular nucleus, POMC mRNA in the anterior pituitary and circulating levels of ACTH (released from the anterior pituitary), and corticosterone (released from the adrenal cortex) are all increased in response to this stimulus. All samples were collected 3 h after injection, demonstrating the continuing stimulation of hormone release despite the relatively short half-life of IL-1β.

Chronic inflammatory stress

Autoimmune diseases in humans and immune-mediated disease models in animals are associated with profound changes in neuroendocrine systems. These systems are crucial for the integrity of the organism. In the absence of a fully functioning HPA axis the prognosis is not good. In humans with rheumatoid arthritis (RA) treated with metyrapone, which prevents synthesis of glucocorticoid steroids, there is an exacerbation of disease activity [136]. Similarly, the severity of streptococcal cell wall-induced arthritis in the Lewis rat can be reduced by treatment with corticosterone at physiological doses and the resistant Fischer strain can be made susceptible if treated with a corticosteroid receptor antagonist [137]. Furthermore, in AA and EAE, surgical removal of the adrenal glands results in an earlier onset of disease, a rapid increase in the severity of the disease (irrespective of the timing of the adrenalectomy) and a fatal outcome if the animals are not treated with steroids [138,139].

In contrast, disruption of neuroendocrine systems at the pituitary level by hypophysectomy, i.e. the surgical removal of the pituitary, prevents the onset of AA. Replacement of growth hormone (GH) and to a lesser extent prolactin (but not other pituitary hormones) can return susceptibility in these animals [140,141]. Of interest in this regard is the recent report concerning a dual-labelling immunocytochemistry study in mice investigating the co-localization of either IL-1 type I or IL-1 type II receptors with specific pituitary cell types. The IL-1 receptors were only found on GH-producing cells [142]. An increase in GH secretion has been reported in the AA rat. Pulsatility was unaffected by the onset of inflammation but the total GH output was increased [143,144]. However, others have reported that plasma GH is decreased in AA [145]. Adding to the confusion growth hormone releasing hormone (GHRH) produces a more potent GH response in AA rats than controls [144], which contrasts with a study on RA patients who have a blunted GH response to GHRH [146]. Hypophysectomy has also been reported to reduce inflammatory responsiveness to thermal injury. This effect could be reversed by treatment with α-helical CRF 10 min before or after exposure to heat [147]. These authors suggested that hypophysectomy may induce a condition whereby endogenous CRF is able to act as an anti-inflammatory agent. Whether CRF is able to operate independently of glucocorticoids is moot. Infusion of CRF through the development of AA in adrenalectomized rats had no effect on the severity of the subsequent inflammation [148].

Regulation of the HPA axis following chronic immune activation

The remaining part of this chapter will concentrate on situations in which there is chronic immune activation. In view of the importance of glucocorticoids as anti-inflammatory agents we will concentrate on alterations in the HPA axis concentrating on the AA rat model. AA is a T-cell-dependent disease model which can be induced in susceptible strains of rat by a single intradermal injection into the base of the tail of a suitable adjuvant such as ground, heat-killed *Mycobacterium butyricum* or the synthetic adjuvant CP20961. Twelve to fourteen days after injection the animals begin to exhibit hind-paw inflammation which subsequently spreads to other joints reaching a maximum severity 21 days after injection [149]. This model has been used extensively for studies on pain, inflammation and arthritis. In addition we will consider changes in other animal disease models and also complement these data with findings in humans with autoimmune disease.

A circadian variation in disease activity has been reported in RA [150] and it is thought that this might be related to the cortisol-influenced circadian variation in the immune system [150,151]. For example, the circadian rhythm of spontaneous natural killer cell activity is lost or demonstrates a change of phase in patients with RA [152]. Disease activity in RA is at a minimum 6 h after the peak in cortisol concentrations. Following i.v. infusion of hydrocortisone the suppression of lymphocytes and monocytes and the stimulation of neutrophils is maximal 4–6 h later, supporting the idea of glucocorticoid-mediated influence on immune system and hence disease activity [150]. RA patients with medium disease activity have been reported to exhibit a shift in maximum and minimum cortisol concentrations to earlier times of the day. However, with high disease activity the circadian rhythm is lost or markedly reduced [153]. These authors found a significant correlation between inflammatory activity of disease and the changes in circadian secretion of cortisol and ACTH. A similar loss of circadian rhythm occurs in rats with AA [154,155]. The normal rhythm is blunted with a shallow peak and trough occurring 6 h earlier than in non-arthritic animals. Corticosterone levels are increased and similar throughout the day to levels normally seen at the circadian peak of secretion.

Investigation of hypothalamic function is not possible in humans with RA. The rat however provides a suitable model for such investigations. In addition to the increase in hormonal activity in AA detailed above, there is also an increase in POMC mRNA in the anterior pituitary [145,149,156]. As noted earlier the only releasing factor known to increase POMC mRNA is CRF [5]. One would therefore predict an increase in CRF mRNA in the PVN and an increase in CRF release into the hypophysial portal blood driving the corticotrophs to increase POMC mRNA and hence increase ACTH release. However, irrespective of the strain of rat or the adjuvant used we have been unable to demonstrate any increase in CRF mRNA in the PVN or release of CRF peptide into the portal blood. Indeed in the Piebald-Viral-Glaxo (PVG) strain of rat [149] and in the Lewis strain [157] there is a paradoxical decrease in CRF mRNA (Fig. 2.3). This decrease is first apparent at day 11 when the first indications of inflammation are apparent and continues to decrease, reaching a nadir 21 days after adjuvant injection when the severity of the disease is at its peak [149]. We have consistently found a good negative correlation between the decrease in CRF mRNA levels in the PVN and the degree of inflammation in the PVG rat.

In contrast to CRF, AVP concentrations in the portal blood and AVP mRNA in the parvocellular cells of the PVN are both increased (Fig. 2.4) suggesting that in the presence of permissive levels of CRF, AVP is able to take over as the major

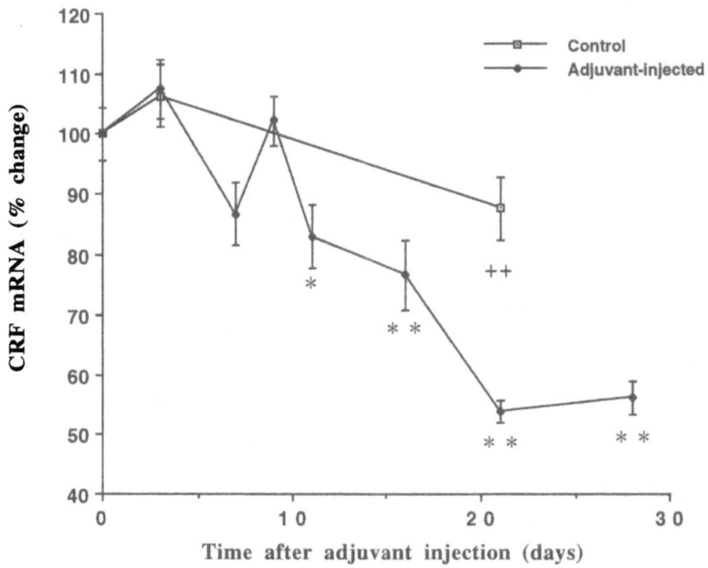

Fig. 2.3. CRF mRNA in the paraventricular nucleus of control, non-arthritic rats and in rats at various times following a single intradermal injection of ground, heat-killed *Mycobacterium butyricum* in paraffin oil given at day 0. In this model the animals become sensitive to handling at around day 11 although obvious inflammation is not usually apparent until day 13/14. Peak clinical symptoms are observed at day 21 and are associated with increased plasma ACTH and corticosterone and increased POMC mRNA in the anterior pituitary. Despite the activation of the hypothalamo-pituitary-adrenal axis CRF mRNA is paradoxically reduced with further reductions associated with increased inflammation. * $P < 0.05$, ** $P < 0.01$ againts day 0 control, ++ $P < 0.01$ againts day 21 control.

stimulator of the HPA axis in the AA rat [149,158]. These data support those from other repeated stress studies which have proposed an increased role for AVP in chronic stress situations as noted above. These data also support the notion that despite the colocalization of AVP within CRF neurons and indeed the co-localization of CRF and AVP within the same secretory vesicles in these neurons they appear to be under independent control and regulation.

The alteration in hypothalamic control mechanisms evidenced by the decrease in CRF mRNA appears to be a common feature in chronic inflammatory disease models. EAE is a T-cell-dependent disease model which has been used as the animal model for multiple sclerosis (MS) research. We have investigated the neuroendocrine changes associated with the development and recovery from disease in an adoptive transfer model of EAE. Rather than injecting myelin basic protein in adjuvant (the standard model), the adoptive transfer model involves injection of activated splenocytes collected from animals with the disease (i.e. no adjuvant is involved). In this model the animals develop peak symptoms after 6–7 days and show full recovery after 11 days. Associated with peak clinical symptoms there is an increase in circulating levels of corticosterone and an increase in

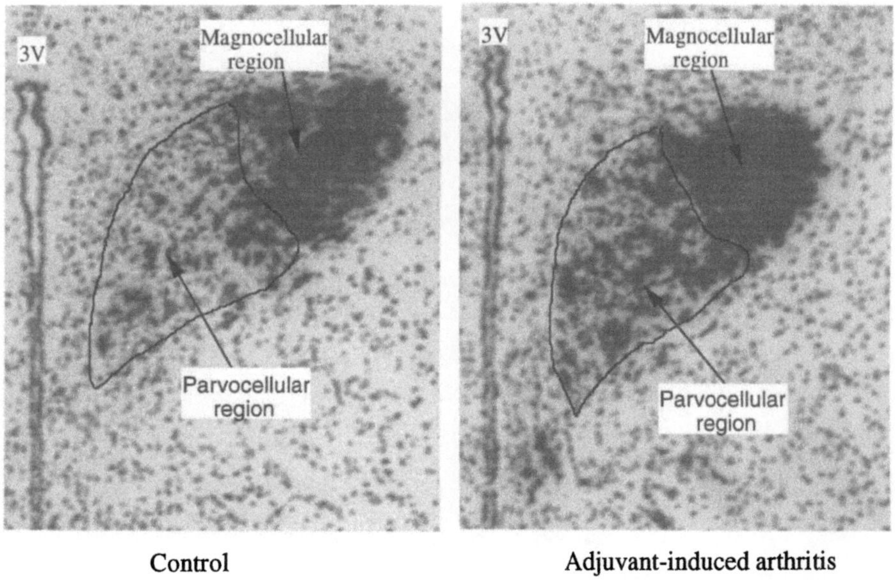

Fig. 2.4. Arginine vasopressin (AVP) mRNA in the paraventricular nucleus (PVN) of control rats and rats 21 days after a single intradermal injection of ground, heat-killed *Mycobacterium butyricum* in paraffin oil (arthritic). The lateral dark staining represents AVP mRNA in the magnocellular cells of the PVN (also known as antidiuretic hormone). The enclosed area is the parvocellular region which is involved in the stress axis but which also contains some magnocellular cells. Probe bound to AVP mRNA in the parvocellular region is increased in the arthritic animals. 3V is the third ventricle.

POMC mRNA in the anterior pituitary. At the level of the PVN there is a paradoxical decrease in CRF mRNA similar to that seen with AA (Fig. 2.5). With recovery the elevated levels of plasma corticosterone and POMC mRNA decline to control levels. Similarly the decreased CRF mRNA levels increase with recovery to levels seen in control animals [159] Increased HPA axis activity has also been reported in humans with MS [160,161]. Post-mortem analysis of MS brains has revealed an increase in the number of CRF-containing cells in the PVN. This increase was found to be due to an increase in CRF neurons co-localizing AVP which rose from 54% in controls to 76% in MS patients [8]. These data support an increased role for AVP in MS. Further support for an increased role for AVP is provided by the observation that MS patients demonstrated an increased relative activity of AVP in the regulation of the HPA axis compared with controls following acute challenge [160].

Eosinophilia-myalgia syndrome (EMS) reached epidemic proportions on the West Coast of the USA in 1989. EMS is an immune-mediated disease characterized by eosinophilia, myalgias, oedema, fasciitis and neuropathies. The source of the epidemic was eventually traced to a rogue batch of impure L-tryptophan (L-Trp) which was being used as a diet supplement and which found its way into a number of preparations. To confirm the agent responsible, Lewis rats were

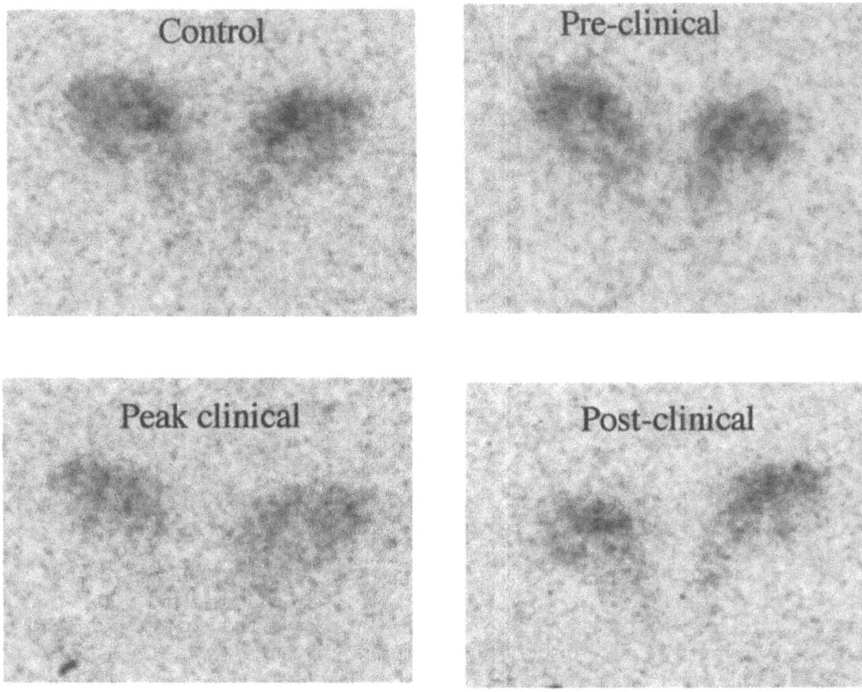

Fig. 2.5. CRF mRNA in the PVN of rats at various times after adoptive transfer of activated splenocytes to induce experimental allergic encephalomyelitis. In this model peak clinical symptoms are observed after 7 days (peak clinical) and the animals have fully recovered by day 11 (post-clinical). Despite increases seen in POMC mRNA in the anterior pituitary and increased circulating corticosterone associated with peak clinical symptoms CRF mRNA in the PVN of these animals is paradoxically reduced.

treated with comparable doses (w/w) of impure L-Trp to those ingested by humans. These animals also developed fasciitis and perimyositis similar to the pathological features of human EMS. The appearance of symptoms of EMS in the rat model was associated with a decrease in CRF mRNA in the PVN [157,162].

Leishmaniasis is a parasitic liver infection caused by the protozoan *Leishmania donovani* which is associated with T-cell activation. This parasite is prevalent in the tropics and some Mediterranean areas. Certain mouse strains are susceptible to infection with the parasite. Following inoculation into a susceptible strain of mouse we were able to demonstrate a decrease in CRF mRNA associated with infection [163]. This finding is of importance as it is the first evidence of a similar alteration in hypothalamic control in a different species. More recently a decrease in CRF mRNA has been noted in a spontaneous mouse model of the autoimmune disease systemic lupus erythematosus (N. Shanks, personal communication).

These data suggest that an alteration in the hypothalamic control of the HPA axis may be part of the adaptive response to chronic immune-mediated disease/inflammatory stressors. This change can be seen in a number of different disease models in different rodent species and may also occur in humans. From the available evidence it would appear that AVP takes over as the major stimulator of the HPA axis while CRF provides a secondary permissive role in chronic inflammatory stress.

Stress responsiveness in disease

CRF is generally considered to be the major corticotrophin-releasing factor responsible for coordinating the response to acute stress. It was therefore of little surprise to find that the ability of rats with AA to respond to acute stress was impaired. These animals were unable to mount an increase of either CRF mRNA in the PVN or plasma corticosterone in response to an intraperitoneal injection of hypertonic saline [26]. Similarly they were unable to respond to acute restraint stress [164]. Seven days after adjuvant injection and prior to the development of inflammation these animals have a hyperresponse of CRF mRNA to this stressor compared with the increase seen in non-arthritic control rats [164]. A similar inability to respond to an acute challenge has been reported in humans with RA. Chikanza and colleagues [165] reported no increase in plasma cortisol in RA patients following surgery. They also found that these patients responded normally to a CRF stimulation test indicating that the pituitary and adrenal responses were intact and further supporting the case for a hypothalamic defect associated with inflammatory disease. As noted above, the inability to mount a glucocorticoid response to either acute or chronic immune stimulation can be life threatening. To determine how the AA rat would respond to an acute immune challenge we injected AA rats with endotoxin. We found that these rats were able to mount a corticosterone response of equal magnitude to that seen in non-arthritic control animals (unpublished observations). These data show that although the ability to respond to acute stress is impaired, the ability of rats with AA to mount a response to acute immune stimulation is intact, suggesting activation of an alternative pathway and providing a further example of differential control of the response to acute stress and acute immune stimulation.

A defect in the regulation of the HPA axis has been proposed as predisposing individuals to autoimmune disease [137,166,167]. The Fischer strain of rat is able to mount a robust HPA axis response to stress and is resistant to EAE and streptococcal cell wall-induced arthritis. In contrast the Lewis rat which mounts a poor response to a variety of immune and psychological challenges is susceptible to these diseases. The defect has been localized to the CRF neurons in the PVN [167] resulting in low basal plasma corticosterone levels and a blunted circadian rhythm. The Lewis rat has smaller adrenal and pituitary glands and a larger thymus than the histocompatible Fischer rat. These differences correlate well with the increased susceptibility to a wide-range of T-cell-mediated autoimmune-like diseases seen in the Lewis rat. While the hyporesponsiveness of the HPA axis may be a factor in the relative susceptibility of this strain, other factors are also likely to be important. The PVG strain of rat is resistant to EAE and and it has been suggested that this resistance is dependent on its hyperresponsiveness to stress [166]. However, we have found the PVG rat to be susceptible to AA despite the robust response to hypertonic saline stress [139].

Impaired HPA axis function has been noted in a number of animal models and also in immune-mediated disease in humans. African sleeping sickness is a potentially lethal parasitic disease in humans caused by the protozoan *Trypanosoma brucei*. In the early acute phase of the disease symptoms can include fever, rash, weight loss, oedema and chronic fatigue. In later stages the parasite may invade the CNS and this can produce a broad spectrum of neurological and psychiatric symptoms. Untreated the disease is eventually fatal. HPA axis function in individuals with sleeping sickness is impaired. They are unable to mount an adrenal response to injected ACTH suggesting adrenal insufficiency [168]. A secondary adrenal insufficiency has also been suggested as they are also unable to mount a suitable ACTH response to injection of CRF. However, it does not appear that these individuals have an adrenal insufficiency which predisposes them to the disease as has been suggested for the Lewis rat. Associated with recovery following antiparasitic treatment ACTH and cortisol responses to CRF return to near normal suggesting the defect is a response to the disease and by inference is not present prior to development.

A similar blunting of the HPA response has been noted in the obese strain of chicken which develops autoimmune thyroiditis; an animal model for Hashimoto disease. These animals have elevated levels of corticosteroid-binding globulin and a corresponding decrease in free circulating corticosterone. They show a blunted corticosterone response to IL-1 and other acute stimuli. The alteration in the response has been localized to the hypothalamo-pituitary part of the axis although the precise locus remains to be determined [169]. The blunted response to challenge is also evident in a number of inbred strains of mice which develop diseases similar to systemic lupus erythematosus (SLE) and Sjögrens syndrome. Parallel to the development of the diseases in susceptible strains there is an age-related decline in plasma corticosterone together with a blunted response to challenge with IL-1 and other stimuli [169].

Ethical considerations have prevented extensive study on alterations in the HPA axis in humans. Some patients with multiple sclerosis (MS) are reported to have elevated levels of plasma cortisol compared with controls [160] while others report no difference in cortisol levels or the diurnal rhythm [170]. Changes in HPA axis responsiveness have been investigated by injection of the hypothalamic-releasing factors CRF and AVP. Michelson and co-workers [160] have suggested that in MS AVP may play a more significant role than CRF in maintaining the activity of the HPA axis. Wei and Lightman [170] found a difference in response to ovine CRF depending on the disease status with a lower cortisol response in patients with secondary progressive MS but a normal response in those with primary progressive MS. Cortisol secretion in patients with RA was also impaired in response to challenge with releasing hormones although in these patients the ACTH response was intact [171]. In contrast patients with newly diagnosed untreated RA showed no difference in plasma ACTH and cortisol to a similar challenge [146]. These authors note, however, that given the presence of inflammation the response might be considered inappropriately low. Together these data suggest an alteration in the responsiveness of the HPA axis associated with autoimmune disease. However, it appears that the mechanism involved is not simply a predisposition to autoimmune disease secondary to a defect in the HPA axis but a more complex interaction between the HPA axis and the disease process.

The effects of stress on disease

The role of stress on inducing disease and influencing the development and severity of disease is complex. Available evidence indicates that negative life events such as death of a spouse, divorce or separation can contribute to the onset or increase the severity of a variety of diseases including RA, insulin-dependent diabetes, Crohn's disease, uveitis, Grave's disease and other disorders [172; 173]. Many patients with arthritis believe psychological stress to exacerbate the severity of their disease [174]. Correlations between stress and both onset of disease and disease flares have also been shown [175]. This provides something of a paradox as the stress-induced increase in cortisol in these patients should act in an anti-inflammatory manner. At the very least, given the inability of patients with arthritis to mount a response to stress [165], one might predict no effect of stress on disease. Instead stress appears to increase disease activity. Numerous studies have been conducted in the rat and shown that a variety of stressful stimuli, e.g. restraint stress [176–178], tail shock and sound stress [179] and conditioned aversive stimulus [180] are able to suppress the development or severity of AA, collagen-induced arthritis and EAE. However, other studies have demonstrated an increase in severity and accelerated onset of both collagen-induced arthritis and EAE in response to noise stress [181], crowding stress [182] and exposure to predator stress [181]. These discrepancies may be due to differences in the stress paradigms employed in the various studies conducted in different laboratories. The effect may differ dependent on the type of stress, its duration and frequency; possibly these and the coping ability of the animals may all be of relevance. In one study the effects of footshock were dependent on the timing of the shocks relative to the onset of initial symptoms [183]. Similarly sound stress was found to only delay the onset of EAE while a more potent stressor (tail shock) reduced the severity [179]. In a recent study a number of predominantly psychological stressors were used in a rotation paradigm to determine the effects of this stress regime on collagen-induced arthritis [184]. This treatment reduced the severity of the disease during the stress regime but 2 days after ceasing stress there was no difference in gross inflammation suggesting a delay in onset. However, examination of the joints revealed the area of tibia invaded by stroma to be reduced in the stressed group confirming a reduction in the severity of the disease.

Less well established are the immune consequences of the stressors used and these will need to be determined before the mechanisms underlying the differences in the effects of stress on disease activity can be better understood. It is likely that the cytokine profile released in response to stress together with the timing and duration of the anti-inflammatory glucocorticoid release may be crucial. For instance, an acute injection of IL-1 will increase the severity of AA;

however, repeated injections of this cytokine have been reported to reduce the inflammation [185]. Treatment with IL-1 receptor antagonist (IL-1ra) has a protective effect in EAE when the animals are treated from day 9 onwards [186]. Gene transfer of IL-1ra directly into ankle joints also suppressed experimental arthritis [187]. These data suggest that treatments which inhibit IL-1 may be beneficial for the treatment of autoimmune/inflammatory disorders.

In AA, associated with inflammation, there is an increase in the splenic content of the immunoneuropeptides CRF, AVP, ACTH and β-endorphin. The increase in ACTH is evident as early as day 3, well before any evidence of inflammation (days 11–14), and prior to increases in CRF and AVP which occur at day 14 suggesting ACTH production in the spleen is independent of CRF and AVP [188]. There is evidence that CRF stimulation of immune POMC is not direct as at the pituitary but instead mediated via IL-1. CRF immunoreactivity and CRF mRNA have been noted in the synovium of patients with RA and animals with AA [189–190]. Elevated levels of CRF-binding protein have been determined in the blood of patients with RA and in septicaemia [191]. Of considerable interest was the finding that in synovial fluid the ligand binding to the CRF-binding protein eluted earlier than endogenous CRF on high performance liquid chromatography (HPLC) suggesting novel peptides in synovial fluid. The recent isolation and characterization of urocortin [103] which has some CRF-like properties may result in a novel direction of research and a novel therapeutic target.

Role of gonadal steroids in disease

The important role of gonadal steroids in autoimmune diseases are evident from the gender discrepancies reported for a variety of diseases. Females (mice, rats and humans), are more prone to autoimmune diseases than are males, e.g. autoimmune thyroiditis (19:1), SLE (9:1), and RA (4:1). The HPA axis may well have a role to play as stress inhibits the hypothalamo-pituitary-gonadal axis at multiple levels and glucocorticoids also inhibit the tissue effects of sex steroids. Sex steroids are also able to modulate the activity of the HPA axis reflecting the bi-directional nature of the gonadal and adrenal systems. The CRF gene contains an oestrogen-responsive element in the 5′ regulatory region suggesting CRF may be involved in the sexually dimorphic response to HPA activation; Females generally have a more active HPA axis response than males. Gonadal competence may also be a factor as young females are many times more susceptible to RA than young males. With increasing age the ratio is reduced. Under the age of 60 the ratio is about 5:1. With increasing age the female to male ratio approaches parity. These changes may reflect the decrease in testosterone associated with increasing age in males and it has been suggested that androgens may have a role in protecting males from developing autoimmune diseases in both humans and in animal models. Serum androgen concentrations are reduced in humans with RA and also in rats with AA [192–194]. Castration increases the incidence, time of onset and severity of SLE [195] streptococcal cell wall-induced arthritis and AA in males [193, 196]. These effects can be reversed by testosterone treatment in AA [193,197] (Fig. 2.6). Treatment with testosterone has also been shown to be effective in male patients [198]. Female NZB/NZW mice usually die following the development of SLE but this can be prevented by treatment with the androgen dihydrotestosterone [199]. Intact males have a greater chance of survival, but removal of testosterone results in increased mortality [200]. In general it appears that in males castration increases the severity of the disease which can be suppressed by androgen replacement.

The role of oestrogens in the female is complex and appears to depend on the type of disease [201]. In SLE oestrogen accelerates the progress of the disease and may be one of the most important contributors to the female preponderance of the disease. In contrast oestrogen suppresses EAE, collagen-induced arthritis and AA. One suggestion for this apparent discrepancy concerns the effects of oestrogen on B-cell and T-cell-mediated immune responses. Oestrogens enhance the antibody response (B-cell), while suppressing cell-mediated immunity (T-cell). EAE, collagen-induced arthritis and AA are all considered to be primarily T-cell-mediated disease models. Testosterone acts to suppress both B-cell and T-cell responses and hence suppresses disease severity.

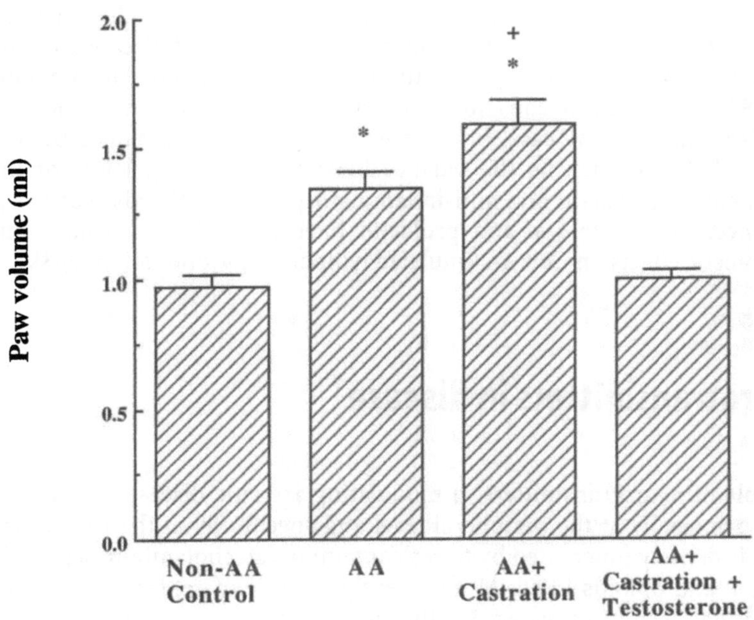

Fig. 2.6. Paw volume (ml) in control, arthritic (AA), arthritic animals that had been castrated (AA + castration) and AA + castration given testosterone replacement (AA + castration + testosterone). Paw volumes were measured 14 days after a single intradermal injection of ground, heat-killed *Mycobacterium butyricum*, in paraffin oil. Values are expressed as mean with a SEM ($n = 6$–8). *$P < 0.01$ compared with control, +$P < 0.05$ AA + castration compared with AA.

The site of action of the gonadal steroids and the mechanism(s) by which these actions are exerted is not fully established. The interactions between the HPA, gonadal and other neuroendocrine systems is complex. Whether the changes reflect the actions of the gonadal steroids themselves or metabolites is not known. Androgen and oestrogen receptors are found on lymphocytes and androgen receptors are also expressed by synovial cells suggesting the possibility of direct effects at the site of inflammation.

Further evidence for the important role of the gonadal steroids comes from observations that changes in activity of autoimmune diseases occur at times of change in gonadal axis function such as puberty, pregnancy, post-partum, menopause [202,203]. At these times the hormonal milieu is dramatically altered, with changes not only in the gonadal axis but also in other hormones such as prolactin and GH which themselves are altered in autoimmune diseases. A pro-inflammatory role for prolactin has been proposed in AA [204]. Removal of the pituitary prevents the development of AA in the rat. Implanting the pituitary under the kidney capsule, which results in hypersecretion of prolactin, or treatment with either exogenous prolactin or GH (but not any other pituitary hormones), reinstates the susceptibility [140,141]. Treatment with bromocriptine, which inhibits prolactin release, also prevents AA [205]. These data do not preclude an indirect effect of prolactin and there are a number of discrepancies in the literature. In RA and AA plasma prolactin concentrations are either not altered from controls [143,143] or have been shown to be increased [206]. There is evidence in AA that treatments which increase the severity of disease are associated with a decrease in circulating prolactin thus opposing the pro-inflammatory view [156,193]. Prolactin concentrations are elevated in SLE and other connective tissue disorders and in many cases have been shown to be accompanied by increased cortisol [207]. It may be the balance between circulating levels of the pro-inflammatory prolactin and anti-inflammatory glucocorticoids that has greater relevance. Corticosterone and prolactin have been shown to have mutually antagonistic effects on Con A-stimulated lymphocyte proliferation [208].

Neurotransmitters in disease

The role of neurotransmitters in modulating arthritic disease has been investigated principally at the peripheral level and specifically at the level of the joint which is densely innervated by fibres containing catecholamines, serotonin, substance P and opioids [209]. More recently the role of central neurotransmitter systems has been investigated. For the purposes of this review we will concentrate on catecholamines and serotonin. At the peripheral level adrenal medullectomy decreases the severity of AA [210,211], which can be reversed by treatment with adrenaline or $\beta 2$-agonists [211]. The roles of a number of catecholamine receptor agonists and antagonists have been investigated in this regard [210,212].

The catecholamines have generally been considered to play a stimulatory role at the hypothalamic level on HPA axis activation to acute stress. There have been relatively few studies on the role of catecholamines in chronic inflammatory stress/ autoimmune disease models. Depletion of endogenous catecholamines by neonatal treatment with 6-hydroxydopamine (6-OHDA) has been shown to increase the severity of disease in experimental autoimmune myasthenia gravis (EAMG) and EAE [213–215]. These treatments would have resulted in a global depletion of catecholamines producing major effects on peripheral sympathetic catecholaminergic systems in addition to central lesions. Irrespective of the mechanism, however, these data demonstrate a protective effect of catecholamine depletion. We noted that noradrenaline concentrations are doubled within the PVN of AA rats suggesting a possible role for endogenous noradrenaline associated with the development of AA [216]. To investigate this further we injected 6-OHDA directly over the PVN to lesion catecholamines locally in this area of AA rats. These lesions produced a 75% reduction in noradrenaline concentrations in these animals. Although we found no effect of this treatment on CRF mRNA levels or on plasma corticosterone concentrations there was a significant increase in the severity of the hind paw inflammation [216], supporting a protective role for central catecholamines and furthermore suggesting the intriguing possibility of treatments affecting central neurotransmitter systems altering the course of disease (Fig. 2.7). These effects do

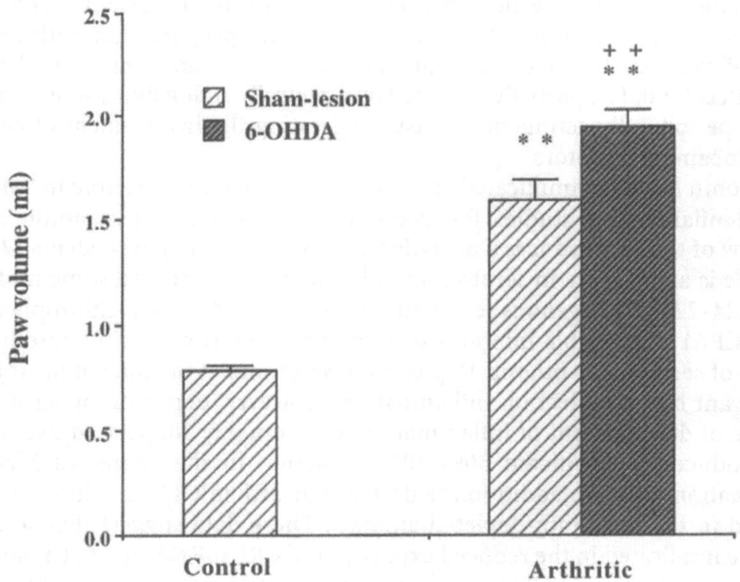

Fig. 2.7. Paw volume (ml) of sham-lesioned arthritic or non-arthritic control (light hatched bars), and 6-hydroxydopamine (6-OHDA) lesioned (dark hatched bars), arthritic rats measured 14 days after a single intradermal injection of ground, heat-killed *Mycobacterium butyricum* in paraffin oil. Bilateral infusion of 6-OHDA directly over the PVN to produce localized depletion of catecholamines took place 2 days prior to induction of arthritis and by day 14 there was a 75% decrease in noradrenaline in the PVN of the lesioned rats. Values are expressed as mean with a SEM ($n = 6$–8). $^{**}P < 0.01$ compared with non-arthritic controls, $^{++}P < 0.01$ compared with sham-lesioned, arthritic rats.

not appear to be mediated via the HPA axis although there remains a high inverse correlation between the levels of CRF mRNA in the PVN and the the severity of the disease. In a monoarthritic rat model tyrosine hydroxylase mRNA levels are increased in the pontine noradrenergic cell groups which project to the spinal cord, suggesting a possible involvement of spinal cord projections in this model [217]. Noradrenaline concentrations are decreased in the lumbar and sacral portions of the spinal cord in chronic relapsing EAE and are decreased with attacks in the cranio-thoracal spinal cord [218]. It has also been demonstrated that agonists directed against the β-adrenergic receptor alleviate EAE [219]. In contrast to these studies demonstrating an increase in severity following depletion of catecholamines, other workers have noted that both central and peripheral depletion of catecholamines can reduce the severity of EAE [220]. These latter data are not easily resolved and require further study. The mechanism by which these actions are mediated is unknown but a recent study demonstrated that immune depression induced by protein calorie malnutrition can be suppressed following central catecholaminergic lesions. These data suggest that central catecholaminergic hyperactivity is one of the mechanisms involved in immunodepression and provided evidence relating central catecholaminergic activity to an improvement of lymphocyte proliferation and IL-1 production [221]. While noradrenaline has received most attention the agents used, e.g. 6-OHDA, would have effects on other catecholamines. In this respect dopamine may also be important. Pergolide (a dopamine receptor agonist), has anti-inflammatory properties in both an acute model of carageenan-induced paw inflammation and in an arthritis model [222]. These effects could be partially inhibited by a centrally acting dopamine antagonist but not peripherally acting antagonists, suggesting the involvement of centrally acting dopamine receptors.

Serotonin has been implicated as having a pro-inflammatory role in a number of acute inflammatory studies. For example, direct injection of serotonin into the hind paw of the rat produces a dose-dependent increase in paw oedema [223]. In AA there is an increase in serotonin in plasma, spinal cord and some mid-brain areas [224–229]. Endogenous serotonin can be depleted using *p*-chlorophenylalanine (PCPA) a reversible inhibitor of tryptophan hydroxylase, the rate-limiting enzyme of serotonin synthesis. Depletion of serotonin at the time of the injection of adjuvant had no effect on inflammation. However, depletion of serotonin at the time of development of inflammation, i.e. from day 10, post-adjuvant injection produced a significant 30%–40% reduction in the degree of hind paw inflammation [230]. In addition the decrease in CRF mRNA seen in AA rats was reversed in the serotonin-depleted animals. These data suggest that serotonin might be implicated in the reduced expression of CRF mRNA in AA. Of particular interest is the possibility that serotonin antagonists may be of clinical benefit in reducing the severity of acute inflammation. However, we were unable to demonstrate any effect of a 5HT2 antagonist in AA in the PVG rat (unpublished observations) and others have demonstrated an increase in severity of streptococcal cell wall-induced arthritis in the Lewis rat treated with a 5HT2 antagonist [137]. Alterations in central 5HT2C receptor mRNA regulation have been noted in the CA1–CA3 hippocampal regions in a monoarthritic rat model suggesting that 5HT2C receptors may be involved in mediating central serotoninergic changes in arthritis [231].

The PCPA depletion paradigm, while reducing total body serotonin by more than 95%, tells us nothing of the site of action of serotonin. It may be acting directly at the level of the joint, it may be altering the immune response via actions on immune tissues, acting within the spine possibly involved in nociception or within the brain. To address the question of a central versus peripheral effect we went on to use 5′7′-dihydroxytryptamine, a toxin of serotonin neurons, injected into the lateral ventricle. These lesions produced a greater reduction in the severity of the disease than seen following total body depletion using PCPA suggesting that alterations in central serotonin concentrations may be more efficacious than alterations to peripheral systems. The mechanisms involved, the central nuclei mediating these effects and the receptor subtypes involved require further studies.

Summary

The interactions of the immune and neuroendocrine systems are likely to remain a major topic of research for many years to come. The complexity of these systems individually is confounding enough. The interactions between these systems, particularly in relation to chronic immune-mediated challenges, much more so. The cell types synthesizing cytokines in the nervous system are being characterized and the neuropeptides in the immune tissues are being determined. The responses of these to a wide variety of stimuli and under many different conditions are being investigated. We can now begin to unravel the mechanisms through which these factors act and begin to understand their role in determining susceptibility or resistance to disease or their ability to modulate the severity of disease.

The barriers which once existed between disciplines separating neuroendocrinologists, psychologists and immunologists are become increasingly blurred. The new discipline of psycho-neuro-endocrin-immunology is attracting increasing numbers of disciples and requires a shift in attitude with a move to a more holistic, less reductionist approach. Individuals need to be broad-minded and willing to explore the interfaces between these disciplines. Given the degree of specialization required for understanding in any one topic it is likely that we will see an increase in collaborative efforts to bridge the gaps in our individual knowledge. The goal of a better understanding of the mechanisms involved in disease processes and of our ability to develop novel therapeutic interventions to combat them is close at hand.

References

1. Harris G (1948) Neural control of the pituitary gland. Physiol Rev 28:139–179
2. Guillemin R, Rosenberg B (1955) Humoral hypothalamic control of anterior pituitary: a study with combined tissue cultures. Endocrinology 57:599–607

3. Saffran M, Schally AV (1955) In vitro bioassay of corticotropin: modification and statistical treatment. Endocrinology 56:523–532
4. Vale W, Spiess J, Rivier C, Rivier J (1981) Characterization of a 41-residue ovine hypothalamic peptide that stimulates secretion of corticotropin and β-endorphin. Science 213:1394–1397
5. Levin N, Blum M, Roberts JL (1989) Modulation of basal and corticotropin-releasing factor-stimulated pro-opiomelanocortin gene expression by vasopressin in rat anterior pituitary. Endocrinology 125:2957–2966
6. Whitnall MH, Mezey EHG (1985) Co-localization of corticotropin-releasing factor and vasopressin in median eminence neurosecretory vesicles. Nature 317:248–250
7. Whitnall MH, Smith D, Gainer H (1987) Vasopressin coexists in half of the corticotropin-releasing factor axons present in the external zone of the median eminence in normal rats. Neuroendocrinology 45:420–424
8. Erkut ZA, Hofman MA, Ravid R, Swaab DF (1995) Increased activity of hypothalamic corticotropin-releasing hormone neurons in multiple sclerosis. J Neuroimmunol 62:27–33
9. Gillies GE, Linton EA, Lowry PJ (1982) Corticotropin-releasing activity of the new CRF is potentiated several times by vasopressin. Nature 299:355–357
10. Turkelson CM, Thomas CR, Arimura A, Chang J, Chang JH, Shimiza M (1982) In vitro potentiation of the activity of synthetic-ovine corticotropin-releasing factor by arginine vasopressin. Peptides 1:111–115
11. Vale W, Vaughan J, Smith M, Yamamoto G, Rivier J, Rivier C (1983) Effects of synthetic ovine CRF, glucocorticoids, catecholamines, neurohypophysial peptides and other substances on cultured corticotropic cells. Endocrinology 113:1121–1131
12. Rivier C, Vale W (1983) Interaction of corticotropin-releasing factor and arginine vasopressin on adrenocorticotropin secretion in vivo. Endocrinology 113:939–942
13. Gibbs DM (1985) Measurement of hypothalamic corticotropin-releasing factors in hypophyseal portal blood. Fed Proc 44:203–206
14. Plotsky PM (1985) Hypophyseotropic regulation of adenohypophyseal adrenocorticotropin secretion. Fed Proc 44:207–213
15. Verbalis JG, Stricker EM, Robinson AG, Hoffman GE (1991) Cholecystokinin activates c-fos expression in hypothalamic oxytocin and corticotropin-releasing hormone neurons. J Neuroendocrinol 3:205–213
16. Pesce G, Guillaume V, Jesova D, Faudon M, Grino M, Oliver C (1990) Epinephrine in rat hypophysial portal blood is derived mainly from the adrenal medulla. Neuroendocrinol. 52:322–327
17. Hökfelt T, Fahrenkrug J, Tatemoto K, Mutt V, Werner S, Hulting A-L, Terenius L, Chang KJ (1983) The PHI (PHI-27)/corticotropin-releasing factor/enkephalin immunoreactive hypothalamic neuron: possible morphological basis for integrated control of prolactin, corticotropin, and growth hormone secretion. PNAS USA 80:858–898
18. Hisano S, Tsuruo Y, Katoh S, Daikoku S, Yanaiharan N, Shibasaki T (1987) Intragranular colocalization of arginine vasopressin and methionine-enkephalin-octapeptide in CRF-axons in the median eminence. Cell Tissue Res 249:497–507
19. Jessop DS, Chowdrey HS, Larsen PJ, Lightman SL (1992) Substance P: multifunctional peptide in the hypothalamo-pituitary system? J Endocrinol 132:331–337
20. Fink G, Dow RC, Casley D, Johnston CI, Bennie J, Carroll S, Dick H (1992) Atrial natriuretic peptide is involved in the ACTH response to stress and glucocorticoid feedback. J Endocrinol 135:37–43
21. Cunningham ET, Sawchenko PE (1988) Anatomical specificity of noradrenergic inputs to the paraventricular and supraoptic nuclei of the rat hypothalamus. J Comp Neurol 274:60–76

22. Liposits Z, Phelix C, Paull WK (1987) Synaptic interaction of serotoninergic axons and corticotropin-releasing factor (CRF) synthesizing neurons in the hypothalamic paraventricular nucleus of the rat. Histochemistry 86:541–549
23. Larsen PJ, Hay-Schmidt A, Vrang N, Mikkelsen JD (1996) Origin of projections from the mid-brain raphe nuclei to the hypothalamic paraventricular nucleus in the rat: a combined retrograde and anterograde tracing study. Neuroscience 70:963–988
24. Imaki T, Shibasaki T, Hotta M, Demura H (1992) Early induction of c-fos precedes increased expression of corticotropin-releasing factor messenger ribonucleic acid in the paraventricular nucleus after immobilization stress. Endocrinology 131:240–246
25. Kononen J, Honkaniemi J, Alho H, Koistinaho J, Iadrola M, Pelto-Huikko M (1992) Fos-like immunoreactivity in the rat hypothalamic-pituitary axis after immobilization stress. Endocrinology 130:3041–3047
26. Harbuz MS, Chalmers J, De Souza L, Lightman SL (1993) Stress-induced activation of CRF and c-fos mRNAs in the paraventricular nucleus are not affected by serotonin depletion. Brain Res 609:167–173
27. Kjaer A, Larsen PJ, Knigge U, Moller M, Warberg J (1994) Histamine stimulates c-fos expression in hypothalamic vasopressin-, oxytocin-, and corticotropin-releasing hormone-containing neurons. Endocrinology 134:482–491
28. Senba E, Umemoto S, Kawai Y, Noguchi K (1994) Differential expression of fos family and jun family mRNAs in the rat hypothalamo-pituitary-adrenal axis after immobilization stress. Mol Brain Res 24:283–294
29. Helmreich DL, Cullinan WE, Watson SJ (1996) The effect of adrenalectomy on stress-induced c-fos mRNA expression in the rat brain. Brain Res 706:137–144
30. Ceccatelli S, Villar MJ, Goldstein M, Hökfelt T (1989) Expression of c-fos immunoreactivity in transmitter-characterized neurons after stress. PNAS USA 86:9569–9573
31. Covenas R, de Leon M, Cintra A, Bjelke B, Gustafsson J-A, Fuxe K (1993) Coexistence of c-fos and glucocorticoid receptor immunoreactivities in the CRF immunoreactive neurons of the paraventricular hypothalamic nucleus of the rat after acute immobilization stress. Neurosci Lett 149:149–152
32. Ma X-M, Levy A, Lightman SL (1997) Rapid changes of heteronuclear RNA for corticotropin-releasing hormone and arginine vasopressin in response to acute stress. J Endocrinol 152:81–89
33. Pezzone MA, Lee W-S, Hoffman GE, Pezzone KM, Rabin BS (1993) Activation of brainstem catecholaminergic neurons by conditioned and unconditioned aversive stimuli as revealed by c-fos immunoreactivity. Brain Res 608:310–318
34. Schreiber SS, Tocco G, Shors TJ, Thompson RF (1991) Activation of immediate early genes after acute stress. NeuroReport 2:17–20
35. Matta SG, Foster CA, Sharp BM (1993) Nicotine stimulates the expression of cfos protein in the parvocellular paraventricular nucleus and brainstem catecholaminergic regions. Endocrinology 132:2149–2156
36. Jingami H, Matsukura S, Numa S, Imura H (1985) Effects of adrenalectomy and dexamethasone administration on the level of prepro-corticotropin-releasing factor messenger ribonucleic acid (mRNA) in the hypothalamus and adrenocorticotropin/B-lipotropin precursor mRNA in the pituitary in rats. Endocrinology 117:1314–1320
37. Bugnon CD, Fellman D, Gouget A (1983) Changes in corticoliberin and vasopressin-like immunoreactivities in the zona externa of the median eminence in adrenalectomized rats. Immunocytochemical study. Neurosci Lett 37:43–49
38. Merchenthaler I, Vigh S, Petrusz P, Schally AV (1983) The paraventriculo-infundibular corticotropin-releasing factor (CRF) pathway as revealed by immunocytochemistry in long-term hypophysectomized or adrenalectomized rats. Regul Pept 5:295–305

39. Paull WK, Gibbs FP (1983) The corticotropin-releasing factor (CRF) neurosecretory system in intact, adrenalectomized, and adrenalectomized-dexamethasone dexamethasone treated rats. Histochemistry 78:303–316
40. Suda T, Tomori N, Tozawa F, Mouri T, Demura H, Shizume K (1983) Effects of bilateral adrenalectomy on immunoreactive corticotropin-releasing factor in the rat median eminence and intermediate-posterior pituitary. Endocrinology 113:1182–1184
41. Swanson LW, P.E. S, Rivier JE, Vale WW (1983) The organization of ovine corticotropin-releasing factor immunoreactive cells and fibers in the rat brain: An immunohistochemical study. Neuroendocrinology 36:165–186
42. Tramu G, Croix C, Pillez A (1983) Ability of the CRF immunoreactive neurons of the paraventricular nucleus to produce a vasopressin-like material. Neuroendocrinology 37:467–469
43. Kiss JZ, Mezey E, Skirboll L (1984) Corticotropin-releasing factor-immunoreactive neurons become vasopressin positive after adrenalectomy. PNAS USA 81:1854–1858
44. Plotsky PM (1987) Regulation of hypophysiotropic factors mediating ACTH secretion. Ann NY Acad Sci 512:205–217
45. Young WS III, Mezey E, Siegel RE (1986) Quantitative in situ hybridization histochemistry reveals increased levels of corticotropin-releasing factor mRNA after adrenalectomy in rats. Neurosci Lett 70:198–203
46. Beyer HS, Matta SG, Sharp BM (1988) Regulation of the messenger ribonucleic acid for corticotropin-releasing factor in the paraventricular nucleus and other brain sites of the rat. Endocrinology 123:2117–2123
47. Lightman SL, Young WS III (1989) Influence of steroids on the hypothalamic corticotropin-releasing factor and preproenkephalin mRNA responses to stress. PNAS USA 86:4306–4310
48. Swanson LW, Simmons DM (1989) Differential steroid hormone and neural influences on peptide mRNA levels in CRH cells of the paraventricular nucleus: a hybridization histochemical study in the rat. J Comp Neurol 285:413–435
49. Wolfson B, Manning RW, Davis LG, Arentzen R, Baldrino FJ (1985) Co-localization of corticotropin releasing factor and vasopressin mRNA in neurones after adrenalectomy. Nature 315:59–61
50. Young WS III, Mezey E, Siegel RE (1986) Vasopressin and oxytocin mRNAs in adrenalectomized and Brattleboro rats: analysis by quantitative in situ hybridization histochemistry. Mol Brain Res 1:231–241
51. Whitnall MH, Key S, Gainer H (1987) Vasopressin-containing and vasopressin-deficient subpopulations of corticotropin-releasing factor axons are differentially affected by adrenalectomy. Endocrinology 120:2180–2182
52. Whitnall MH (1988) Distributions of pro-vasopressin expressing and pro-vasopressin deficient CRH neurons in the paraventricular hypothalamic nucleus of colchicine-treated normal and adrenalectomized rats. J Comp Neurol 275:13–28
53. Kovacs KJ, Mezey E (1987) Dexamethasone inhibits corticotropin-releasing factor gene expression in the rat paraventricular nucleus. Neuroendocrinology 46:365–368
54. Harbuz MS, Lightman SL (1989) Glucocorticoid inhibition of stress-induced changes in hypothalamic corticotrophin-releasing factor messenger RNA and proenkephalin A messenger RNA. Neuropeptides 14:17–20
55. Sawchenko PE (1987) Evidence for a local site of action for glucocorticoids in inhibiting CRF and vasopressin expression in the paraventricular nucleus. Brain Res 403:213–224
56. Kovacs KJ, Makara GB (1988) Corticosterone and dexamethasone act at different brain sites to inhibit adrenalectomy-induced adrenocorticotropin hypersecretion. Brain Res 474:205–210
57. Hu S-B, Tannahill LA, Biswas S, Lightman SL (1992) Release of corticotropin-releasing factor-41, arginine vasopressin and oxytocin from rat foetal hypothalamic cells in

culture: response to activation of intracellular second messengers and to corticosteroids. J Endocrinol 132:57–65
58. Herman JP, Wiegand SJ, Watson SJ (1990) Regulation of basal corticotropin-releasing hormone and arginine vasopressin messenger ribonucleic acid expression in the paraventricular nucleus: effects of selective hypothalamic deafferentations. Endocrinology 127:2408–2417
59. Selye H (1936) A syndrome produced by diverse nocuous agents. Nature 138:32
60. Harbuz MS, Lightman SL (1992) Stress and the hypothalamo-pituitary-adrenal axis: acute, chronic and immunological activation. J Endocrinol 134:327–339
61. Romero LM, Sapolsky RM (1996) Patterns of ACTH secretagog secretion in response to psychological stimuli. J Neuroendocrinol 8:243–258
62. Lightman SL, III. YWS (1988) Corticotrophin-releasing factor, vasopressin and pro-opiomelanocortin mRNA responses to stress and opiates in the rat. J Physiol 403:511–523
63. Harbuz MS, Lightman SL (1989) Responses of hypothalamic and pituitary mRNA to physical and psychological stress in the rat. J Endocrinol 122:705–711
64. Lightman SL, Young WSI (1987) Changes in hypothalamic preproenkephalin A mRNA following stress and opiate withdrawal. Nature 328:643–645
65. Harbuz MS, Chowdrey HS, Jessop DS, Biswas S, Lightman SL (1991) Role of catecholamines in mediating messenger RNA and hormonal responses to stress. Brain Res 551:52–57
66. Harbuz MS, Russell JA, Sumner BEH, Kawata M, Lightman SL (1991) Rapid changes in the content of proenkephalin A and corticotrophin-releasing hormone mRNAs in the paraventricular nucleus during morphine withdrawal in urethane-anaesthetized rats. Mol Brain Res 9:285–291
67. Watts AG (1991) Ether anesthesia differentially affects the content of preprocorticotropin-releasing hormone, prepro-neurotensin/neuromedin N and preproenkephalin mRNAs in the hypothalamic paraventricular nucleus of the rat. Brain Res 544:353–357
68. Linton EA, Tilders FJH, Hodgkinson S, Berkenbosch F, Vermes I, Lowry PJ (1985) Stress-induced secretion of adrenocorticotropin in rats is inhibited by administration of antisera to ovine corticotropin-releasing factor and vasopressin. Endocrinology 116:966–970
69. Harbuz MS, Nicholson SA, Gillham B, Lightman SL (1990) Stress responsiveness of hypothalamic corticotrophin-releasing factor and pituitary proopiomelanocortin mRNAs following high-dose glucocorticoid treatment and withdrawal in the rat. J Endocrinol 127:407–415
70. Dallman MF, Jones MT (1973) Corticosteroid feedback control of ACTH secretion: effects of stress-induced corticosterone secretion on subsequent stress responses in the rat. Endocrinology 92:1367–1375
71. Schmidt ED, Binnekade R, Janszen AWJW, Tilders FJH (1996) Short stressor induced long-lasting increases of vasopressin stores in hypothalamic corticotropin-releasing hormone (CRH) neurons in adult rats. J Neuroendocrinol 8:703–712
72. Schmidt ED, Janszen AWJW, Wouterlood F, Tilders FJH (1995) Interleukin-1-induced long-lasting changes in hypothalamic corticotropin-releasing hormone (CRH)-neurons and hyperresponsiveness of the hypothalomo-pituitary-adrenal axis. J Neurosci 15:7417–7426
73. Kant GJ, Eggleston T, Landman-Roberts L, Kenion CC, Driver GC, Meyerhoff JL (1985) Habituation to repeated stress is stressor specific. Pharmacol Biochem Behav 22:631–634
74. Hashimoto K, Suemaru S, Takao T, Sugarwara M, Makino S, Ota S (1988) Corticotropin-releasing hormone and pituitary-adrenocortical responses in chronically stressed rats. Regul Pept 23:117–126

75. Spencer RL, McEwen BS (1990) Adaptation of the hypothalamo-pituitary-adrenal axis to chronic ethanol stress. Neuroendocrinology 52:481–489
76. Daniels-Severs A, Goodwin A, Keil LC, Vernikos-Danellis J (1973) Effect of chronic crowding and cold on the pituitary-adrenal system: responsiveness to an acute stimulus during chronic stress. Pharmacology 9:348–356
77. Sakellaris PC, Vernikos-Danellis J (1975) Increased rate of response of the pituitary-adrenal system in rats adapted to chronic stress. Endocrinology 97:597–602
78. Vernikos J, Dallman MF, Bonner C, Katzen A, Shinsako J (1982) Pituitary-adrenal function in rats chronically exposed to cold. Endocrinology 110:413–420
79. Scribner KA, Walker C-D, Cascio CS, Dallman MF (1991) Chronic streptozotocin diabetes in rats facilitates the acute stress response without altering pituitary or adrenal responsiveness to secretagogues. Endocrinology 129:99–108
80. Scaccianoce S, Muscolo LAA, Cigliana G, Navarra D, Nicolai R, Angelucci L (1991) Evidence for a specific role of vasopressin in sustaining pituitary-adrenocortical stress response in the rat. Endocrinology 128:3138–3143
81. De Goeij DCE, Kvetnansky R, Whitnall MH, Jesova D, Berkenbosch F, Tilders FJH (1991) Repeated stress-induced activation of corticotropin-releasing factor neurons enhances vasopressin stores and colocalization with corticotropin-releasing factor in the median eminence of rats. Neuroendocrinology 53:150–159
82. Hauger RL, Lorang M, Irwin M, Aguilera G (1990) CRF receptor regulation and sensitization of ACTH responses to acute ether stress during chronic intermittent immobilization stress. Brain Res 532:34–40
83. Khansari DN, Murgo AJ, Faith RE (1990) Effects of stress on the immune system. Immunol Today 11:170–175
84. Bertini R, Bianchi M, Ghezzi P (1988) Adrenalectomy sensitizes mice to the lethal effects of interleukin 1 and tumour necrosis factor. J Exp Med 167:1708–1712
85. Riley V (1981) Psychoneuroendocrine influences on immuno-competence and neoplasia. Science 212:100–1109
86. Keller SE, Weiss JM, Schleiffer SJ, Miller NE, Stein M (1983) Stress-induced suppression of immunity in adrenalectomized rats. Science 221:1301–1304
87. Jankovic BD (1989) Neuroimmunomodulation: facts and dilemmas. Immunol Lett 21:101–118
88. Meickle WA, Dorchuck RW, Araneo BA, Stringham JD, Evans TG, Spruance SL, Daynes RA (1991) The presence of a dehydroepiandrosterone-specific receptor binding complex in murine T cells. J Steroid Biochem Molec Biol 42:293–304
89. Blauer KL, Poth M, Rogers WM, Bernton EW (1991) Dehydroepiandrosterone antagonizes the suppressive effects of dexamethasone on lymphocyte proliferation. Endocrinology 129:3174–3179
90. Weksler ME (1993) Immune senesence and adrenal steroids-immune dysregulation and the action of dehydroepiandrosterone (DHEA) in old animals. Eur J Clin Pharmacol 45:21–23
91. Watson RR, Huls A, Araghiniuam M, Chung SB (1996) Dehydroepiandrosterone and diseases of aging. Drugs Aging 9:274–291
92. Khorram O, Vu L, Yen SSC (1997) Activation of immune function by dehydroepiandrosterone (DHEA) in age-advanced men. J Gerontol (A) 52:M1–M7
93. Blalock JE (1994) The syntax of immune-neuroendocrine communication. Immunol Today 15:504–511
94. Labeur MS, Arzt E, Wiegers GJ, Holsboer F, Reul JMHM (1995) Long-term intracerebroventricular corticotropin-releasing hormone administration induces distinct changes in rat splenocyte activation and cytokine expression. Endocrinology 136:2678–2688
95. Meyer WJI III, Smith EM, Richards GE, Cavallo A, Morrill AC, Blalock JE (1987) In vivo immunoreactive adrenocorticotropin (ACTH) production by human mononu-

clear leukocytes from normal and ACTH-deficient individuals. J Clin Endocrinol Metab 64:98–105
96. Jain R, Zwickler D, Hollander CS, Brand H, Saperstein A, Hutchinson B, Brown C, Audhya T (1991) Corticotropin-releasing factor modulates the immune response to stress in the rat. Endocrinology 128:1329–1336
97. Serda SM, Wei ET (1992) Epinephrine-induced pulmonary oedema in rats is inhibited by corticotropin-releasing factor. Pharmacol Res 26:85–91
98. Wei ET, Thomas HA (1994) Correlation of neuroendocrine and anti-edema activities of alanine-corticotropin-releasing factor analogs. Eur J Pharmacol 263:319–321
99. Berkenbosch F, Wolvers DAW, Derijk R (1991) Neuroendocrine and immunological mechanisms in stress-induced immunomodulation. J Steroid Biochem Mol Biol 40:639–647
100. Sundar SK, Cierpial MA, Kilts C, Ritchie JC, Weiss JM (1990) Brain IL-1-induced immunosuppression occurs through activation of both pituitary-adrenal axis and sympathetic nervous system by corticotropin-releasing factor. J Neurosci 10:3701–3706
101. Karalis K, Sano H, Redwine J, Listwak S, Wilder RL, G.P. C (1991) Autocrine or paracrine inflammatory actions of corticotropin-releasing hormone in vivo. Science 254:421–423
102. Stephanou A, Jessop DS, Knight RA, Lightman SL (1990) Corticotrophin-releasing factor-like immunoreactivity and mRNA in human leukocytes. Brain Behav Immun 4:67–73
103. Vaughan J, Donaldson C, Bittencourt J, Perrin MH, Lewis K, Sutton S, Chan R, Turnbull AV, Lovejoy D, Rivier C, Rivier J, Sawchenko PE, Vale W (1995) Urocortin, a mammalian neuropeptide related to fish urocortin I and to corticotrophin-releasing factor. Nature 378:287–292
104. Behan DP, Khongsaly O, Ling N, De Souza EB (1996) Urocortin interaction with corticotropin-releasing factor (CRF) binding protein (CRF-BP): a novel mechanism for elevating 'free' CRF levels in human brain. Brain Res 725:263–267
105. Webster EL, De Souza ED (1988) Corticotropin-releasing factor receptors in mouse spleen: identification, autoradiographic localization and regulation by divalent cations and guanine nucleotides. Endocrinology 122:609–617
106. Weidenfeld J, Abramsky O, Ovadia H (1989) Effect of interleukin-1 on ACTH and corticosterone secretion in dexamethasone and adrenalectomized pretreated male rats. Neuroendocrinology 50:650–654
107. Berkenbosch F, Van Oers J, Del Rey A, Tilders F, Besedovsky H (1987) Corticotropin-releasing factor-producing neurons in the rat activated by interleukin-1. Science 238:524–526
108. Sapolsky R, Rivier C, Yamamoto G, Plotsky P, Vale W (1987) Interleukin-1 stimulates the secretion of hypothalamic corticotropin-releasing factor. Science 238:522–524
109. Watanabe T, Morimoto A, Sakata Y, Murakami N (1990) ACTH response induced by interleukin-1 is mediated by CRF secretion stimulated by hypothalamic PGE. Experientia 46:481–484
110. Berkenbosch F, De Goeij DEC, Del Rey A, Besedovsky HO (1989) Neuroendocrine, sympathetic and metabolic responses induced by interleukin-1. Neuroendocrinology 50:570–576
111. Suda T, Tozawa F, Ushiyama T, Sumitomo T, Yamada M, Demura H (1990) Interleukin-1 stimulates corticotropin-releasing factor gene expression in rat hypothalamus. Endocrinology 126:1223–1228
112. Cambronero JC, Borrell J, Guaza C (1989) Glucocorticoids modulate rat hypothalamic corticotrophin-releasing factor release induced by interleukin-1. J Neurosci Res 24:470–476

113. Tsagarakis S, Gillies G, Rees LH, Besser M, Grossman A (1989) Interleukin-1 directly stimulates the release of corticotrophin releasing factor from rat hypothalamus. Neuroendocrinology 49:98–101
114. Navarra P, Tsagarakis S, Faria MS, Rees LH, Besser GM, Grossman AB (1991) Interleukins-1 and -6 stimulate the release of corticotropin-releasing hormone-41 from rat hypothalamus in vitro via the eicosanoid cyclooxygenase pathway. Endocrinology 128:37–44
115. Chover-Gonzalez AJ, Harbuz MS, Lightman SL (1993) Effect of adrenalectomy and stress on interleukin-1β mediated activation of hypothalamic corticotropin releasing factor mRNA. J Neuroimmunol 42:155–160
116. Ju G, Zhang X, Jin B-Q, Huang CS (1991) Activation of corticotropin-releasing factor-containing neurons in the paraventricular nucleus of the hypothalamus by interleukin-1 in the rat. Neurosci Lett 132:151–154
117. Wan W, Janz L, Vriend CY, Sorensen CM, Greenberg AH, Nance DW (1993) Differential induction of c-fos immunoreactivity in hypothalamus and brain stem nuclei following central and peripheral administration of endotoxin. Brain Res Bull 32:581–587
118. Elmquist JK, Ackermann MR, Register KB, Rimler RB, Ross LR, Jacobson CD (1993) Induction of fos-like immunoreactivity in the rat brain following *Pasteurella multicida* endotoxin administration. Endocrinology 133:3054–3057
119. Hare AS, Clarke G, Tolchard S (1995) Bacterial lipopolysaccharide-induced changes in fos protein expression in the rat brain:correlation with thermoregulatory changes and plasma corticosterone. J Neuroendocrinol 7:791–799
120. Day HEW, Akil H (1996) Differential pattern of c-fos mRNA in rat brain following central and systemic administration of interleukin-1beta: implications for mechanism of action. Neuroendocrinology 63:207–218
121. Cunningham ET, DeSouza E (1994) Interleukin-1 receptors in the brain and endocrine tissues. Immunol Today 14:161–176
122. Ericsson A, Liu C, Hart RP, Sawchenko PE (1995) Type I interleukin-1 receptor in the rat brain: distribution, regulation, and relationship to sites of IL-1-induced cellular activation. J Comp Neurol 361:681–698
123. Plotkin SR, Banks WA, Kastin AJ (1996) Comparison of saturable transport and extracellular pathways in the passage of interleukin-1a across the blood–brain barrier. J Neuroimmunol 67:41–47
124. Banks WA, Kastin AJ, Broadwell RD (1995) Passage of cytokines across the blood–brain barrier. Neuroimmunomodulation 2:241–248
125. Burrought M, Cabellos C, Prasad S, Tuomanen E (1992) Bacterial components and the pathophysiology of injury to the blood–brain barrier:does cell wall add to the effects of endotoxin in gram-negative meningitis. J Infect Dis 165 (Suppl 1):S82–S85
126. deVries HE, Blom-Roosmalen MCM, vanOosten M, deBoer AG, vanBerkel TJC, Briemer DD, Kuiper J (1996) The influence of cytokines on the integrity of the blood-brain barrier in vitro. J Neuroimmunol 64:37–43
127. Fleshner M, Goehler LE, Hermann J, Relton JK, Maier SF, Watkins LR (1995) Interleukin-1β induced corticosterone elevation and hypothalamic NE depletion is vagally mediated. Brain Res Bull 37:605–610
128. Gaykema RPA, Dijkstra I, Tilders FJH (1995) Subdiaphragmatic vagotomy suppresses endotoxin-induced activation of hypothalamic corticotropin-releasing hormone neurons and ACTH secretion. Endocrinology 136:4717–4720
129. Kapcala LP, He JR, Gao Y, Pieper JO, DeTolla LJ (1996) Subdiaphagmatic vagotomy inhibits intra-abdominal interleukin-1β stimulation of adrenocorticotropin secretion. Brain Res 728:247–254
130. Hori T, Katafuchi T, Take S, Shimizu N, Niijima A (1995) The autonomic nervous system as a communication channel between the brain and the immune system. Neuroimmunomodulation 2:203–215

131. Naito Y, Fukata J, Nakaishi S, Nakai Y, Tamai S, Mori K, Imura H (1990) Chronic effects of interleukin-1 on hypothalamus, pituitary and adrenal glands in rat. Neuroendocrinology 51:637–641
132. Mengozzi M, Ghezzi P (1991) Defective tolerance to the toxic and metabolic effects of interleukin-1. Endocrinology 128:1668–1672
133. Sweep CGJ, Van Der Meer MJM, Hermus ARMM, Smals AGH, Van Der Meer JWM, Pesman GJ, Willemsen SJ, Benraad TJ, Kloppenborg PWC (1992) Chronic stimulation of the pituitary-adrenal axis in rats by interleukin-1β infusion: in vivo and in vitro studies. Endocrinology 130:1153–1164
134. Hadid R, Spinedi E, Daneva T, Grau G, Gaillard RC (1995) Repeated endotoxin treatment decreases immune and hypothalamo-pituitary-adrenal axis responses: effects of orchidectomy and testosterone therapy. Neuroendocrinology 62:348–355
135. Hadid R, Spinedi E, Giovambattista A, Chautrad T, Gaillard RC (1996) Decreased hypothalamo-pituitary-adrenal axis response to neuroendocrine challenge under repeated endotoxemia. Neuroimmunomodulation 3:62–68
136. Panayi GS (1992) Neuroendocrine modulation of disease expression in rheumatoid arthritis. Eular Congr Rep 2:2–12
137. Sternberg EM, Hill JM, Chrousos GP, Kamilaris T, Listwak SJ, Gold PW, Wilder RL (1989) Inflammatory mediator-induced hypothalamic-pituitary-adrenal axis activation is defective in streptococcal cell wall arthritis susceptible Lewis rats. PNAS USA 86:2374–2378
138. Mason D, MacPhee I, Antoni F (1990) The role of the neuroendocrine system in determining genetic susceptibility to experimental encephalomyelitis in the rat. Immunology 70:1–5
139. Harbuz MS, Rees RG, Lightman SL (1993) Hypothalamo-pituitary responses to acute stress and changes in circulating glucocorticoids during chronic adjuvant-induced arthritis in the rat. Am J Physiol 264:R179–R185
140. Nagy E, Berczi I, Friesen HG (1983) Regulation of immunity in rats by lactogenic and growth hormones. Acta Endocrinol 102:351–357
141. Berczi I, Nagy E, Asa SL, Kovacs K (1984) The influence of pituitary hormones on adjuvant arthritis. Arthritis Rheum 27:682–688
142. French RA, Zachary JF, Dantzer R, Frawley LS, Chizzonite R, Parnet P, Kelley KW (1996) Dual expression of p80 type I and p68 type II interleukin-1 receptors on anterior pituitary cells synthesizing GH. Endocrinology 137:4027–4036
143. Calvino B, Besson J-M, Mounier F, Kordon C, Bluet-Pajot M-T (1992) Chronic pain induces a paradoxical increase in growth hormone secretion without affecting other hormones related to acute stress in the rat. Pain 49:27–32
144. Bluet-Pajot MT, Mournier F, Slama A, Videau C, Kordon C, Epelbaum J, Calvino B (1996) The increase in growth-hormone secretion in experimentally induced arthritic rats is an adaptive process involved in the regulation of inflammation. Neuroendocrinology 63:85–92
145. Neidhart M, Fluckiger EW (1992) Hyperprolactinaemia in hypophysectomized or intact male rats and the development of adjuvant arthritis. Immunology 77:449–455
146. Templ E, Koeller M, Riedl M, Wagner O, Graninger W, Luger ANA (1996) Anterior pituitary function in patients with newly-diagnosed rheumatoid arthritis. Br J Rheumatol 35:350–356
147. Wei ET, Wong JC, Kiang JG (1990) Decreased inflammatory responsiveness of hypophysectomized rats to heat is reversed by a corticotropin-releasing factor (CRF) antagonist. Regul Pept 27:317–323
148. Harbuz MS, Chowdrey HS, Lightman SL, Wei ET, Jessop DS (1996) An investigation into the effects of chronic infusion of corticotrophin-releasing factor on hind paw inflammation in adjuvant-induced arthritis. Stress 1:105–111

149. Harbuz MS, Rees RG, Eckland D, Jessop DS, Brewerton D, Lightman SL (1992) Paradoxical responses of hypothalamic CRF mRNA and CRF-41 peptide and adenohypophyseal POMC mRNA during chronic inflammatory stress. Endocrinology 130:1394–1400
150. Harkness JAL, Richter MB, Panayi GS, Van de Pette K, Unger A, Pownall R (1982) Circadian variation in disease activity in reheumatoid arthritis. Br Med J 284:551–554
151. Kirkham BW, Panayi GS (1989) Diurnal periodicity of cortisol secretion, immune reactivity and disease activity in rheumatoid arthritis: implications for steroid treatment. Br J Rheumatol 28:154–157
152. Masera RG, Carignola R, Staurenghi AH, Sartori ML, Lazzero A, Griot G, Angeli A (1994) Altered circadian-rhythms of natural-killer (NK) cell-activity in patients with autoimmune rheumatic diseases. Chronobiologica 21:127–132
153. Neeck G, Federlin K, Graef V, Rusch D, Schmidt KL (1990) Adrenal secretion of cortisol in patients with rheumatoid arthritis. J Rheumatol 17:24–29
154. Persellin RH, Kittinger GW, Kendall JW (1972) Adrenal response to experimental arthritis in the rat. Am J Physiol 222:1545–1549
155. Sarlis NJ, Chowdrey HS, Stephanou A, Lightman SL (1992) Chronic activation of the hypothalamo-pituitary-adrenal axis and loss of circadian rhythm during adjuvant-induced arthritis in the rat. Endocrinology 130:1775–1779
156. Stephanou A, Sarlis NJ, Knight RA, Lightman SL, Chowdrey HS (1992) Glucocorticoid mediated responses of plasma ACTH and anterior pituitary proopiomelanocortin, growth hormone and prolactin mRNAs during adjuvant-induced arthritis in the rat. J Mol Endocrinol 9:273–281
157. Brady LS, Page SW, Thomas FS, Rader JL, Lynn AB, Misiewicz-Poltorak B, Zelazowski E, Crofford LJ, Zelazowski P, Smith C, Raybourne RB, Love LA, Gold PW, Sternberg EM (1994) 1′1-Ethylidenebis[L-tryptophan], a contaminant implicated in L-tryptophan eosinophilia myalgia syndrome, suppresses mRNA expression of hypothalamic corticotropin-releasing hormone in Lewis (LEW/N) rat brain. Neuroimmunomodulation 1:59–65
158. Chowdrey HS, Larsen PJ, Harbuz MS, Jessop DS, Aguilera G, Eckland DJA, Lightman SL (1995) Evidence for arginine vasopressin as the primary activator of the HPA axis during adjuvant-induced arthritis. Br J Pharmacol 116:2417–2424
159. Harbuz MS, Leonard JP, Lightman SL, Cuzner ML (1993) Changes in hypothalamic corticotrophin releasing factor (CRF) and pituitary proopiomelanocortin (POMC) messenger RNA during the course of experimental allergic encephalomyelitis (EAE). J Neuroimmunol 45:127–132
160. Michelson D, Stone L, Galliven E, Magiakou MA, Chrousos GP, Sternberg EM, Gold PW (1994) Multiple sclerosis is associated with alterations in hypothalamic-pituitary-adrenal axis function. J Clin Endocrinol Metab 79:848–853
161. Reder AT, Mackowiec RL, Lowy MT (1994) Adrenal size is increased in multiple sclerosis. Arch Neurol 51:151–154
162. Crofford LJ, Rader JI, Dalakas MC, Hill RH, Page SW, Needham LL, Brady LS, Heyes MP, Wilder RL, Gold PW, Illa I, Smith C, Sternberg EM (1990) L-tryptophan implicated in human eosinophilia-myalgia syndrome causes fasciitis and perimyositis in the Lewis rat. J Clin Invest 86:1757–1763
163. Harbuz MS, Jessop DS, Chowdrey HS, Blackwell JM, Larsen PJ, Lightman SL (1995) Evidence for altered control of hypothalamic CRF in immune-mediated diseases. Ann N Y Acad Sci 771:449–458
164. Aguilera G, Jessop DS, Harbuz MS, Kiss A, Lightman SL (1997) Biphasic regulation of hypothalamic-pituitary corticotropin releasing hormone receptors during development of adjuvant-induced arthritis in the rat. J Endocrinol 153:185–191

165. Chikanza IC, Petrou P, Chrousos GP, Kingsley G, Panayi GS (1992) Defective hypothalamic response to immune/inflammatory stimuli in patients with rheumatoid arthritis. Arthritis Rheum 35:1281–1288
166. Mason D (1991) Genetic variation in the stress response; susceptibility to experimental allergic encephalomyelitis and implications for human inflammatory disease. Immunol Today 12:57–60
167. Sternberg EM, Young WS, Bernardini R, Calogero AE, Chrousos GP, Gold PW, Wilder RL (1989) A central nervous system defect in biosynthesis of corticotropin-releasing hormone is associated with the susceptibility to streptococcal cell wall-induced arthritis in Lewis rats. PNAS USA 86:4771–4775
168. Reincke M, Heppner C, Petske F, Allolio B, Arlt W, Mbulamberi D, Siekmann L, Vollmer D, Winkelmann W, Chrousos GP (1994) Impairment of adrenocortical function associated with increased plasma tumour necrosis factor-alpha and interleukin-6 concentrations in African trypanosomiasis. Neuroimmunomodulation 1:14–22
169. Hu Y, Dietrich H, Herold M, Heinrich PC, Wick G (1993) Disturbed immuno-endocrine communication via the hypothalamo-pituitary-adrenal axis in autoimmue disease. Int Arch Allergy Immunol 102:232–241
170. Wei TCM, Lightman SL (1997) The neuroendocrine axis in mulitple sclerosis. Brain (in press)
171. Gudbjornsson B, Skogseid B, Oberg K, Wide L, Hallgren R (1996) Intact adrenocorticotropic hormone-secretion but impaired cortisol response in patients with active rheumatoid arthritis-effect of glucocorticoids. J Rheumatol 23:596–602
172. Dorian B, Garfinkel PE (1987) Stress, immunity, and illness-a review. Psychol Med 17:393–407
173. Thomason BT, Brantley PJ, Jones GN, Dyer HR, Morris JL (1992) The relation between stress and disease activity in rheumatoid arthritis. J Behav Med 15:215–220
174. Affleck G, Pfeiffer C, Tennen H, Fifield J (1987) Attributional processes in rheumatoid arthritis. Arthritis Rheum 30:927–931
175. Rimon R, Laasko R (1985) Life stress and rheumatoid arthritis. Psychother Psychosom 43:38–43
176. Levine S, Strebel R, Wenk EJ, Harman PJ (1962) Suppression of experimental allergic encephalomyelitis by stress. Proc Soc Exp Biol Med 109:294–298
177. Kuroda Y, Mori T, Hori T (1994) Retsraint stress suppresses experimental allergic encephalomyelitis in Lewis rats. Brain Res Bull 34:15–17
178. Levine S, Saltzman A (1987) Nonspecific stress prevents relapses of experimental allergic encephalomyelitis in rats. Brain Behav Immun 1:336–341
179. Bukilica M, Djordjevic S, Maric I, Dimitrijevic M, Markovic BM, Jankovic BD (1991) Stress-induced suppression of experimental allergic encephalomyelitis in the rat. Int J Neurosci 59:167–175
180. Lysle DT, Luecken LJ, Maslonek KA (1992) Suppression of the development of adjuvant-induced arthritis by a conditioned aversive stimulus. Brain Behav Immun 6:64–73
181. Rogers MP, Trentham DE, McCune WJ, Ginsberg BI, Rennke HG, Reich P, David JR (1980) Effects of psychological stress on the induction of arthritis in rats. Arthritis Rheum 23:1337–1342
182. Amkraut AA, Solomon GF, Kraemer HC (1971) Stress, early experience and adjuvant-induced arthritis in the rat. Psychol Med 33:203–214
183. Rabin BS, Cohen S, Ganguli R, Lysle DT, Cunnick JE (1989) Bidirectional interaction between the central nervous system and the immune system. Crit Rev Immunol 9:279–312
184. Miller SC, Rapier SH, Holtsclaw LI, Turner BB (1995) Effects of psychological stress on joint inflammation and adrenal function during induction of arthritis in the Lewis rat. Neuroimmunomodulation 2:329–338

185. Jacobs C, Young D, Tyler S, Callis G, Gillis S, Conlon PJ (1988) In vivo treatment with IL-1 reduces the severity and duration of antigen-induced arthritis. J Immunol 141:2967–2974
186. Martin D, Near SL (1995) Protective effect of the interleukin-1 receptor antagonist (IL-1ra) on experimental allergic encephalomyelitis in rats. J Neuroimmunol 61:241–245
187. Makarov SS, Olsen JC, Johnston WN, Anderle SK, Brown RR, Baldwin AS, Haskill JS, Scwaab JH (1996) Suppression of experimental arthritis by gene transfer of interleukin 1 receptor antagonist cDNA. PNAS USA 93:402–406
188. Jessop DS, Renshaw D, Lightman SL, Harbuz MS (1995) Changes in ACTH and β-endorphin immunoreactivity in immune tissues during a chronic inflammatory stress are not correlated with changes in corticotropin-releasing hormone and arginine vasopressin. J Neuroimmunol 60:29–35
189. Crofford LJ, Sano H, Karalsi K, Webster EL, Goldmuntz EA, Chrousos GP, Wilder RL (1992) Local secretion of corticotrophin-releasing hormone in the joints of Lewis rats with inflammatory arthritis. J Clin Invest 90:2555–2564
190. Crofford LJ, Sano H, Karalis K, Friedman TC, Epps HR, Remmers EF, Mathern P, Chrousos GP, Wilder RL (1993) Corticotrophin-releasing hormone in synovial fluids and tissues of patients with rheumatoid arthritis and osteoarthritis. J Immunol 151:1587–1596
191. Woods RJ, David J, Baigent S, Gibbins J, Lowry PJ (1996) Elevated levels of corticotrophin-releasing factor binding protein in the blood of patients suffering from arthritis and septicaemia and the presence of novel ligands in synovial fluid. Br J Rheumatol 35:120–124
192. Martens HE, Sheets PK, Tenover JS, Dugowson CE, Bremner WJ, Starkebaum G (1994) Decreased testosterone levels in men with rheumatoid arthritis: effect of low dose prednisone therapy. J Rheumatol 21:1427–1431
193. Harbuz MS, Perveen-Gill Z, Lightman SL, Jessop DS (1995) A protective role for testosterone in adjuvant-induced arthritis. Br J Rheumatol 34:1117–1122
194. Rivier C (1995) Luteinizing-hormone-releasing hormone, gonadotropins, and gonadal steroids in stress. Ann NY Acad Sci 771:187–191
195. Roubinian J, Talal N, Siiteri PK, Sadakian JA (1979) Sex hormone modulation of autoimmunity in NZB/NZW mice. Arthritis Rheum 22:1162–1165
196. Allen JB, Blatter D, Calandra GB, Wilder RL (1983) Sex hormonal effects on the severity of streptococcal cell wall-induced polyarthritis in the rat. Arthritis Rheum 26:560–563
197. Steward A, Bayley DL (1992) Effects of androgens in models of rheumatoid arthritis. Agents Actions 35:268–272
198. Cutolo M, Ballearri E, Giusti M, Intra E, Accardo S (1991) Androgen replacement therapy in male patients with rheumatoid arthritis. Arthritis Rheum 34:1–5
199. Homo-Delarche F, Fitzpatrick F, Christeff N, Nunez EA (1991) Sex steroids, glucocorticoids, stress and autoimmunity. J Steroid Biochem Mol Biol 40:619–637
200. Walker SE, Besch-Williford CL, Keisler DH (1994) Accelerated deaths from systemic lupus erythematosus in NZB × NZW F1 mice treated with the testosterone blocking drug flutamide. J Lab Clin Med 124:401–407
201. Schuurs AHWM, Verheul HAM (1989) Sex hormones and autoimmune disease. Br J Rheumatol 28(Suppl I):59–61
202. Wilder RL (1995) Neuroendocrine-immune system interactions and autoimmunity. Ann Rev Immunol 13:307–338
203. Wilder RL (1996) Adrenal and gonadal-steroid hormone deficiency in the etiopathogenesis of rheumatoid arthritis. J Rheumatol 23:10–12
204. Chikanza IC, Panayi GS (1991) Hypothalamic-pituitary mediated modulation of immune function: prolactin as a neuroimmune peptide. Br J Rheumatol 30:203–207

205. Neidhart M (1989) Bromocriptine microcapsules inhibit ornithine decarboxylase activity induced by Freund's complete adjuvant in lymphoid tissues of male rats. Endocrinology 125:2846–2852
206. Jorgensen C, Bressot N, Bologna C, Sany J (1995) Dysregulation of the hypothalamo-pituitary axis in rheumatoid arthritis. J Rheumatol 22:1829–1833
207. Neidhart M (1996) Elevated serum prolactin or elevated proplactin/cortisol ratio are associated with autoimmune processes in systemic lupus erythematosus and other connective tissue diseases. J Rheumatol 23:476–481
208. Sandi C, Cambronero JC, Borrell J, Guaza C (1992) Mutually antagonistic effects of corticosterone and prolactin on rat lymphocyte proliferation. Neuroendocrinology 56:574–581
209. Levine JD, Fields HL, Basbaum AI (1993) Peptides and the primary afferent nociceptor. J Neurosci 13:2273–2286
210. Levine JD, Coderre TJ, Helms C, Basbaum AI (1988) β2-adrenergic mechanisms in experimental arthritis. PNAS USA 85:4553–4556
211. Coderre TJ, Basbaum AI, Dallman MF, Helms C, Levine JD (1990) Epinephrine exacerbates arthritis by an action at presynaptic β2-adrenoceptors. Neuroscience 34:521–523
212. Coderre TJ, Chan AK, Helms C, Basbaum AI, Levine JD (1991) Increasing sympathetic nerve terminal-dependent plasma extravasation correlates with decreased arthritic joint injury in rats. Neuroscience 40:185–189
213. Agius MA, Checinski ME, Richman DP, Chelmicka-Schorr E (1987) Sympathectomy enhances the severity of experimental autoimmune myasthenia gravis (EAMG). J Neuroimmunol 16:11–12
214. Chelmicka-Schoor E, Checinski ME, Arnason BGW (1988) Chemical sympathectomy augments the severity of experimental allergic encephalomyelitis in Lewis rats. J Neuroimmunol 17:347–350
215. Chelmicka-Schoor E, Kwasniewski MN, Wollmann RL (1992) Sympathectomy augments adoptively transferred experimental allergic encephalomyelitis. J Neuroimmunol 37:99–103
216. Harbuz MS, Chover-Gonzalez AJ, Biswas S, Lightman SL, Chowdrey HS (1994) Role of central catecholamines in the modulation of corticotrophin-releasing factor mRNA during adjuvant-induced arthritis in the rat. Br J Rheumatol 33:205–209
217. Cho H-J, Lee H-S, Bae M-A, Joo K (1995) Chronic arthritis increases tyrosine hydroxylase mRNA levels in the pontine noradrenergic cell groups. Brain Res 695:96–99
218. Krenger W, Honegger CG, Feurer C, Cammiuli S (1986) Changes of neurotransmitter systems in chronic relapsing experimental allergic encephalomyelitis in rat brain and spinal cord. J Neurochem 47:1247–1254
219. Weigmann K, Muthyala S, Kim DH, Arnason BGW, Chelmicka- Schorr E (1995) β-adrenergic agonists suppress chronic/relapsing experimental allergic encephalomyelitis (CREAE) in Lewis rats. J Neuroimmunol 56:201–206
220. Leonard JP, MacKenzie FJ, Patel HA, Cuzner ML (1991) Hypothalamic noradrenergic pathways influence neuroendocrine and clinical status in experimental allergic encephalomyelitis. Brain Behav Immun 5:328–338
221. Schlesinger L, Arevalo M, Simon V, Lopez M, Munoz C, Hernandez A, Carreno P, Belmar J, White A, Haffnercavaillon N (1995) Immune depression induced by protein-calorie malnutrition can be suppressed by lesioning central noradrenaline systems. J Neuroimmunol 57:1–7
222. Bendele AM, Spaethe SM, Benslay DN, Bryant HU (1991) Anti-inflammatory activity of pergolide, a dopamine receptor agonist. J Pharmacol Ther 259:169–175
223. Sufka KJ, Schomburg FM, Giordano J (1992) Receptor mediation of 5-HT induced inflammation and nociception in rats. Pharmacol Biochem Behav 41:53–56

224. Weil-Fugazza J, Godefroy F, Besson JM (1979) Changes in brain and spinal tryptophan and 5-hydroxyindoleacetic acid levels following acute morphine administration in normal and arthritic rats. Brain Res 175:291–301
225. Garzon J, Lerida M, Sanchez-Blazquez P (1990) Effect of intrathecal injection of pertussis toxin on substance P, norepinephrine and serotonin contents in various neural structures of arthritic rats. Life Sci 47:1915–1923
226. Marlier L, Poulat P, Rajaofetra N (1991) Modifications of serotonin-, substance P- and calcitonin gene-related peptide-like immunoreactivities in the dorsal horn of the spinal cord of arthritic rats: a quantitative immunocytochemical study. Exp Brain Res 85:482–490
227. Pertsch M, Krause E, Hirschelmann R (1993) A comparison of serotonin (5-HT) blood levels and activity of 5-HT2 antagonists in adjuvant arthritic Lewis and Wistar rats. Agents Actions 38:C98–101
228. Sofia RD, Vassar HB (1974) Changes in serotonin (5HT) concentrations in brain tissue of rats with adjuvant-induced polyarthritis. Arch Int Pharmacodyn 211:74–79
229. Godefroy F, Weil-Fugazza J, Besson J-M (1987) Complex temporal changes in 5-hydroxytryptamine synthesis in the central nervous system induced by experimental polyarthritis in the rat. Pain 28:223–238
230. Harbuz MS, Perveen-Gill Z, Lalies MD, Jessop DS, Lightman SL, Chowdrey HS (1996) The role of endogenous serotonin in adjuvant-induced arthritis in the rat. Br J Rheumatol 35:112–116
231. Holmes MC, French KL, Seckl JR (1995) Modulation of serotonin and corticosteroid receptor gene expression in the rat hippocampus with circadian rhythm and stress. Mol Brain Res 28:186–192

Chapter 3

The role of the innervation of lymphoid tissue in the regulation of the Th1/Th2 dichotomy

Barbara Kruszewska, Jan A. Moynihan and David L. Felten

Neural-immune interactions

Immune responses can be modulated by input from the sympathetic nervous system (SNS) [reviewed in 1–3]. There is ample evidence that the components necessary for such an interaction are present in a normal animal. Histological studies clearly show the presence of nerve fibres within synaptic distance of cells in many lymphoid tissues, including spleen, lymph nodes, thymus and bone marrow [4,5]. These postganglionic sympathetic nerve fibres release norepinephrine (NE) as the major neurotransmitter, and are visualized easily with immunocytochemical methods utilizing antibodies to tyrosine hydroxylase, the rate-limiting enzyme in NE synthesis. Nerve fibres staining for numerous neuropeptides also have been found in these lymphoid organs; a role for these other neurotransmitters (such as neuropeptide Y) has not been elucidated fully. Norepinephrine is the most prominent neurotransmitter candidate in sympathetic postganglionic nerve fibres and therefore has been the focus of extensive research.

Immune activation: intracellular events

The activation of a lymphocyte is triggered by a specific signal from an antigen and is mediated by cell surface receptors. This signal, however, is subject to modulation by events triggered by other receptors on the lymphocyte surface. A simplified diagram of the intracellular events associated with such an interaction is shown in Fig. 3.1. The cell has on its surface a T-cell receptor which can be activated by a specific ligand, usually a peptide major histocompatibility complex (MHC). Activation of this receptor triggers a cascade of events, mediated by CD3, G-protein and inositol trisphosphate (IP_3), which leads to a release of calcium from intracellular stores. Increased calcium levels in turn trigger activation of protein kinase C and also bind with calmodulin; both of these signals can regulate gene transcription. Increased mRNA for interferon gamma (IFN-γ) and interleukin 2 (IL-2) can be detected within 1 hour of T-cell activation [6]; transient activation of genes for these and many other proteins leads to production of proteins needed for T-cell mitosis and functions, including clonal proliferation.

Modulation by adrenergic receptors

T-cells also have adrenergic receptors on their surface, activation of which can interact with the events accompanying T-cell receptor activation. If the receptor is a β1-adrenergic receptor, as shown in Fig. 3.1, activation with norepinephrine triggers a cascade of G-protein-mediated events, leading to enhanced adenylate cyclase activity and intracellular cyclic adenosine monophosphate (cAMP) and protein kinase A activity. A wide range of biochemical events and intracellular processes are known to be affected by cAMP. One that is directly linked with T-cell activation is cytoskeleton-associated actin assembly [7]. One of the earliest events following T-cell activation by an antigen is the formation of pseudopodia, a process that is dependent on the formation and assembly of F-actin and which

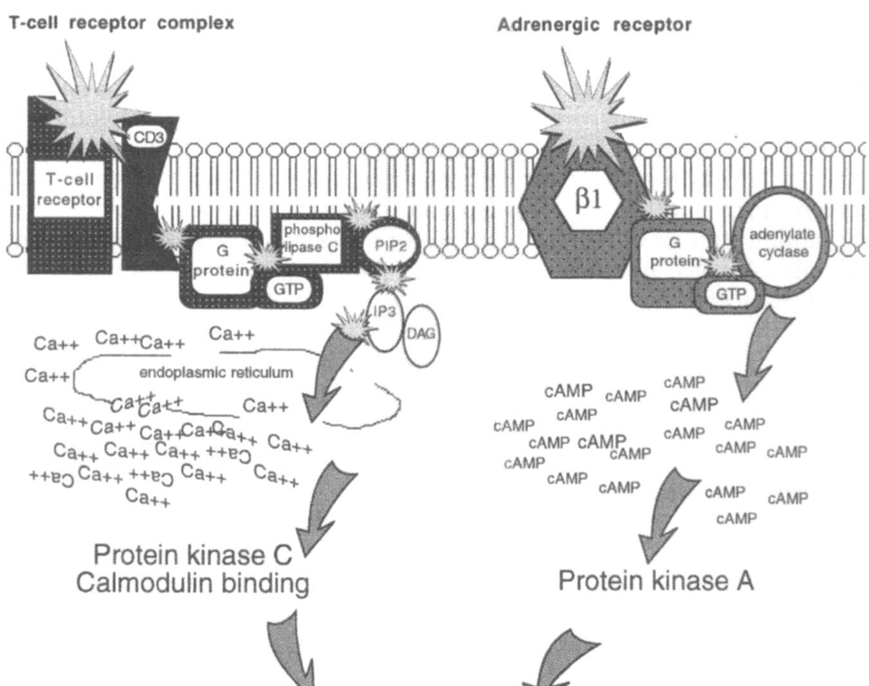

Fig. 3.1. Second messenger cascades triggered by activation of the T-cell receptor and a β-adrenergic receptor. PIP_2, phosphatidylinositol-4,5-bisphosphate; DAG, 1,2-diacylglycerol; IP_3, inositol-1,4,5-triphosphate.

enhances the interaction of the T-cell receptor with antigen and co-stimulatory signals which drive proliferation. Selliah et al. [8] have shown that in T-cells this process is blocked by the β-adrenergic-mediated increase of cAMP. In general, high intracellular cAMP concentrations are associated with decreased cell proliferation [9]. Activation of a β1-adrenergic receptor therefore works *against* the events triggered by activation of the T-cell receptor.

Other adrenergic receptor subtypes trigger different second messenger cascades. The α2-receptor is linked to an inhibitory G-protein, which reduces cAMP levels [10]. Its effect is therefore opposite to that of a β-receptor. Changes in both calcium and cAMP have been reported for α1-adrenergic activation. There is general agreement that in most cells α1-receptor activation is coupled to the phosphatidylinositol/IP_3 pathway [10]. This is the same pathway activated by the T-cell receptor complex; thus activation of this adrenergic receptor at the time of T-cell activation should result in a synergistic increase in intracellular calcium. However, there are also reports of a different α1-transduction mechanism leading to cAMP accumulation [11], which would be expected to have an anti-proliferative effect. Calcium and cAMP concentrations may be reciprocally regulated; some forms of cAMP phosphodiesterase (an enzyme which inactivates cAMP) are regulated by the calcium-binding protein calmodulin [12].

Although the precise intracellular mechanisms have not been worked out, it is probable that the ultimate response of a T-lymphocyte depends upon the sum of the signals received at its surface, including those triggered by catecholamine interactions with adrenoceptors. A direct example of such intracellular "crosstalk" has been demonstrated [13,14]. Resting human T-cells show a rise in cAMP levels when stimulated with isoproterenol, a β-adrenergic agonist, but not phytohaemagglutinin (PHA), a T-cell mitogen. When cells are exposed to both PHA and isoproterenol, there is a synergistic rise in cAMP that is 2- to 10-fold greater than that seen with isoproterenol alone. This effect is blocked by propranolol, a β-adrenergic antagonist. The same effect is observed using norepinephrine instead of isoproterenol for adrenergic activation.

Evidence for presence of adrenergic receptors on immune cells

Radioligand studies as well as functional studies have confirmed that various immune cell types (T-cells, B-cells, macrophages, neutrophils, eosinophils) express adrenergic receptors on their cell surface. The presence of the β-adrenergic subtype has been well established by radioligand binding studies [15–25] and confirmed by functional and pharmacological studies [26–30].

Several binding studies also indicate the presence of the α-subtype in a lymphoblastoid cell line [31], guinea-pig spleen cells [32] and human lymphocytes [33]; but α-adrenergic effects have been mainly implied from pharmacological studies [29,34–38]. A recent autoradiographic study [39] found that the distribution of α2-adrenoceptors in the lymph node correlates with previously reported noradrenergic innervation of the lymph node. In any case, the cascade of events triggered by activation of any of these receptor subtypes would be expected to interact with those triggered by the T-cell receptor.

The Th1/Th2 paradigm

In recent years a classification of T helper ($CD4^+$) cells has emerged based on the pattern of cytokines they secrete [40–42]. T helper-1 (Th1) cells are defined as those which secrete primarily IFN-γ and IL-2. T helper-2 (Th2) cells are those which secrete IL-4, IL-5, IL-6 and IL-10. The Th1- or Th2-cell profile is believed to emerge from a common precursor cell, the Th0-subset, which is capable of secreting both IFN-γ and IL-4 [42,43]. A Th0-cell can be driven into either subset with the appropriate signal; however, once it becomes committed to either phenotype it does not switch, since it has not been possible to convert a clone of one subtype into the other. Additional cytokine secretion patterns which do not fit these two subsets also have been reported [44].

This classification system originally was defined in murine T-cell clones but has since become correlated with different effector functions in vivo. IL-2 and IFN-γ production enhance macrophage activity, linking the Th1 pattern of cytokine synthesis with increased immunity against intracellular pathogens [45], such as those associated with viral, fungal or bacterial infections. On the other hand, Th2-cell cytokines such as IL-4 are essential for B-cell differentiation and antibody production, associating this subset with humoral immunity, useful protection against extracellular microbes.

A classic example of this is seen in the case of *Leishmania major*, a protozoan parasite which establishes itself in macrophages [46]. Most strains of mice infected with this pathogen develop a protective immune response that eliminates the pathogen. T-cell clones from these infected animals produce IFNγ and do not produce IL-4. In the BALB/cJ strain, however, infection with *L. major* is usually fatal. Cells taken from animals of this strain have been shown to produce mainly IL-4, which drives a non-protective humoral immune response

The cytokine products of one subset are often inhibitory for the other subset [42]. For example, IL-4 stimulates Th2-cell expansion while suppressing Th1-cell differentiation and macrophage activation [47–48]. On the other hand, IFN-γ produced by Th1-cells inhibits Th2-cell function [49,50].

Denervation studies: review of results

Given that immune cells receive sympathetic innervation, and that catecholamines are capable of modulating immune responses, removal of sympathetic input can be a useful way to assess the effect of the nervous system on immune function. Sympathectomy can be accomplished in several ways. One is by surgically cutting the sympathetic nerve supply to a particular lymphoid organ. This is a cumbersome method and in many cases an impractical one, because prolonged immune suppression from invasive surgery occurs as a superimposed problem. A more commonly used technique is chemical sympathectomy, in which nerve terminals are destroyed by administration of 6-hydroxydopamine (6-OHDA). 6-OHDA is a neurotoxin which selectively destroys noradrenergic nerve terminals, without destroying the cell bodies [51]. SNS input also can be blocked at the receptor level, utilizing various adrenergic receptor antagonists. This process of receptor blockade does not evoke physical degeneration and withdrawal of nerve terminals, but pharmacologically evokes systemic actions.

The majority of denervation studies to date have utilized chemical sympathectomy to examine the effect of removal of sympathetic input. One of the earliest studies [26] showed that sympathectomy (coupled with adrenalectomy, to ensure NE depletion) increased the antibody response to sheep red blood cells (SRBC) in rats. This, together with data showing decreasing NE levels in the splenic pulp during the immune response, led the authors to propose that "interruption of sympathetic nervous innervation removes an important suppressor of the immune response".

Studies since that time have demonstrated that the assessment of immune function following chemical sympathectomy requires much more complex interpretation. As Madden & Felten [2] point out, "it was initially hoped that a unifying theme for catecholamine-immune system interactions could be established." Instead, new studies continue to demonstrate the diversity of responses. A partial review of the literature (summarized in Table 3.1) shows that both enhancement and suppression of various parameters have been reported, in addition to studies where no effect was observed (not included in this table). Although there is no doubt that chemical sympathectomy is invaluable in assessing neural-immune interactions, it is crucial that interpretation of such data take into account many factors as discussed below.

Parameter measured

The immune response of an organism to a pathogen is a complex orchestration of a series of events. It involves many cell types (T lymphocytes, B lymphocytes, macrophages), numerous cytokines and other chemical signals, and takes place in various lymphoid organs in a precisely organized temporal sequence. The type of immune response elicited (humoral or cell-mediated) varies with the nature of antigen. The lymphoid tissue examined also will be a factor, partly because of temporal differences and partly because, as Swain et al. [47] have shown, lymphokine secretion is different in different lymphoid organs. This is further complicated by the fact that there is continuous migration of immune cells between lymphoid organs and the circulatory pool. At any point in the immune cascade, activity of some components will be suppressed while that of others may be enhanced. Since it is seldom feasible to examine *all* of the parameters involved in an immune response, it is important to acknowledge that the results of changes in any one parameter show only a partial picture.

Compartment effects and other effects of 6-OHDA

6-OHDA does not cross the blood–brain barrier in a normal animal; therefore, when it is injected peripherally into an adult animal its effects are limited to the

Table 3.1. Changes in immune responses following chemical sympathectomy

Strain & Species	Cell type[a]	D[b]	Antigen in		Enhancement	Suppression	Ref.
			vivo	vitro			
C3H/He mice	S	1	SRBC	SRBC		↓ 1° Ab (PFC)	57
C3H mice (newborn)	S	8	SRBC	SRBC	↑ PFC response		4
A/J mice	S	10	TNP, PC PC-KLH		↑ PFC response to thymus-independent Ag only		59
C3H/Hej, BALB/cJ mice	LN	1	SRBC	SRBC		↓ PFC response	56
A/J mice	S	10	None	None	↑ β-adrenergic receptor density	↓ B-cell number by 25%	20
A/J mice	S	10	None	None	↑ Thy-1.2+ and Lyt-2 cells		21
CBA/J × A/J F1 (newborn)	S	10	None	None		↓ Lyt-2 cells	
C3H, BALB/c DBA/2, C57Bl/6 mice	S, LN	1–2	SRBC			↓ 1° and 2° PFC response	58
B6C3F1 mice	S	1	None	SRBC	↑ β-adrenergic receptor density	↓ PFC response	24
BALB/cByJ mice	L	1	TNCB			↓ IL-2, CTL, DTH	60
DBA/2 & C57B/6 mice	S	2	None	Con A	↑ T-cell proliferation in DBA/2 only		61
BALB/cByJ mice	S, LN	1	None	Con A	↑ IFN-γ in LN	↓ IFN-γ in spleen ↓ Thy-1+ and CD4+ T cells ↓ T-cell proliferation ↓ IL-2 in spleen	54
				LPS	↑ B-cell proliferation in LN ↑ polyclonal IgG in LN	↓ B cell proliferation in spleen ↓ Polyclonal IgM in LN	
BALB/cJ & C57Bl/6J mice	S	1	KLH	KLH	↑ IL-4, IL-2, proliferation ↑ IFN-γ, ↑ Ig in C57 only		65
C3H/OuJiCo mice	T	2	LPS	Con A		↓ T cell proliferation	62
	S	2	LPS	PHA	↑ proliferation & NK cell activity		

[a] S, spleen; LN, lymph nodes; T, thymus.
[b] D, denervation protocol (number of i.p. 6-OHDA injections).

periphery. Chemical sympathectomy studies where 6-OHDA is injected into the cisterna magna [i.e. 52,53] cannot be compared to those with peripheral injection since the effects are not limited to the periphery. Additionally, since the blood–brain barrier is not completely formed in newborn animals, neonatal sympathectomy [4,21] also will require adjustments in interpretation. Finally, a direct effect of 6-OHDA on immune cells should be eliminated by the use of an appropriate control, such as pre-exposure to the catecholamine uptake blocker desipramine. However, it is possible that prolonged exposure of lymphoid tissue to desipramine itself can enhance and prolong the availability of NE for interactions with surface receptors on immunocytes.

Extent of denervation

The chemical denervation protocol is based on the assumption that sympathectomy depletes the majority of adrenergic fibres. Measurement of catecholamine levels following sympathectomy shows that it is possible to achieve up to 90% denervation with the proper dosage [54]. However, it has become clear that the amount of 6-OHDA needed for complete denervation varies with the species and strain of animal. For example, BALB/cJ mice require only 100 mg of 6-OHDA per kg of body weight to achieve 90% denervation of the spleen, whereas the C57B1/6J mouse strain requires two and a half times that amount (250 mg/kg) to achieve the same level of denervation. Several early studies were done before this species dose difference was known, resulting in reports of no effect of sympathectomy with the C57B1/6J strain [55,56] which may have been due to use of suboptimal doses of 6-OHDA. The extent of denervation can be assessed by measurement of catecholamine levels using high-performance liquid chromatography (HPLC) with electrochemical detection.

Of those studies where 6-OHDA was injected peripherally, denervation protocols varied from a single injection to one injection per week for 10 weeks. Chronic sympathectomy may have different effects from acute withdrawal of input.

Immune state of animal

As noted earlier, the cytokine profiles of immunized animals are very different from animals which have not been exposed to antigen. Mosmann et al. [44] point out that "the Th1 and Th2 phenotypes are found most easily after vigorous immunization of the animals used to derive the T-cell clones" As can be seen in Table 3.1, half of the studies have been done on animals primed with an antigen in vivo; the rest used naive animals. Removal of SNS input to naive T-cells may have different consequences from removal of input to cells which are actively involved in an immune response. The choice of antigen, whether for immunization in vivo or stimulation in vitro, also will determine the pathway triggered. This is obvious for antigens which drive the immune response in a humoral versus cell-mediated direction, but may also be true in more subtle situations. For example, Li et al. [27] find that while lipopolysaccharide (LPS) and anti-mouse μ-chain antibodies both activate mouse B-cells, modulation of this activation by β-adrenergic agonists is different for the two antigens.

Cell-mediated versus humoral immune responses

The majority of sympathectomy studies to date examined parameters which are associated with humoral immunity, such as measurement of serum antibody titres and plaque-forming cells (PFC). Of six studies which measured PFC responses in the spleen, two report enhancement while four report suppression. The antibody response was suppressed in SRBC-primed mice of the C3H/He, BALB/cJ, DBA/2 and C57B1/6 strains [56–58] and in naive B6C3F1 mice [24]. It was enhanced in immune A/J mice [59] and newborn C3H mice [4]. As pointed out earlier, it is difficult to make any generalizations from these studies since they involved five different strains of mice, denervation protocols that ranged from one to ten 6-OHDA injections in neonates or adults, and immune and non-immune animals.

Much less has been done with parameters associated with cell-mediated immunity, such as the delayed-type hypersensitivity (DTH) response and cytotoxic T-cell response (CTL). Only one study [60] has measured CTL and DTH responses following denervation. In BALB/cByJ and C3H/HeJ mice, the DTH response (ear swelling) to a contact sensitizing agent was decreased following chemical sympathectomy. There was also a decrease in the CTL response of sensitized lymph node cells which was accompanied by a suppression of IL-2 production in vitro.

Madden et al. [54] have shown that chemical sympathectomy of non-immune BALB/cByJ mice alters the in vitro activity of spleen and lymph node cells. Con A or LPS-induced activity by spleen cells (IL-2 and IFN-γ production, T- and B-cell proliferation) was decreased in all cases, whereas both specific enhancement (IFN-γ, B-cell proliferation, polyclonal IgG secretion) and specific suppression (T-cell proliferation, polyclonal IgM secretion) were observed with lymph node cells. Pharmacological studies [29] also suggest a balance between inhibitory effects mediated by the α-adrenergic receptor and enhancing effects mediated by a β-adrenergic receptor, in the generation of a CTL response in BALB/c and C57B1/6 mice.

Both enhancement [61] and suppression [62] of T-cell proliferation have been reported following chemical sympathectomy. Given that T helper cells participate in both humoral and cell-mediated responses, measurement of proliferation in a heterogeneous cell population does not allow classification into either a humoral or cell-mediated category.

Effect of denervation on two mouse strains dominant for two types of immune responses

In light of the reciprocal regulation of cytokines associated with humoral and cell-mediated immunity, the possibility exists that the conflicting effects of sympathectomy may be at least partly due to differential regulation of the two types of immune responses. That is, it is possible that humoral (Th2-driven) responses are modulated differently from cell-mediated (Th1-driven) responses. Such differential modulation of immune responses by nervous system input is suggested by the work of Sheridan et al. [63], who showed that cellular and humoral immune responses are differentially affected by restraint stress, and that at least one component of this effect is mediated by an adrenal-independent (non-glucocorticoid) mechanism [64].

Given the difficulty of comparing studies which utilize various sympathectomy protocols, strains and antigens, the strategy in our laboratory has been to focus on two inbred strains of mice, BALB/cJ and C57Bl/6J, which are dominant for the two types of immune responses. As discussed earlier, the BALB/cJ strain responds to pathogens with primarily a humoral immune response. If the pathogen is intracellular, for example, *Leishmania*, the resulting immune response will be non-protective, making the animal susceptible to attack from this antigen. On the other hand, the C57Bl/6J mouse is able to mount an effective cell-mediated response and is not susceptible to this disease. Mice of the BALB/cJ strain characteristically produce IL-4 and IL-10, cytokines produced by Th2-cells, whereas C57Bl/6J mice produce predominantly IL-2 and IFN-γ, cytokines associated with Th1-type cells. This clear difference in immune dominance makes these two strains useful in examining the role that neural modulation plays on different immune responses. We have therefore embarked on a systematic examination of various immune parameters following denervation in these two strains of mice.

Experimental protocol

Our standard experimental protocol has been reported in detail elsewhere [65] and is described briefly here. Animals were denervated with an i.p. injection of 6-OHDA 2 days prior to immunization. The dose of 6-OHDA used was specific for the strain: 100 mg/kg body weight for BALB/cJ mice, 250 mg/kg for C57Bl/6J. C57Bl/6J mice require a higher dose to achieve approximately 90% denervation. Control animals received an ascorbate saline vehicle. Two days later animals were injected i.p. with 100 μg keyhole limpet haemocyanin (KLH), a protein antigen which elicits a primarily humoral immune response. Six days later animals were killed by decapitation and the spleen was removed. The extent of denervation was verified by HPLC analysis of catecholamine levels from a segment of the spleen. Trunk blood was collected and serum was stored for analysis of antibody titres. Spleen cells were cultured in the presence of KLH and supernatants were collected at 24-h intervals; 24 h and 48 h supernatants were assayed by ELISA for IL-2; 48 h and 72 h supernatants were assayed for IL-4 and IFN-γ.

Results

Removal of sympathetic input under these conditions resulted in a general enhancement of the immune parameters measured (Fig. 3.2, Table 3.2). Both IL-2

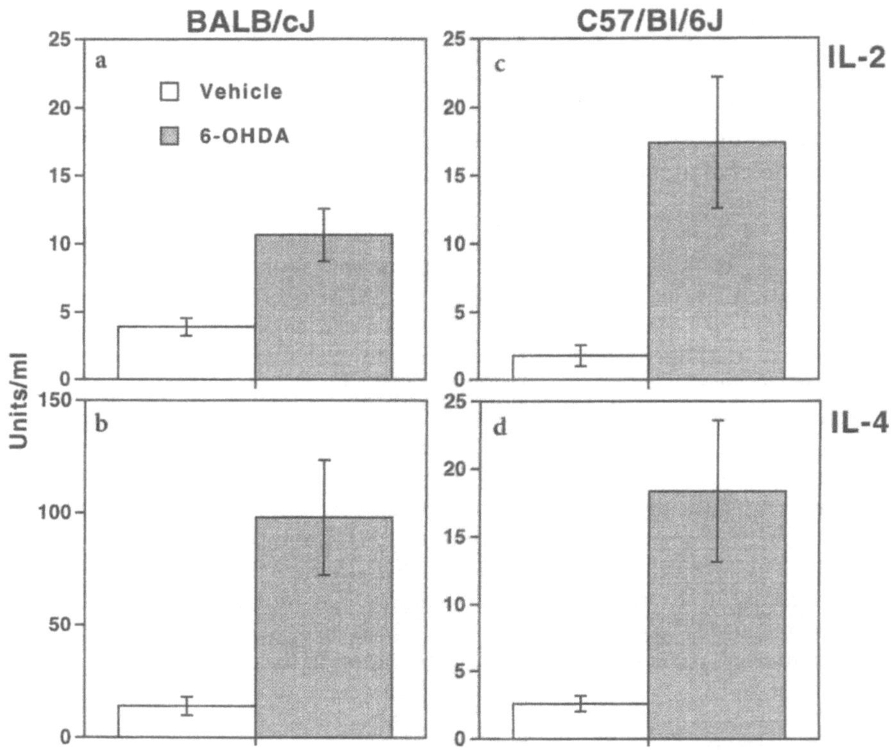

Fig. 3.2. Keyhole limpet hemocyanin (KLH)-stimulated cytokine production in vitro by splenocytes from BALB/cJ (**a**,**b**) and C57Bl/6J (**c**,**d**) mice. **a–c**, IL-2; **b–d**, IL-4. Animals were denervated with 6-hydroxydopamine (6-OHDA), immunized 2 days later with KLH and killed 6 days after immunization. Spleen cells were cultured with 160 μg/ml KLH for 24 h (IL-2) or 72 h (IL-4). Results are expressed as mean ± SEM of 8–12 animals per group. (Adapted from *J Immunology* 155: 4613–4620, 1995.) Copyright 1995. The American Association of Immunologists.

Table 3.2. Effect of chemical sympathectomy on two strains of mice

	BALB/cJ	*C57Bl/6J*
In vitro		
IL-2	Increased	Increased
IL-4	Increased	Increased
IFN-γ	No change	Increased
Proliferation	Increased	Increased
Serum Ig		
IgM	No change	Increased
IgG	No change	Increased
IgG1	No change	Increased
IgG2a	No change	Increased

(Adapted from: J Immunol 155: 4613–4620, 1995.)

and IL-4 production were greatly increased in both strains of denervated mice (Fig. 3.2). Additionally, IFN-γ was increased in the C57B1/6J strain (not shown). The increased cytokine production was from cells which were KLH-specific, since very little cytokine was produced by cells grown without KLH in culture. Most interestingly, there was a substantial increase in serum antibody titres in the C57B1/6J animals, with levels of IgM, IgG, IgG1 and IgG2a all increasing following sympathectomy (Fig. 3.3). Except for a slight enhancement of IgG1 on day 18, there were no comparable increases in antibody levels in the BALB/cJ strain (Fig. 3.4).

This general enhancement of the humoral immune response is consistent with a tonic inhibitory mechanism of sympathetic input. In this model, sympathetic

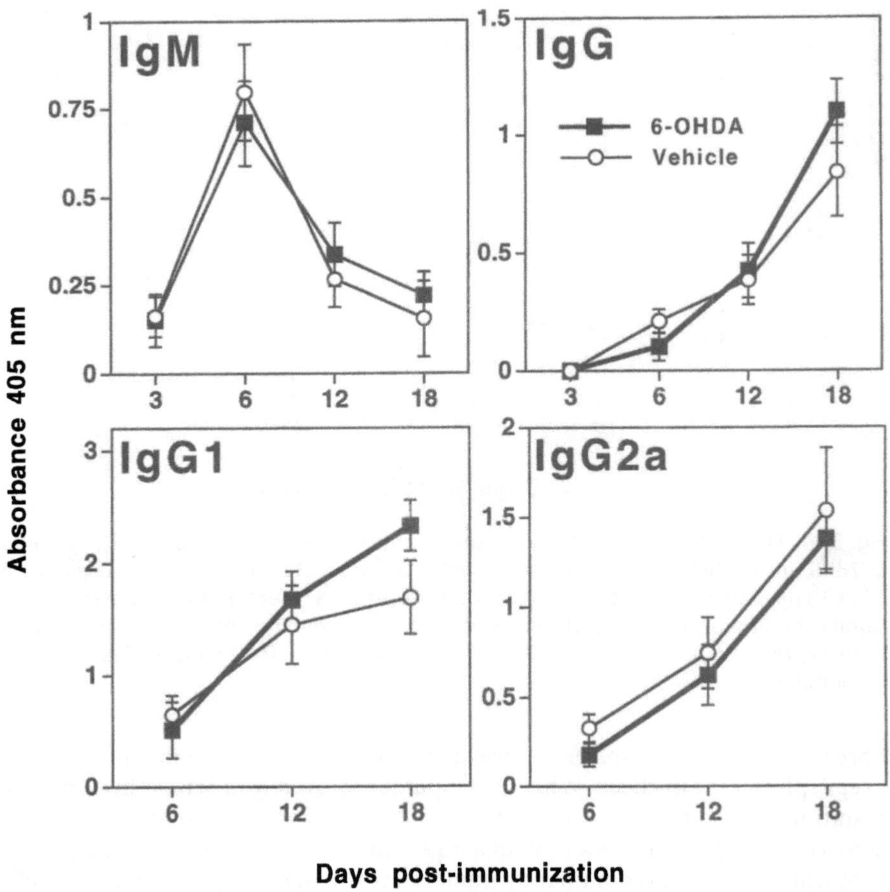

Fig. 3.3. Effect of sympathectomy on serum antibody levels (IgM, IgG, IgG1, IgG2a) in BALB/cJ mice. Animals were immunized with KLH 2 days after denervation, and killed 3, 6, 12 or 18 days after immunization. Results are expressed as mean ± SEM of 8–12 animals per group. (Adapted from *J Immunology* 155: 4613–4620, 1995.) Copyright 1995. The American Association of Immunologists.

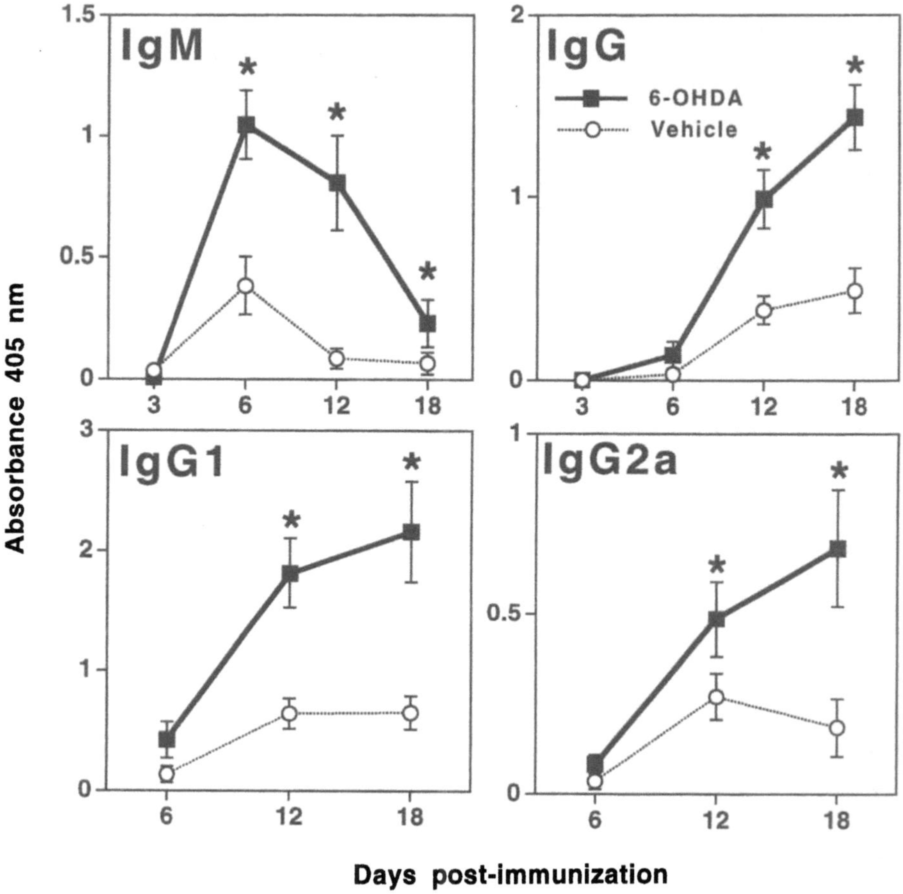

Fig. 3.4. Effect of sympathectomy on serum antibody levels (IgM, IgG, IgG1, IgG2a) in C57Bl/6J mice. Animals were immunized with KLH 2 days after denervation and killed 3, 6, 12 or 18 days after immunization. Results are expressed as mean ± SEM; asterisks indicate significant differences ($P < 0.05$) compared with corresponding vehicle control. (Adapted from *J Immunology* 155: 4613–4620, 1995.) Copyright 1995. The American Association of Immunologists.

fibres innervating the spleen release a steady stream of norepinephrine which keeps spleen cells in check. When this input is removed with sympathectomy, the resulting release from inhibition is manifested as increased humoral immune activity. Given that KLH is a protein antigen, the immune response elicited in this protocol should be predominantly humoral, involving IL-4 producing Th2-cells. This type of response is characteristic of the BALB/cJ strain, so it is not surprising that a release from inhibition increases IL-4 in this strain. The lack of an increase in antibody titres in this strain may indicate that it is not possible to further enhance antibody production in a strain that is already biased for this process. It is also possible that antibody production is suppressed by the increased IL-2 production. As discussed earlier, there is evidence that Th2-like responses (i.e. antibody production) can be suppressed by products of the Th1 subset (i.e. IL-2) [49–50].

The accompanying increase in IL-2 production was unexpected, but may be due to several factors. Immunization with KLH may trigger some component of cell-mediated immunity, and both Th1 and Th2 populations are released from inhibition by sympathectomy. Alternatively, cells other than Th1/Th2 (for example, the Th0 subset which is known to secrete IL-2 [44]) may be released from inhibition as well. A Th subset which secretes both IL-2 and IL-4 has been reported [66] although it is not well characterized.

The C57B1/6J strain, which normally does not mount an efficient humoral immune response, is able to do so when sympathetic input is removed. In this strain, there is enhancement of not only its dominant cytokines (IL-2 and IFN-γ) but also of the cytokine associated with humoral immunity, IL-4. The presence of IL-4 drives T-cells into a Th2 mode [67–69]. The C57B1/6J strain normally produces very little IL-4. If sympathectomy lifts from inhibition cells which enhance IL-4 production, the result could lead to an increase in antibody production. The accompanying increase in serum antibody titres therefore correlates with enhanced IL-4 levels. The Th1 subset also plays a role in humoral immunity in that IFN-γ induces heavy-chain switching to the IgG2a isotype.

The ability of an animal biased towards one type of immune response to switch to a different response has been reported previously. A Th1-like response can be induced in *Leishmania*-infected BALB/cJ mice by blocking the action of IL-4 [70]. Conversely, expression of an IL-4 transgene in resistant mice rendered them susceptible [71]. Antigen dose alone is sufficient to direct the Th phenotype, with mid-range doses eliciting secretion of Th1-associated cytokines, whereas doses at the very high or very low end of the range trigger a Th2-like profile [72]. High levels of Th1-type cytokines can be elicited from BALB/c mice with sublethal doses of herpes virus [73].

Differential modulation: possible mechanisms

Cellular mechanisms offer many points at which differential modulation could take place (Table 3.3). The type and density of receptor plays a crucial role in a cell's ability to respond to sympathetic input. Adrenergic receptors are not distributed randomly on immune cells. Radioligand binding studies show that receptor density on the immune cell surface varies not only among different immune cells, but even within the T-cell subsets. Mouse spleen B-cells have a

Table 3.3. Differential modulation of immune responses by sympathetic nervous system input: possible mechanisms

Differences in	Example
Cell heterogeneity	Th1, Th2
Adrenergic receptor subtypes	$\alpha1$, $\alpha2$, $\beta1$, $\beta2$
Second messenger cascades	PIP_2, IP_3, Ca, cAMP
Receptor density	750–2900 per cell
Affinity of neurotransmitter binding	

higher number than mouse spleen T-cells [20,21]. Khan et al. [74] found 750 receptors per cell on T helper cells, 2900 on T suppressor cells, and 1800 on cytotoxic T-cells. Intriguing evidence for the heterogeneity of receptor type on Th-subtypes has been shown by the work of Sanders et al. [25], who found that Th1 clones do not have β-adrenergic receptors, while Th2 clones do. Within the Th2 population, receptor numbers ranged from 200 to 513 per cell. Furthermore, receptor density is not a static feature. Exposure of the cell to antigen, mitogen or neurotransmitter can alter receptor expression and sensitivity. For example, high concentrations of NE agonist can down-regulate receptors, whereas adrenergic deprivation can lead to up-regulation and/or supersensitivity [reviewed in 75].

As discussed previously, at least four major classes of adrenergic receptor subtypes have been identified. Although the $\beta 2$-subtype is though to be the primary subtype in human and mouse T-cells, evidence is accumulating for the existence of the other subtypes as well. Different subtypes are linked to different second messenger cascades, some with opposing effects.

Summary

Given that both humoral (Th2-driven) and cell-mediated (Th1-driven) immune responses are complex events involving multiple pathways and feedback loops, the modulation of these responses by the nervous system is likely to be a complex process as well. It is not possible, with our current state of knowledge, to generalize whether immune responses are enhanced or suppressed by SNS input. The ultimate response of a lymphoid cell will depend upon the sum total of signals it receives from its surface receptors. The Th1 and Th2 subsets of T-helper cells may have different sensitivity to neural input. It has become clear in recent years that cytokines such as IL-2, IL-4 and IL-12 play a crucial role in directing the immune response. Since these cytokines are both produced by, and act on, lymphoid cells, it is reasonable to assume that any perturbation of the system, such as that triggered by removal of sympathetic input, would have an impact on the nature of the immune response. Our current protocol utilizing two strains of mice, each dominant for one type of immune response, offers a good model in which to examine systematically the effect of neural input on a diversity of immune parameters.

References

1. Ader R, Felten DL, Cohen N (1991) Psychoneuroimmunology, 2nd edn. Academic Press, San Diego
2. Madden KS, Felten DL (1995) Experimental basis for neural-immune interactions. Physiol Rev 75:77–106

3. Madden KS, Sanders VM, Felten DL (1995) Catecholamine influences and sympathetic neural modulation of immune responsiveness. Annu Rev Pharmacol Toxicol 35:417–448
4. Williams JM, Peterson RG, Shea PA, Schmedtje JF, Bauer DC, Felten DL (1981) Sympathetic innervation of murine thymus and spleen: evidence for a functional link between the nervous and immune systems. Brain Res Bull 6:83–94
5. Felten SY, Felten DL (1991) Innervation of lymphoid tissue. In: Ader R, Felten DL, Cohen N (eds.) Psychoneuroimmunology, 2nd edn. Academic Press, San Diego, pp 27–69
6. Crabtree GR (1989) Contingent genetic regulatory events in T lymphocyte activation. Science 243:355–361
7. Parsey MV, Lewis GK (1993) Actin polymerization and pseudopod reorganization accompany anti-CD3-induced growth arrest in Jurkat T cells. J Immunol 151:1881–1893
8. Selliah N, Bartik MM, Carlson SL, Brooks WH, Roszman TL (1995) cAMP accumulation in T-cells inhibits anti-CD3 monoclonal antibody-induced actin polymerization. J Neuroimmunol 56:107–112
9. Kammer GM (1988) The adenylate cyclase-cAMP-protein kinase: a pathway and regulation of the immune response. Immunol Today 9:222–229
10. Exton JH (1985) Mechanisms involved in α-adrenergic phenomena. Am J Physiol 248:E633–E647
11. Morgan NG, Waynick LE, Exton JH (1983) Characterisation of the α-1-adrenergic control of hepatic cAMP in male rats. Eur J Pharmacol 96:1–10
12. Weishaar RE (1987) Multiple molecular forms of phosphodiesterase: an overview. J Cyclic Nucleotide Protein Phosphor Res 11:463–472
13. Carlson SL, Brooks WH, Roszman TL (1989) Neurotransmitter-lymphocyte interactions: dual receptor modulation of lymphocyte proliferation and cAMP production. J Neuroimmunol 24:155–162
14. Roszman TL, Brooks WH (1992) Signaling pathways of the neuroendocrine-immune network. Chem Immunol 52:170–190
15. Bishopric NH, Cohen HJ, Lefkowitz RJ (1980) Beta-adrenergic receptors in lymphocyte subpopulations. J Allergy Clin Immunol 65:29–33
16. Johnson DL, Gordon MA (1980) Characteristics of adrenergic binding sites associated with murine lymphocytes isolated from spleen. J Immunopharmacol 2:435–452
17. Brodde O-E, Engel G, Hoyer D, Bock KD, Weber F (1981) The β-adrenergic receptor in human lymphocytes: subclassification by the use of a new radio-ligand ($\pm$)-^{125}iodocyanopindolol. Life Sci 29:2189–2198
18. Landmann RMA, Bürgisser E, West M, Bühler FR (1984) Beta-adrenergic receptors are different in subpopulations of human circulating lymphocytes. J Recept Res 4:37–50
19. Loveland BE, Jarrot B, McKenzie IFC (1981) The detection of β-adrenoceptors on murine lymphocytes. Int J Immunopharmacol 3:45–55
20. Miles K, Atweh S, Otten G, Arnason BGW, Chelmicka-Schorr E (1984) β-adrenergic receptors on splenic lymphocytes from axotomized mice. Int J Immunopharmacol 6:171– 177
21. Miles K, Chelmicka-Schorr E, Atweh S, Otten G, Arnason BGW (1985) Sympathetic ablation alters lymphocyte membrane properties. J Immunol 135:797s–801s
22. Pochet R, Delespesse G, Gausset PW, Collet H (1979) Distribution of beta-adrenergic receptors on human lymphocyte subpopulations. Clin Exp Immunol 38:578–584
23. Williams LT, Snyderman R, Lefkowitz RJ (1976) Identification of β-adrenergic receptors in human lymphocytes by (-) [^{3}H] alprenolol binding. J Clin Invest 57:149–155
24. Fuchs BA, Campbell KS, Munson AE (1988) Norepinephrine and serotonin content of the murine spleen: its relationship to lymphocyte β-adrenergic receptor density and the humoral immune response in vivo and in vitro. Cellular Immunol 117:339–351
25. Sanders VM, Street NE, Fuchs BA (1994) Differential expression of the β-adrenoceptor by subsets of T-helper lymphocytes. FASEB J 8:A114

26. Besedovsky HO, Del Rey A, Sorkin E, Da Prada M, Keller HH (1979) Immunoregulation mediated by the sympathetic nervous system. Cellular Immunol 48:346–355
27. Li YS, Kouassi E, Revillard JP (1990) Differential regulation of mouse B-cell activation by β-adrenoceptor stimulation depending on type of mitogens. Immunology 69:367–372
28. Sanders VM, Munson AE (1985) Norepinephrine and the antibody response. Pharmacol Rev 37:229–248
29. Hatfield SM, Petersen BH, DiMicco JA (1986) Beta-adrenoceptor modulation of the generation of murine cytotoxic T lymphocytes in vitro. J Pharmacol Exp Ther 239:460–466
30. Feldman RD, Hunninghake GW, McArdle WL (1987) β-adrenergic receptor-mediated suppression of interleukin 2 receptors in human lymphocytes. J Immunol 139: 3355–3359
31. Borda ES, de Bracco MME, Leirós CP, Sterin-Borda L (1990) Expression of α-adrenoceptors in a human transformed lymphoblastoid cell line. J Neuroimmunol 29:165–172
32. McPherson GA, Summers RJ (1982) Characteristics and localization of [^{3}H]-clonidine binding in membranes prepared from guinea-pig spleen. Clin Exp Pharmacol Physiol 9:77–87
33. Titinchi S, Clark B (1984) Alpha2-adrenoceptors in human lymphocytes: direct characterisation by [^{3}H]yohimbine binding. Biochem Biophys Res Commun 121:1–7
34. Hadden JW, Hadden EM, Middleton E Jr (1970) Lymphocyte blast transformation. I. Demonstration of adrenergic receptors in human peripheral lymphocytes. Cell Immunol 1:583–595
35. Sanders VM, Munson AE (1985) Role of alpha-adrenoceptor activation in modulating the murine primary antibody response in vitro. J Pharmacol Exp Ther 232:395–400
36. Heilig M, Irwin M, Grewal I, Sercarz E (1993) Sympathetic regulation of T-helper cell function. Brain Behav Immun 7:154–163
37. Felsner P, Hofer D, Rinner I, Mangge H, Gruber M, Korsatko W, Schauenstein K (1992) Continuous in vivo treatment with catecholamines suppresses in vitro reactivity of rat peripheral blood T-lymphocytes via α-mediated mechanisms. J Neuroimmunol 37:47–57
38. Felsner P, Hofer D, Rinner I, Porta S, Korsatko W, Schauenstein K (1995) Adrenergic suppression of peripheral blood T-cell reactivity in the rat is due to activation of peripheral α2-receptors. J Neuroimmunol 57:27–34
39. Fernández-López A, Pazos A (1994) Identification of α2-adrenoceptors in rat lymph node and spleen: an autoradiographic study. Eur J Pharmacol 252:333–336
40. Mosmann TR, Cherwinski H, Bond MW, Giedlin MA, Coffman RL (1986) Two types of murine helper T cell clone. I. Definition according to profiles of lymphokine activities and secreted proteins. J Immunol 136:2348–2357
41. Mosmann TR, Coffman RL (1989) Heterogeneity of cytokine secretion patterns and functions of helper T cells. Adv Immunol 46:111–147
42. O'Garra A, Murphy K (1994) Role of cytokines in determining T-lymphocyte function. Curr Opin Immunol 6:458–466
43. Coffman RL, Varkila K, Scott P, Chatelain R (1991) Role of cytokines in the differentiation of $CD4^+$ T-cell subsets in vivo. Immunol Rev 123:189–207
44. Mosmann TR, Schumacher JH, Street NF, Budd R, O'Garra A, Fong TAT, Bond MW, Moore KWM, Sher A, Fiorentino DF (1991) Diversity of cytokine synthesis and function of mouse $CD4^+$ T cells. Immunol Rev 123:209–229
45. Cher DJ, Mosmann TR (1987) Two types of murine helper T cell clone. II. Delayed-type hypersensitivity is mediated by Th1 clones. J Immunol 138:3688–3694
46. Howard JG (1986) Immunological regulation and control of experimental leishmaniasis. Int Rev Exp Pathol 28:79–116

47. Swain SL, Bradley LM, Croft M, Tonkonogy S, Atkins G, Weinberg AD et al. (1991) Helper T-cell subsets: phenotype, function and the role of lymphokines in regulating their development. Immunol Rev 123:114–144
48. Sher A, Gazzinelli RT, Oswald IP, Clerici M, Kullberg M, Pearce EJ et al. (1992) Role of T-cell derived cytokines in the downregulation of immune responses in parasitic and retroviral infection. Immunol Rev 127:183–204
49. Fernandez-Botran R, Sanders VM, Mosmann TR, Vitetta ES (1988) Lymphokine-mediated regulation of the proliferative response of clones of T helper 1 and T helper 2 cells. J Exp Med 168:543–558
50. Gajewski TF, Fitch FW (1988) Anti-proliferative effect of IFN-γ in immune regulation. I. IFN-γ inhibits the proliferation of Th2 but not Th1 murine helper T lymphocyte clones. J Immunol 140:4245–4252
51. Kostrzewa RM, Jacobowitz DM (1974) Pharmacological actions of 6-hydroxy-dopamine. Pharmacol Rev 26:199–288
52. Cross RJ, Jackson JC, Brooks WH, Sparks DL, Markesbery WR, Roszman TL (1986) Neuroimmunomodulation: impairment of humoral immune responsiveness by 6-hydroxydopamine treatment. Immunology 57:145–152
53. Cross RJ, Roszman TL (1988) Central catecholamine depletion impairs in vivo immunity but not in vitro lymphocyte activation. J Neuroimmunol 19:33–45
54. Madden KS, Moynihan JA, Brenner GJ, Felten SY, Felten DL, Livnat L (1994) Sympathetic nervous system modulation of the immune system. III. Alterations in T and B cell proliferation and differentiation in vitro following chemical sympathectomy. J Neuroimmunol 49:77–87
55. Hall NR, McClure JE, Hu S-K, Tare NS, Seals CM, Goldstein AL (1982) Effects of 6-hydroxydopamine upon primary and secondary thymus-dependent immune responses. Immunopharmacology 5:39–48
56. Felten DL, Livnat S, Felten SY, Carlson SL, Bellinger DL, Yeh P (1984) Sympathetic innervation of lymph nodes in mice. Brain Res Bull 13:693–699
57. Kasahara K, Tanaka S, Ito T, Hamashima Y (1977) Suppression of the primary immune response by chemical sympathectomy. Res Comm Chem Pathol Pharmacol 16:687–694
58. Livnat S, Felten SY, Carlson, SL, Bellinger DL, Felten DL (1985) Involvement of peripheral and central catecholamine systems in neural-immune interactions. J Neuroimmunol 10:5–30
59. Miles K, Quintáns J, Chelmicka-Schorr E, Arnason BGW (1981) The sympathetic nervous system modulates antibody response to thymus-independent antigens. J Neuroimmunol 1:101–105
60. Madden KS, Felten SY, Felten DL, Sundaresan PR, Livnat S (1989) Sympathetic neural modulation of the immune system. I. Depression of T cell immunity in vivo and in vitro following chemical sympathectomy. Brain Behav Immun 3:72–89
61. Lyte M, Ernst S, Driemeyer J, Baissa B (1991) Strain-specific enhancement of splenic T cell mitogenesis and macrophage phagocytosis following peripheral axotomy. J Neuroimmunol 31:1–8
62. Delrue-Perollet C, Li K-S, Vitiello S, Neveu PJ (1995) Peripheral catecholamines are involved in the neuroendocrine and immune effects of LPS. Brain Behav Immun 9:149–162
63. Sheridan JF, Feng N, Bonneau RH, Allen CM, Huneycutt BS, Glaser R (1991) Restraint stress differentially affects anti-viral cellular and humoral immune responses in mice. J Neuroimmunol 31:245–255
64. Bonneau RH, Sheridan JF, Feng N, Glaser R (1993) Stress-induced modulation of the primary cellular immune response to herpes simplex virus infection is mediated by both adrenal-dependent and independent mechanisms. J Neuroimmunol 42:167–176

65. Kruszewska B, Felten SY, Moynihan JA (1995) Alterations in cytokine and antibody production following chemical sympathectomy in two strains of mice. J Immunol 155:4613- 4620
66. Street NE, Schumacher JH, Fong TAT, Bass H, Fiorentino DF, Leverah, JA, Mosmann TR (1990) Heterogeneity of mouse helper T cells: evidence from bulk cultures and limiting dilution cloning for precursors of Th1 and Th2 cells. J Immunol 144:1629–1639
67. Swain SL, Weinberg AD, English M, Huston G (1990) IL-4 directs the development of Th2-like helper effectors. J Immunol 145:3796–3806
68. Hsieh C-S, Heimberger AB, Gold JS, O'Garra A, Murphy KM (1992) Differential regulation of T helper phenotype development by interleukins 4 and 10 in an $\alpha\beta$-TCR transgenic system. PNAS USA 89:6065–6069
69. Seder RA, Paul WE, Davis MM, Fazekas de St. Groth B (1992) The presence of interleukin 4 during in vitro priming determines the lymphokine-producing potential of CD4+ T cells from T cell receptor transgenic mice. J Exp Med 176:1091–1098
70. Chatelain R, Varkila K, Coffman RL (1992) IL-4 induces a Th2 response in *Leishmania major*-infected mice. J Immunol 148:1182–1187
71. Noben-Trauth N, Kropf P, Muller I (1996) Susceptibility to *Leishmania major* infection in interleukin-4-deficient mice. Science 271:987–990
72. Hosken NA, Shibuya K, Heath AW, Murphy KM, O'Garra A (1995) The effect of antigen dose on $CD4^+$ T helper cell phenotype development in a T cell receptor-$\alpha\beta$-transgenic model. J Exp Med 182:1579–1584
73. Brenner GJ, Cohen N, Moynihan JA (1994) Similar immune response to nonlethal infection with herpes simplex virus-1 in sensitive (BALB/c) and resistant (C57B1/6) strains of mice. Cell Immunol 157:510–524
74. Khan MM, Sansoni P, Silverman ED, Engleman EG, Melmon KL (1986) β-adrenergic receptors on human suppressor, helper and cytotoxic lymphocytes. Biochem Pharmacol 35:1137–1142
75. Kobilka B (1992) Adrenergic receptors as models for G protein-coupled receptors. Annu Rev Neurosci 15:87–114

Chapter 4

Modulation of glucocorticoid activity by metabolism of steroids in non-lymphoid organs

Brian R. Walker

Introduction

The adrenal cortex secretes three classes of steroid hormones: mineralocorticoids, androgens and glucocorticoids [1]. In rodents the principal glucocorticoid is corticosterone, but in humans cortisol (i.e. 17α-OH-corticosterone) is more abundant because 17-hydroxylase is expressed in adrenal fasciculata/reticularis cells. This enzyme also accounts for greater production of adrenal androgens (dehydroepiandrosterone and androstenedione) in humans than in rodents. A wide variety of synthetic glucocorticoids is available, in most of which affinity for the glucocorticoid receptor is increased by inclusion of a halogen group in the 9α-position, as in beclomethasone or dexamethasone.

The only consistent physiological stimulus to the adrenal cortex which enhances glucocorticoid secretion is adrenocorticotrophin (ACTH). Secretion of cortisol in response to other stimuli, including gastrointestinal peptide [2], other peptides derived from the ACTH precursor proopiomelanocortin (POMC) [3], and transforming growth factor-β (TGF-β) [4], may not be physiological.

The actions of glucocorticoids are numerous and diverse, and are mediated by activation of corticosteroid type 2, or glucocorticoid, receptors. Glucocorticoid receptors are expressed in almost all mammalian cells. In addition, some glucocorticoids interact with other receptors. For example, cortisol (but not dexamethasone) has the same affinity as aldosterone for the corticosteroid type 1, or mineralocorticoid, receptor [5]. Cortisol has also been shown to bind to putative cell membrane receptors [6,7]. As a result of these diverse actions, excessive secretion of glucocorticoids, exemplified in clinical Cushing's syndrome [8,9], causes hypertension (mediated at least in part by enhanced vasoconstriction [10] and sodium retention [11]), opposes the actions of insulin causing obesity and glucose intolerance [12,13], enhances protein catabolism causing muscle wasting, has profound neuropsychiatric effects causing depression, cognitive impairment and/or psychosis [14], and impairs numerous aspects of the immune response (see elsewhere in this volume). By contrast, inadequate cortisol secretion, exemplified in Addison's disease [15], is associated with hypotension, dilutional hyponatraemia, hypoglycemia, depression and death. This absolute requirement for glucocorticoids is increased at times of stress, when the principal role of glucocorticoids may be to limit potential deleterious effects of the acute phase reaction. In this regard, the time course of glucocorticoid action, dependent on intracellular receptor activation and gene transcription and maximal after 6 h, is significantly slower than most responses to stress, and allows glucocorticoids to act as a physiological "brake" on the stress response.

The purpose of this chapter is to review factors which can influence glucocorticoid activity. Other chapters will address factors which influence ACTH secretion and control cortisol secretion rates. This chapter will deal with factors which have been recognized more recently to influence glucocorticoid activity, and which exert their influence on glucocorticoid-responsive target organs.

The normal response to glucocorticoids

The glucocorticoid receptor is one of a family of intracellular receptors which includes all the steroid receptors (mineralocorticoid, androgen, oestrogen, progesterone and vitamin D), the retinoid receptors, the thyroid hormone receptor, and many more "orphan" receptors for which endogenous ligands have yet to be identified. With some important differences, these receptors share similar modes of action [16]. Key elements of this action are represented in Fig. 4.1. Much of the work identifying these steps has been performed in transfected transformed cell

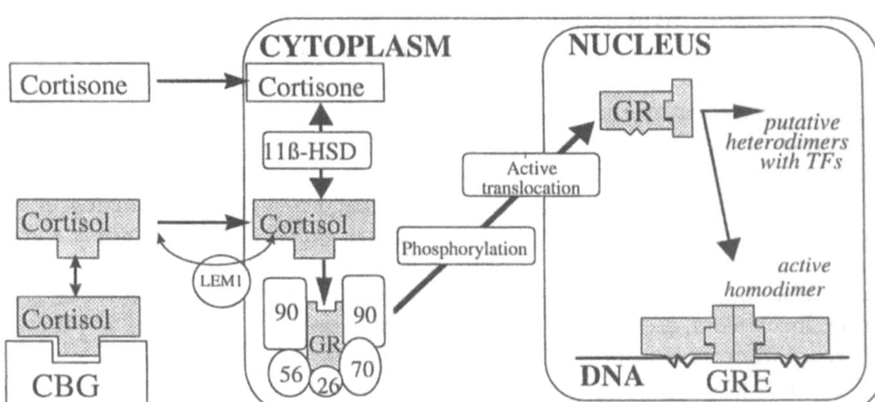

Fig. 4.1. Factors influencing tissue sensitivity to cortisol. 11β-HSD, 11β-hydroxysteroid dehydrogenases; GR, glucocorticoid receptor; ‾\/\/‾, zinc fingers, which bind to DNA; 90, 70, 56, and 26 indicate heat shock proteins of these molecular weights, in kDa, which dissociate from the activated receptor; TF, transcription factor, e.g. AP-1; GRE, glucocorticoid response element, typically a palindromic sequence to which the receptors bind; LEM1 is a yeast transmembrane transporter; CBG, corticosteroid-binding globulin.

lines, and for some factors the relevance to normal cells is uncertain. What is clear is that all of these steps may differ according to cell type, and many of them could be subject to regulation.

Access of ligand to intracellular receptors

Cortisol and corticosterone both circulate in complexes with corticosteroid-binding globulin (CBG). Only the small proportion (~5%) of free steroid is able to diffuse into interstitial fluid and across the cell membrane. If the dissociation of glucocorticoid from its binding protein were uniform, then the regulation of free cortisol concentrations could occur via hypothalamic-pituitary feedback independently of the bound element of circulating cortisol. However, recent evidence suggests that the dissociation from CBG may be influenced by the local environment. For example, cortisol is dissociated from CBG by neutrophil lysosymal enzyme release [17]. Moreover, CBG may not always be limited to the plasma compartment and its leakage into interstitial fluid may enhance the delivery of cortisol [18,19]. Finally, the movement of free steroid across the cell membrane may be subject to active transport, as has been observed with the LEM-1 transporter in yeast cell membrane [20], and this will in turn result in dissociation of further free steroid from its binding protein.

Once in the target cell, cortisol is not certain to reach the glucocorticoid receptor. It may bind to other proteins, for example the β-isoform of the glucocorticoid receptor which, by virtue of a truncated C-terminus, does not induce gene transcription [21]. Alternatively, cortisol may be metabolized by a variety of enzymes, including 11β-dehydrogenase which is described in detail below. It may also be actively exported out of the cell by proteins such as LEM1. However, by the same token, the concentration of intracellular cortisol can be enhanced, especially by the conversion of its most abundant circulating metabolite, cortisone, into cortisol by the enzyme 11β-reductase (see below).

Another important variable is the availability of adequate numbers of receptors for ligand binding. However, glucocorticoid receptors have a relatively low affinity for their ligand (K_d for cortisol 40 nM compared with free cortisol concentrations ~10–50 nM) [22] but are expressed in abundance in most tissues. They are therefore rarely saturated. For that reason the regulation of their expression, for example, down-regulation by ligand exposure [23], may have surprisingly little influence on absolute receptor occupancy and biological response.

Activation of intracellular receptors

By contrast with other members of the intracellular receptor family which are exclusively intranuclear, the glucocorticoid receptor is found in both cytoplasm and nucleus and there remains some controversy about where ligand binding takes place [24]. In its non-activated state in cytosol the receptor is ~94 kDa and is associated with several heat shock proteins (hsp). These hsps can influence the structure of the non-activated receptor complex such that ligand-binding affinity is altered [25]. This, or the dependence of the formation of the hsp-receptor complex on adenosine triphosphate (ATP), might explain the reversible change in receptor affinity for dexamethasone which has been observed in leukocytes from patients with steroid-resistant asthma. This phenomenon appears to depend on other humoral factors, since ligand-binding affinity is normalized after 48 h culture of these cells but not if the patient's serum is included in the culture medium [26]. Ligand binding results in dissociation of the receptor from the hsps, phosphorylation of serines, and the initiation of active translocation to the nucleus. Once in the nucleus, glucocorticoid receptors need to dimerize before they bind to DNA, but they may dimerize with a number of proteins, thus forming homodimers with each other or heterodimers with transcription factors (including *jun* transcription factor and the *fos-jun* heterodimer, AP-1) [16]. The availability of these potential "molecular chaperones" for the receptor may, at least in transfected cells, dictate the extent to which activated glucocorticoid receptors bind to DNA. For example, activated glucocorticoid receptors may be "squelched" by excessive levels of AP-1, calreticulin, or NF-κB [16].

Dimerized glucocorticoid receptors bind to a limited array of response elements on DNA. A common conformation is the palindromic repeat sequence which allows binding of a glucocorticoid receptor homodimer. This is one of the commonest motifs observed in the promoter regions of human genes. Once bound, the receptor interacts with other transcription factors either to induce or repress gene transcription. Binding to DNA is influenced, in a cell-specific manner, by the chromatin conformation and the presence of other gene-specific promoters which allow gene transcription to occur. Some of these effects may oppose, and indeed overcome, glucocorticoid action, as in the "dominant negative" effect of AP-1 or NF-κB. Finally, the transcripts produced or repressed in response to activated glucocorticoid receptors will clearly have different actions according to the cellular environment in which they are expressed.

Lessons from abnormal tissue responsiveness to glucocorticoids in humans

From the above, it is clear that there is a bewildering array of variables which influence the responses to glucocorticoids, either in transfected cells or in theory. To establish which of these is important in humans it is helpful to consider clinical syndromes in which dysfunctional glucocorticoid signalling has a demonstrable clinical impact. In the population, inter-individual variability in sensitivity to synthetic glucocorticoids has been reported to be a clinical problem in the treatment of asthma and has been observed in the skin. Only rarely, however, do syndromes of profound alterations in sensitivity to endogenous glucocorticoids present spontaneously.

Dysfunction of the glucocorticoid receptor

Congenital resistance to glucocorticoids has been observed in several families, and is associated with a variety of mutations in the glucocorticoid receptor or in hsp90 [16,27,28]. Some cases remain unexplained by molecular investigation. The clinical presentation is predictable, with impaired negative feedback resulting in enhanced ACTH and cortisol levels but no features of Cushing's syndrome. This biochemical response seems to overcome the resistance such that clinical consequences of glucocorticoid deficiency are rarely encountered. However, the enhanced ACTH drive also stimulates adrenal androgens and the precursor mineralocorticoid 11-deoxycorticosterone, which may produce amenorrhoea/hirsutism and hypokalaemic hypertension, respectively.

Enhanced sensitivity of the glucocorticoid receptor leading to Cushing's syndrome has only been reported in one case [29]. The molecular defect has not been identified.

Alterations in cortisol metabolism: 11β-dehydrogenase deficiency

An alternative explanation for variable tissue sensitivity to cortisol is variability in the local metabolism in the target tissue. This possibility is also exemplified by a rare congenital syndrome, but in this case the subsequent clinical investigation opened a new area of biology.

Between 1972 and 1984, fewer than 20 children and one adult had been reported with the "syndrome of apparent mineralocorticoid excess" [30–33]. Inheritance followed an autosomal recessive pattern. These patients had profound hypertension accompanied by hallmarks of excessive activation of mineralocorticoid receptors, including hypokalaemia. However, no known mineralocorticoid hormone could be detected in their serum. Another unexplained but consistent feature of the syndrome was an abnormality of cortisol metabolism in which the metabolites of cortisol (i.e. tetrahydrocortisols and cortols) are present in urine in much greater abundance than those of cortisone (tetrahydrocortisone and cortolones). Normally, these metabolites are represented in approximately equal proportions. Subsequent detailed clinical investigation showed that it was cortisol which was responsible for stimulation of mineralocorticoid receptors in these patients [32,34], and that the altered cortisol metabolites could be accounted for by impaired activity of 11β-dehydrogenase, the enzyme which converts cortisol to cortisone (Fig. 4.2). However, the link between these two observations remained unclear. Importantly, despite impaired clearance of cortisol because of impaired 11β-dehydrogenase activity, these patients had intact hypothalamic-pituitary negative feedback so that normal circulating cortisol concentrations were achieved as a result of lower cortisol secretion rates. Thus, any contribution of cortisol to the aetiology of the condition must be mediated by a change in peripheral sensitivity rather than a change in secretion.

In parallel with these clinical observations, surprising evidence was emerging concerning the properties of the mineralocorticoid receptor. The mineralocorticoid receptor, by contrast with the glucocorticoid receptor [35], has a restricted

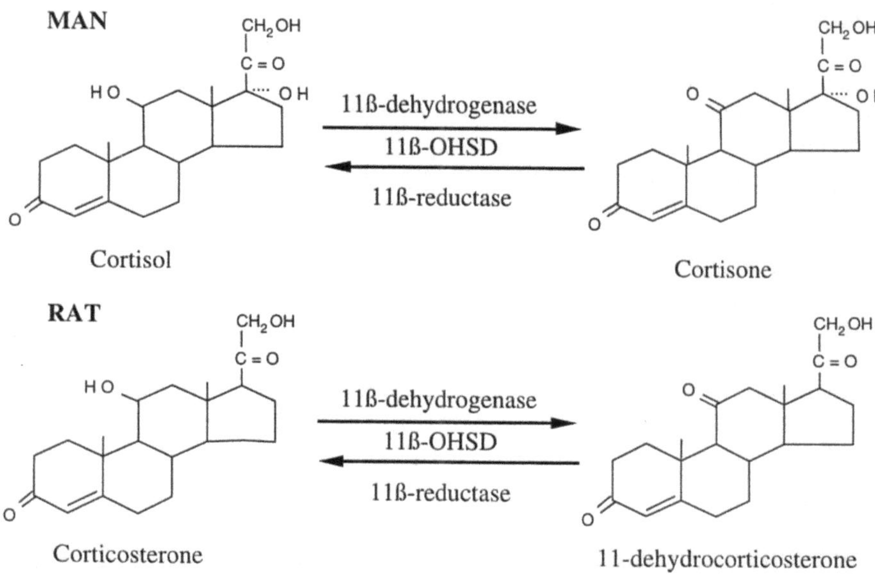

Fig. 4.2. Reactions catalysed by 11β-HSDs in man and rat.

localization in sites where aldosterone modulates electrolyte transport, such as distal nephron, colon, salivary and sweat glands [36–38]. In these sites it had been documented in vivo that mineralocorticoid receptors bind aldosterone but not cortisol [39]. However, in other sites, such as hippocampus, mineralocorticoid receptors are expressed but they bind glucocorticoids as well as mineralocorticoids [5,40]. Moreover, when the mineralocorticoid receptor was cloned and expressed, it was found to have the same affinity for cortisol as for aldosterone in vitro [5,40]. How then could it avoid binding cortisol in the kidney in vivo when cortisol concentrations are 100-fold higher than those of aldosterone? The syndrome of apparent mineralocorticoid excess seemed to be a circumstance in which this selectivity was lost and mineralocorticoid receptors were flooded with cortisol.

The next important step came again from clinical observations. It had been know since the 1940s that liquorice, when taken habitually in excess or used in the therapy of peptic ulceration, induces a syndrome of hypokalaemia, sodium retention, and hypertension [41–43]. Most interestingly, there was also evidence that liquorice-induced mineralocorticoid excess was dependent on the presence of cortisol, since it could be reversed by dexamethasone administration [44] and was absent in patients with Addison's disease [45]. In a landmark study, Stewart and Edwards [46] demonstrated that the mode of action of liquorice, or its principal active constituent glycyrrhetinic acid (Fig. 4.3) [47], was to inhibit 11β-dehydrogenase activity [46,48,49]. In rat kidney, they showed that mineralocorticoid receptors do not normally bind glucocorticoids, but they can be induced to do so by administration of glycyrrhetinic acid and inhibition of 11β-dehydrogenase [50,51]. Thus, the physiological role of 11β-dehydrogenase is to protect mineralocorticoid receptors from cortisol and allow them to bind aldosterone.

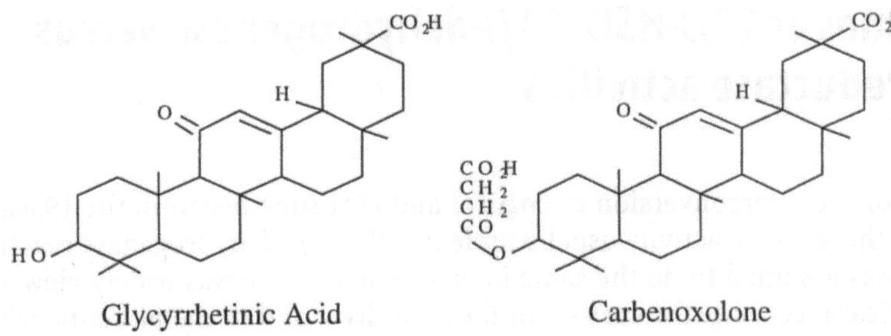

Fig. 4.3. Structure of inhibitors of 11β-HSD. Glycyrrhetinic acid is the principal active constituent of liquorice, and carbenoxolone is its hemisuccinate. Both have been used in the treatment of peptic ulceration [41,153,154].

Since then, most cases of apparent mineralocorticoid excess syndrome have been attributed to mutations in the 11β-dehydrogenase gene on the long arm of chromosome 16 (11β-HSD2) [52–55]. In addition, impaired conversion of cortisol to cortisone has been invoked as the explanation for the hypokalaemia which characterizes the ectopic ACTH syndrome [56–58], and as a contributor to the poorly understood aetiology of essential hypertension [59,60].

The ubiquitous role of pre-receptor metabolism

The importance of pre-receptor metabolism of ligands for corticosteroid receptors should, perhaps, have come as no surprise. Pre-receptor metabolism is important for many of the other members of the intracellular receptor family [61]. Usually, these conversions serve to activate a circulating pool of inert hormone in the target tissues. For example, thyroxine is converted to tri-iodothyronine by 5′-monodeiodinase, testosterone is converted to 5α-dihydrotestosterone by 5α-reductase, etc. In fact, because cortisone is not bound to CBG and has no diurnal rhythm, its *free* circulating concentrations are similar, or in excess of, those of cortisol, especially during the nocturnal nadir of cortisol secretion [56]. Thus, it is not only inactivation of cortisol which is important to corticosteroid receptor activation, but reactivation of cortisone may also be important.

The relevance of the equilibrium between cortisol and cortisone to corticosteroid receptor activation has now been explored in animal experiments and in transformed cells. Pre-receptor metabolism has been observed to influence sensitivity to glucocorticoids in a wide variety of sites, and in species as distant as mammals and toads. In addition, two enzymes responsible for 11β-dehydrogenase and 11β-reductase activity have been cloned and characterized. Finally, the effects of targeted disruption of one of these genes, 11β-HSD1 which encodes 11β-reductase, have been observed in a transgenic mouse. This chapter does not seek to provide a comprehensive review of this substantial literature, but some examples illustrate the important principles.

Isozymes of 11β-HSD: 11β-dehydrogenase versus 11β-reductase activities

Studies of the interconversion of cortisol and cortisone began in the 1950s [62]. Because this enzyme activity usually preferred the 11β-dehydrogenase reaction in vitro it was assumed to do the same in vivo. The enzyme was widely viewed as a simple clearance route for cortisol in the adult liver and in the placenta, where it might protect the fetus from excessive exposure to maternal cortisol [63,64]. However, when Monder and colleagues purified the enzyme from rat liver microsomes [65], they observed that the 11β-reductase and 11β-dehydrogenase activities of the protein could be dissociated in vitro [66,67]. When Monder's purified 11β-HSD1 enzyme was cloned and expressed in vitro [68] its activity was dependent upon the cell type in which it was expressed, being a dehydrogenase in CHO-cells [69] but a reductase in COS-7 cells [70]. In hepatocytes [71] and hippocampal cells [72] in primary culture 11β-HSD1 is also a reductase. Moreover, whole organ perfusion experiments suggested that the 11β-reductase activity predominates in liver [73]. Similar observations have been made in human liver [56]. However, there remains the possibility that 11β-HSD1 catalyses 11β-dehydrogenase activity in some tissues, for example in the lung [74] and in vascular smooth muscle (see below). The relative contribution of each tissue to turnover of cortisol and cortisone has not been studied in detail. In human venous blood obtained by selective catheterization in vivo, the only organs from which venous effluent contained substantially different cortisol/cortisone ratios from arterial blood were kidney, liver, and adrenal [56].

When the role of 11β-dehydrogenase in the kidney was first described [50,51], it was assumed that the renal enzyme was the same as Monder's 11β-HSD1. However, numerous kinetic differences, differences in regulation, and discrepant results with antibodies and cDNA probes generated for 11β-HSD1 led to the prolonged search for, and eventual cloning of, 11β-HSD2 by several groups [75–78]. Differences between the two isozymes are summarized in Table 4.1. 11β-HSD2 is a high-affinity exclusive 11β-dehydrogenase under all circumstances studied. Its limited distribution is in keeping with a specific role in protecting mineralocorticoid receptors from cortisol and protecting the fetus from maternal cortisol.

11β-Reductase deficiency

The understanding that 11β-dehydrogenase and 11β-reductase activities are catalysed by different enzymes allowed us to understand a group of clinical conditions in which 11β-reductase activity is deficient. In 1985, a syndrome was described in which the ratio of cortisol : cortisone metabolites was low [79]. This turned out to be due, not to enhanced 11β-dehydrogenase activity, but to impaired conversion of cortisone to cortisol by the enzyme 11β-reductase. The resultant increase in cortisol clearance induced enhanced ACTH secretion and a

Table 4.1. Characteristics of the isoenzymes of 11β-HSD

		11β-HSD1	Homology	11β-HSD2
Amino acid sequence		34 kDa 287–292 amino acids [68,69,156–159]	20%	44 kDa 405–427 amino acids [75,76,78,160]
cDNA		1200 bp	14%	1900 bp
Cofactor affinity [88,161–164]		NADP(H)		NAD
Substrate affinity				
	Corticosterone	1.8 μM		5 nM
	Cortisol	17 μM		50 nM
	Dexamethasone	negligible		500 μM
Favoured reaction		Reductase		Dehydrogenase
Sites expressed		Liver, etc.		Distal nephron, placenta, etc.

syndrome of adrenal androgen excess. A similar abnormality has been reported in patients with polycystic ovarian syndrome [80]. No defect in the 11β-reductase gene (11β-HSD1) has yet been reported in these patients [81]. In addition, a group of patients have been described with cortisol-dependent mineralocorticoid excess and impaired renal 11β-dehydrogenase activity in whom cortisol/cortisone ratios are normal [82–86]. This has been called the "syndrome of apparent mineralocorticoid excess type 2". It has emerged that they have simultaneous impairment of 11β-dehydrogenase in kidney and 11β-reductase in liver. The molecular basis of this disorder remains to be established, but it cannot be attributed to a mutation in 11β-HSD1 or 11β-HSD2 alone and may reflect an abnormality of post-transcriptional modification or protein localization which is common to both enzymes.

One important clinical difference between 11β-HSD1 and 11β-HSD2 is their metabolism of synthetic glucocorticoids. For example, dexamethasone is avidly converted to 11-dehydrodexamethasone by 11β-HSD2 [75] but is not metabolized by 11β-HSD1 even under conditions which permit 11β-dehydrogenase activity [87]. By contrast, only 11β-HSD1 converts 11-dehydrodexamethasone back to dexamethasone. Thus, in sites where 11β-dehydrogenase is abundant, this will protect all receptors from cortisol, but will only protect them from dexamethasone if 11β-HSD2 is responsible for the activity.

Modulation of corticosteroid receptor activation by 11β-HSDs in multiple sites

Although 11β-HSDs are expressed in many tissues, there are few sites in which a physiological role has been demonstrated convincingly in vivo. In *distal nephron*, it is now accepted that mineralocorticoid receptors are protected from cortisol by 11β-HSD2. This is especially true in human kidney, in which 11β-HSD1 is not expressed significantly [88].

In the *placenta*, 11β-HSD2 is expressed in abundance and is an exclusive dehydrogenase in perfused tissue [89], homogenates, cell lines (eg JEG-3) and cells in primary culture. Increased exposure of the fetus to glucocorticoids results in growth retardation and in offspring with higher blood pressure [90] and glucose intolerance [91]. Administration of carbenoxolone to pregnant rats produces the same results in the offspring [92], but it is difficult to establish whether this results from enhanced transplacental transfer of maternal corticosterone, or from effects of carbenoxolone on the fetus. However, in humans and rats there is a direct relationship between 11β-dehydrogenase activity in the placenta and birth weight [89,90,93], and relationships between birth weight and adult blood pressure and glucose tolerance [94,95], so that 11β-HSD2 could provide a placental barrier which, when impaired, programmes subsequent cardiovascular risk in the offspring.

In the *liver*, only 11β-HSD1 is expressed in all species. The liver also expresses glucocorticoid receptors in abundance but does not express mineralocorticoid receptors. Since cortisone does not bind to CBG and circulates at concentrations similar to those of free cortisol [56,96,97], this observation suggests that the conversion of cortisone to cortisol could enhance local glucocorticoid receptor activation, thus ensuring adequate exposure of these low affinity receptors to their endogenous ligand in a tissue-specific manner (Fig. 4.4). This suggestion is borne out by multiple lines of evidence. In cultured hepatic cell lines and primary cultured cells inhibition of 11β-reductase activity results in impaired glucocorticoid activity [71], by contrast with the enhanced glucocorticoid activity associated with inhibition of 11β-dehydrogenase in other cell lines [98]. In rats, carbenoxolone

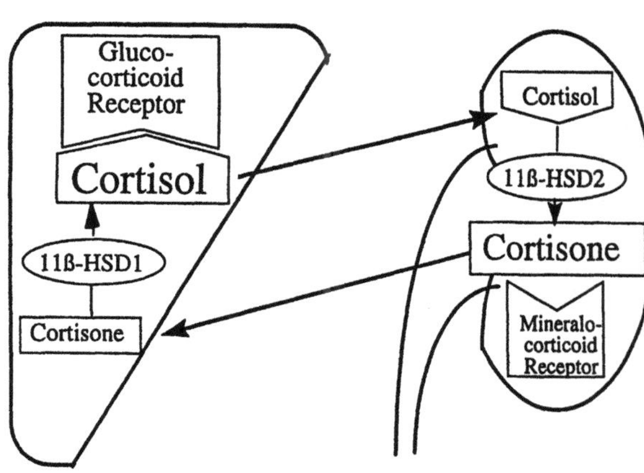

Fig. 4.4. Contrasting direction of 11β-HSD reactions in liver (left) and kidney (right). Predominant conversion of cortisol to cortisone by the dehydrogenase 11β-HSD2 in kidney results in protection of local mineralocorticoid receptors. Predominant conversion of cortisone to cortisol by the reductase 11β-HSD1 in liver results in enhanced activation of glucocorticoid receptors.

inhibits hepatic 11β-HSD type 1 and produces a dose-dependent reduction in fasting glucose concentration (P.M. Jamieson et al., unpublished work). We have also employed the selective down-regulation of hepatic 11β-reductase expression which is observed with oestrogen administration in rats [99]. Oestrogen increases the expression of phosphoenolpyruvate carboxykinase (PEP-CK; the rate-limiting step in gluconeogenesis and principal site of action for glucocorticoids and insulin in the liver) in adrenalectomized animals, but decreases its expression in non-adrenalectomized rats [100]. This glucocorticoid-dependent effect of oestrogen is consistent with a fall in intra-hepatic glucocorticoid concentrations consequent on inhibition of 11β-reductase activity. In mice, a transgenic strain with selective knockout of 11β-HSD type 1 has recently been bred. In the fasting state, these animals have relative hypoglycaemia, and lower hepatic PEP-CK and glucose-6-phosphatase mRNA expression and activity, changes which are consistent with lower intra-hepatic glucocorticoid levels and enhanced insulin sensitivity [101]. Finally in humans, carbenoxolone inhibits conversion of cortisone to cortisol in the liver (Fig. 4.5) [48,102], and results in an increase in sensitivity to insulin which was not attributable to enhanced peripheral insulin sensitivity (Fig. 4.6) [103]. This suggests that hepatic insulin sensitivity was increased as a consequence of lowering intra-hepatic cortisol concentrations.

In the *brain*, 11β-HSD1 is the predominant enzyme and is expressed in a limited range of sites [104–107]. Its role is unclear, but experiments in primary cultured rat hippocampal cells have shown that the enzyme is a reductase, and that its activity enhances the neurotoxic effects of glucocorticoids [72]. However, experiments to address the functional importance of 11β-HSD in the central nervous system in vivo have produced conflicting results [108–110].

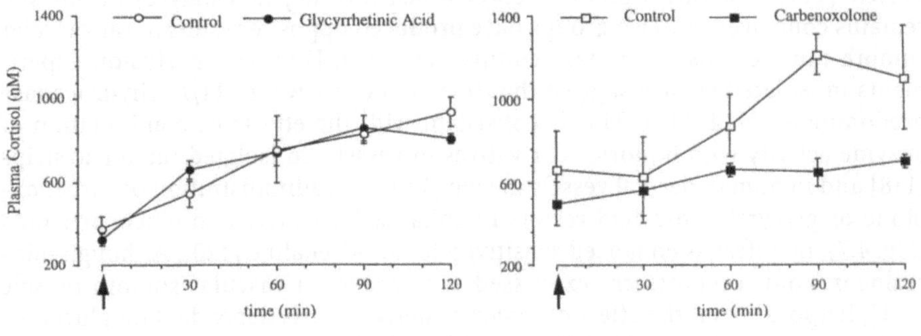

Fig. 4.5. Inhibition of hepatic 11β-reductase by carbenoxolone but not by glycyrrhetinic acid in man. Two groups of three healthy subjects were studied. They fasted at 0900 h before and after therapy with glycyrrhetinic acid (170 mg 8 hourly for 8 days) or carbenoxolone (100 mg 8 hourly for 8 days). At time 0 they took 25 mg of cortisone acetate orally (arrowhead). Blood was withdrawn during the following 2 h for measurement of plasma cortisol by direct radioimmunoassay. Bars are SEM. Comparison of the curves by 2-way repeated measure analysis of variance confirms that glycyrrhetinic acid had no effect but carbenoxolone reduced the conversion of cortisone to cortisol ($P < 0.04$) [102].

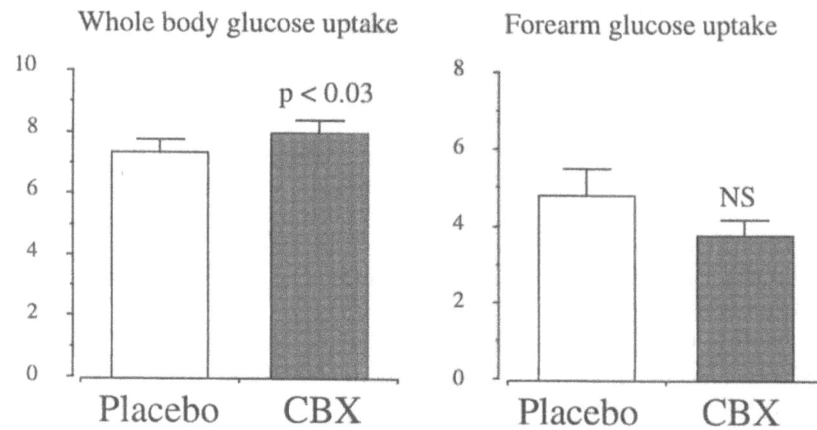

Fig. 4.6. Influence of 11β-HSD on insulin sensitivity in man. Seven healthy males participated in a double-blind cross-over study comparing carbenoxolone (100 mg 8 hourly for 8 days) with placebo. Euglycaemic hyperinsulinaemic clamps were performed with measurement of forearm glucose uptake by plethysmography and collection of arterialized and deep forearm vein samples. Whole body insulin sensitivity is represented by the M value (in $\mu mol.kg^{-1}.min^{-1}$), or rate of dextrose infusion to maintain euglycaemia in the face of constant insulin infusion. Forearm insulin sensitivity is represented as forearm glucose uptake in $\mu mol.100\ ml^{-1}.min^{-1}$. Bars are SEM. Carbenoxolone increased whole-body insulin sensitivity without affecting peripheral insulin sensitivity, consistent with lowering intra-hepatic cortisol concentrations as a result of inhibition of 11β-reductase activity of 11β-HSD1 (Fig. 4.5) [155].

In *blood vessels*, 11 β-HSD1 but not 11β-HSD2 is expressed in vascular smooth muscle [111]. The direction of reaction which the enzyme catalyses in this site remains controversial. Two groups have produced opposite results in rat vascular smooth muscle cells in primary culture [112,113]. However, perfusion experiments in isolated vessels suggest that this is one site where 11β-dehydrogenase predominates [114–117]. This is consistent with the effects of manipulation of enzyme activity with liquorice derivatives in vessels. In isolated rat aortic strips [118] and in human dermal vessels in vivo [119,120], administration of carbenoxolone or glycyrrhetinic acid results in enhanced sensitivity to glucocorticoids (Fig. 4.7), manifest as enhanced sensitivity to noradrenaline [120]. Although mineralocorticoid receptors are expressed at low level in vascular smooth muscle [121], it appears that the effect of liquorice derivatives is dependent on glucocorticoid receptor activation [118,122,123]. Thus, 11β-HSD1 appears, in this tissue at least, to function as a dehydrogenase and protect glucocorticoid receptors from cortisol. Vascular 11β-HSD1 expression and activity are reduced in some hypertensive rat strains [114–117]. Defects in this protection may therefore contribute to enhanced vasoconstrictor sensitivity to glucocorticoids in patients with essential hypertension [124]. The enzyme is also expressed in *cardiac myocytes* [111,125] where it could modulate mineralocorticoid receptor-induced fibrosis [126].

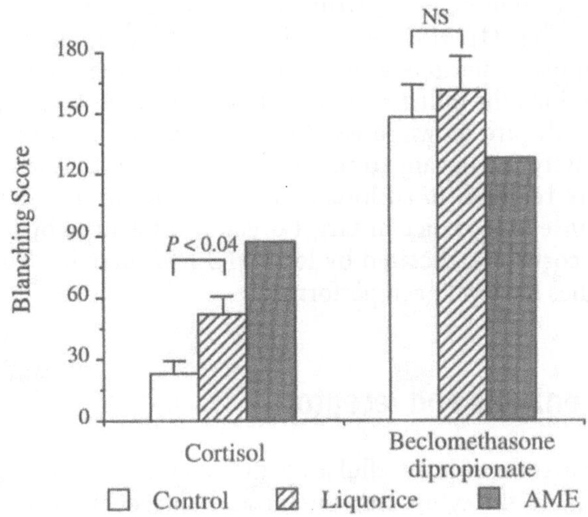

Fig. 4.7. Influence of 11β-HSD on vascular sensitivity to glucocorticoids. Six healthy volunteers were studied before and after open administration of liquorice (300 g daily for 7 days). Vasoconstriction was assessed in the skin on a visual scale after overnight occluded topical application of cortisol or beclomethasone dipropionate, a synthetic glucocorticoid which is not metabolized by 11β-HSD2. A patient with apparent mineralocorticoid excess (AME) type 1 [32], in whom a defect in 11β-HSD2 gene has not yet been identified, was studied in the same way. Bars are SEM. Liquorice potentiated the blanching response to cortisol but not to beclomethasone dipropionate, and a similar pattern was observed in the patient with congenital 11β-dehydrogenase impairment [120]. This suggests that 11β-HSD1, which is the predominant enzyme in vascular smooth muscle, protects vascular glucocorticoid receptors from cortisol. NS = not significant.

In the *testis*, early experiments also suggested that 11β-HSD1 acts as a dehydrogenase. Its expression is markedly enhanced during pubertal development, and administration of carbenoxolone permitted cortisol to suppress LH-induced testosterone release in cultured Leydig cells [98,127]. This protective mechanism appeared to be under physiological control and might contribute to stress-induced down-regulation of the hypothalamic-pituitary-gonadal axis [128]. However, these results have been contradicted by recent data from our laboratory which are more consistent with 11β-HSD1 acting as a reductase in Leydig cells (C.M. Leckie et al., unpublished work). 11β-HSD1 is also expressed in the *ovary* [129]. 11β-HSD1 in human granulosa-lutein cells obtained during oocyte harvesting for in vitro fertilization acts as a dehydrogenase, and protects the cells from cortisol-dependent inhibition of steroidogenesis [130]. Interestingly, the presence of 11β-HSD1 in these cells is associated with a lower chance of successful pregnancy [131].

In addition to these sites, 11β-HSDs have been described and manipulated in a wide variety of systems, as diverse as the *inner ear* [132] and *toad bladder mucosa* cell lines [133]. However, the relevance of 11β-HSD activity to *immune function* has not been investigated until the important work by Daynes and colleagues.

Their results in the lymph node stromal cell are described in Chapter 5. Disappointingly, neither 11β-HSD isozyme is expressed in circulating leukocytes. Glucocorticoids induce a peripheral neutrophilia. In very crude experiments in man, we have examined the influence of liquorice derivatives on peripheral white cell counts. The results are shown in Fig. 4.8. These data do not provide evidence that 11β-HSD activity is relevant to immune function, but clearly much more detailed studies are required to address this. An important question is whether the immune response in a particular target organ is influenced by the prevailing ratio of cortisol : cortisone, dictated by local 11β-HSD activity. To our knowledge, no such studies have yet been performed.

Localization of enzyme and receptor

Although the modulation of intracellular cortisol concentrations is portrayed in Fig. 4.1 as occurring in the cytoplasm, the intracellular localization of 11β-HSD, rather like the localization of the glucocorticoid receptor, remains to be established. In the placenta, the hypothesis that 11β-HSD2 protects the fetus from maternal cortisol suggests that the enzyme plays an endocrine role in protecting distant organs. Similarly, 11β-HSD1 and 11β-HSD2 are expressed in the adrenal cortex, such that there is substantial adrenal secretion of cortisone [56,96, 134–136], which has a potential endocrine action in sites where it is converted to

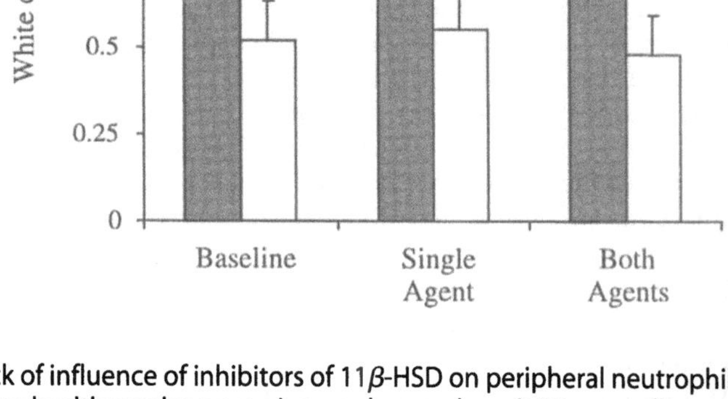

Fig. 4.8. Lack of influence of inhibitors of 11β-HSD on peripheral neutrophil counts. Two groups of three healthy males were given carbenoxolone (100 mg orally every 8 hours) or glycyrrhetinic acid (170 mg orally every 8 hours) for 2 days, followed by both drugs in combination for a further 2 days. White cell counts ($10^9.1^{-1}$) were measured each morning in venous whole blood. Increased glucocorticoid activity in the bone marrow results in a rapid and maintained neutrophilia. However, no significant differences were observed in this small study. Bars are SEM.

cortisol. Original studies in the kidney used the probes available for 11β-HSD1 and showed that it was expressed in different cells from the mineralocorticoid receptors it was thought to protect [50]. A paracrine protection of receptors was therefore suggested. However, it rapidly became clear that 11β-HSD2 is expressed in the same cells as the mineralocorticoid receptor [75,77,137,138], and that the mechanism was likely to be autocrine. The most recent studies using con-focal microscopy suggest that 11β-HSD2 is localized in the endoplasmic reticulum [139]. Nevertheless, it remains possible that 11β-HSDs play a paracrine role in other tissues. For example, Daynes et al. have shown expression of 11β-HSD2 in stromal cells of lymph nodes, which are probably not the principal target cells for glucocorticoids in this site (see Chapter 5).

Regulation of pre-receptor metabolism

Regulation of 11β-HSD expression

Following the discovery of enzyme-mediated receptor protection of renal mineralocorticoid receptors, a large number of experiments was performed to examine regulation of enzyme expression in the rat. Unfortunately, these experiments used tools generated against Monder's 11β-HSD1. Effects on 11β-HSD2 can only be inferred indirectly, and few experiments have been repeated with newly available tools for 11β-HSD2. The results are summarized in Table 4.2. In summary, 11β-HSD2 is subject to dramatic changes during development in utero, but remains constitutively expressed in the adult. By contrast, 11β-HSD1 in the adult is subject to regulation by glucocorticoids, sex steroids, thyroid hormones, and growth hormone. Moreover, chronic stress induced by adjuvant arthritis results in chronic induction of 11β-HSD1 expression [140]. The mediators of this effect, specifically the influence of cytokines on 11β-HSD expression, have not been investigated.

Table 4.2. Regulation of 11β-HSDs

	11β-HSD1	11β-HSD2
Up-regulation	Glucocorticoids [140,145,165,166]	Oestradiol [167]
	Thyroxine [168–172]	
Down-regulation	Insulin [71,166]	ACTH [56]
	Oestradiol [99,167,173]	
	Growth Hormone [99]	

In addition to these animal experiments, human studies have addressed the influence of ACTH on peripheral cortisol inactivation. These studies arose from the surprising observation that patients with Cushing's syndrome are more likely to have evidence of mineralocorticoid excess (i.e. hypokalaemia and fluid retention) if they have ectopic ACTH production than pituitary-dependent Cushing's syndrome [141–143]. This used to be attributed to excessive secretion of deoxycorticosterone, but in fact there is a poor relationship between levels of this mineralocorticoid and the degree of hypokalaemia. An alternative explanation, confirmed in several studies, is that the inactivation of cortisol by 11β-HSD2 is less effective in the ectopic ACTH syndrome [56,57,144]. The reason for this is unclear, but one possibility is that the higher levels of ACTH observed in ectopic production are responsible. In healthy humans, we showed that conversion of cortisol to cortisone is indeed impaired when ACTH is present compared with when ACTH is suppressed (Fig. 4.9) [56]. In rats, this effect was not observed in adrenalectomized animals [145] raising the possibility that an ACTH-dependent adrenal product is responsible. However, the mechanism of the effect remains elusive, and no ACTH-dependent inhibitory activity has yet been identified [146]. Importantly, it suggests that the enhanced reactivation of glucocorticoid as a result of induction of 11β-HSD1 in chronic stress may be further exacerbated by impaired inactivation of cortisol by an effect of ACTH. The net result is that activation of the hypothalamic-pituitary-adrenal axis results in both enhanced cortisol secretion and enhanced tissue sensitivity to cortisol.

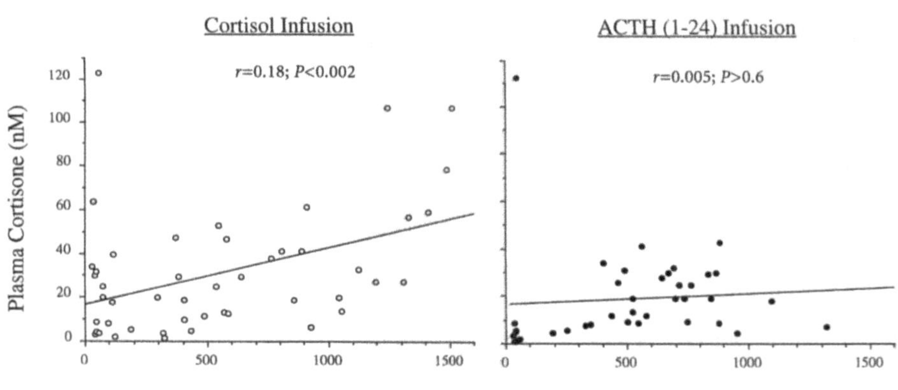

Fig. 4.9. Contrasting conversion of cortisol to cortisone in the presence and absence of ACTH in man. Five healthy males were studied on two occasions in random order. On both occasions they took 1 mg dexamethasone orally at 2300 h to suppress endogenous ACTH and cortisol. Between 0900 h and 1100 h they were infused with incrementally increasing concentrations of cortisol (on one occasion) or ACTH (on the other occasion) to produce a similar range of plasma cortisol concentrations. Plasma cortisone was measured by radioimmunoassay and rose with increasing cortisol concentrations during cortisol infusion, i.e. in the absence of ACTH, demonstrating that 11β-dehydrogenase activity could not be saturated. However, in the presence of ACTH, cortisone failed to rise, consistent with ACTH-dependent suppression of 11β-dehydrogenase activity [56].

Competitive inhibitors of 11β-HSD activity

Of course there is more to enzyme activity than just the amount of protein expressed. From the above, it is clear that tissue specific factors dictate the equilibrium between dehydrogenase and reductase activities. The nature of these is unknown. They may include variations in glycosylation of the enzyme, a phenomenon which has been shown to influence enzyme kinetics in transfected cells [69]. In addition, a lot of work has addressed the possibility that there are endogenous inhibitors of 11β-HSD activity. Two approaches have been taken: to screen known compounds for inhibitory activity in vitro; or to measure inhibitory activity in urine and subsequently seek to identify it.

Screening of compounds "off the shelf" has yielded a long list of inhibitors of 11β-dehydrogenase activity (Table 4.3). However, none of the endogenous compounds on this list have been shown to exert an important influence on glucocorticoid action. Potentially more interesting has been the investigation of inhibitory activity which has been detected in human urine. This activity, which Morris and colleagues attributed to "glycyrrhetinic acid like factors" (GALFs; Fig. 4.10), is increased during pregnancy [147]. Initial reports [148] that GALF activity was enhanced in hypertensive subjects were not confirmed [146,149]. However, GALFs turn out to be one of few inhibitors which display selectivity such that they inhibit 11β-HSD1 and 11β-HSD2 to a variable degree [150]. Thus, inhibitory activity for 11β-HSD2 may indeed be increased in patients with low-renin essential hypertension [151]. In order to understand the importance of GALFs, we urgently need to identify them. Unfortunately, this has proved impossible to date.

Table 4.3. Inhibitors of 11β-HSDs

Endogenous	Exogenous	Unidentified
11-OH-progesterone [174]	Glycyrrhetinic acid [46,49]	GALFs [146–149, 151,175]
3α, 5β-tetrahydroprogesterone	Glycyrrhizic acid	ACTH-dependent inhibitors [56]
3α, 5β-tetrahydro-11-deoxy-corticosterone	Carbenoxolone [48]	
11-epicortisol	CHAPS [176]	
Chenodeoxycholic acid	Ketoconazole	
Cholic acid [177]	Saiboku-To [178]	
	Gossypol [179]	
	Metyrapone [180]	
	11-epiprednisolone	

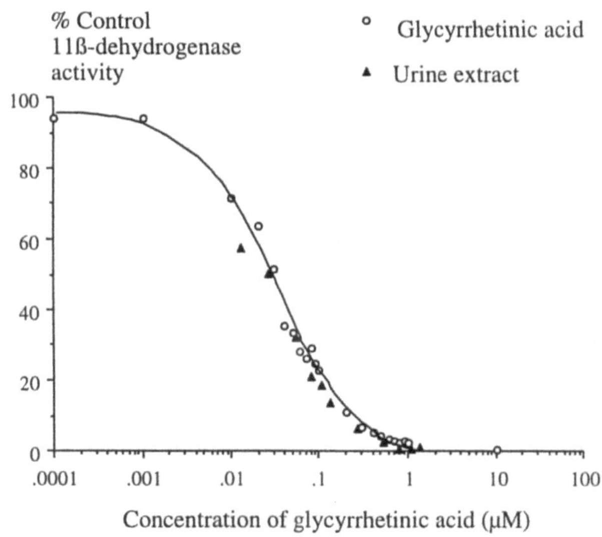

Fig. 4.10. Urinary glycyrrhetinic acid like factors (GALFs). Inhibition of 11β-dehydrogenase activity, measured as the conversion of ^{3}H-corticosterone to ^{3}H-11-dehydrocorticosterone in the presence of NADP in rat liver microsomes, with increasing concentrations of glycyrrhetinic acid (open circles) or with serial dilution of a urinary eluate from a healthy male volunteer (black triangles). The inhibition by neat urinary eluate was equivalent to that by 0.15 μM glycyrrhetinic acid. This data point was used to locate the curve for varying concentrations of urinary eluate, which was then found to lie parallel to the curve for serial dilution of glycyrrhetinic acid [146].

Cortisol metabolism and cortisol secretion

In the above discussions, it is assumed that any alteration in peripheral metabolism of cortisol will not change the circulating concentration of cortisol because hypothalamic-pituitary feedback will result in adjustment of ACTH and cortisol secretion. For example, in the syndrome of apparent mineralocorticoid excess due to 11β-dehydrogenase deficiency, decreased cortisol clearance is associated with lower cortisol secretion rates and normal circulating cortisol concentrations [32]. Similarly, enhanced cortisol clearance due to impaired 11β-reductase activity is associated with enhanced cortisol secretion to maintain normal circulating levels [79]. However, this appropriate homeostatic adjustment may have consequences for the stress response. Chronic suppression of the hypothalamic-pituitary-adrenal axis, for example with exogenous synthetic glucocorticoids, is associated with impaired responses to stress and may be associated with "hypoadrenal crisis" and death. In chronic stress, cortisol clearance may be impaired due to induction of 11β-reductase activity. This could explain paradoxical shrinkage of the adrenal cortex, for example in tuberculosis, and render patients susceptible to hypoadrenal crisis during acute-on-chronic stress. Experimental evidence in support of this hypothesis is presented in Chapter 9.

Another link between cortisol metabolism and secretion arises from the expression of 11β-HSD1 in hippocampus, hypothalamus, and anterior pituitary. In rat hippocampal cells in primary culture, 11β-reductase activity converts 11-dehydrocorticosterone to corticosterone and thereby enhances glucocorticoid action [72]. It remains to be established whether the same activity makes an important contribution to negative feedback regulation of corticotrophin-releasing hormone (CRH) and ACTH secretion [109].

Summary and future perspectives

In summary, the conventional view of the activity of glucocorticoids being regulated exclusively by the hypothalamic-pituitary-adrenal axis has been challenged by evidence that tissue sensitivity to glucocorticoids is also regulated, and that this regulation is integrated in the response to stress. However, despite these advances in the last 10 years, much remains to be explored in the mechanism whereby 11β-HSD activities influence corticosteroid receptor activation. Key cell biology areas at present include the determinants of the intracellular localization of the enzymes and their relationship with pools of non-activated receptors, the regulators of equilibrium between dehydrogenase and reductase activities of 11β-HSD1, and investigation of the promoters of both 11β-HSD genes. Important physiology experiments are needed to establish the regulation of 11β-HSD2 in vivo and to establish the identity and influence of endogenous inhibitors of enzyme activity. The extent to which variations in 11β-HSD activity can explain other aspects of variable glucocorticoid activity, including interactions with dihydroepiandrosterone [152], remains to be investigated. Most importantly, there may be value in manipulating cortisol metabolism in the treatment of hypertension or insulin resistance. This field has been dominated by investigators with a background in the adrenal cortex and cardiovascular disease. There may be important principles to be applied by neuroscientists and immunologists in their own fields.

Acknowledgements

Aspects of this work have been supported by grants from the British Heart Foundation, Medical Research Council, Scottish Office, and Wellcome Trust.

References

1. James VHT (1992) The adrenal gland, 2nd ed. Raven Press, New York

2. Lacroix A, Bolte E, Tremblay J, Dupre J, Poitras P, Fournier H et al. (1992) Gastric inhibitory polypeptide-dependent cortisol hypersecretion: a new cause for Cushing's syndrome. N Engl J Med 327:974–980
3. Al Dujaili EAS, Williams BC, Edwards CRW, Salacinski PR, Lowry PJ. (1982) Human gamma-MSH precursor potentiates ACTH-induced adrenal steroidogenesis by stimulating mRNA synthesis. Biochem J 204:301–305
4. Perrin A, Pascal O, Defarge G, Feige J-J, Chambaz EM (1990) Transforming growth factor β is a negative regulator of steroid 17α-hydroxylase expression in bovine adrenocortical cells. Endocrinology 128:357–362
5. Arriza JL, Weinberger C, Cerelli G (1987) Cloning of human mineralocorticoid receptor complementary DNA; structural and functional kinship with the glucocorticoid receptor. Science 237:268–275
6. Orchinik M, Murray TF, Moore FL (1991) A corticosteroid receptor in neural membranes. Science 252:1848–1851
7. McEwan BS (1991) Non-genomic and genomic effects of steroids on neural activity. Trends Pharmacol Sci 12:141–147
8. Cushing H. (1912) The pituitary body and its disorders. Lippincott, Philadelphia
9. Walker BR, Edwards CRW (1992) Cushing's syndrome. In: James VHT (ed) The adrenal gland. Raven Press, New York, pp 289–318
10. Walker BR, Williams BC. (1992) Corticosteroids and vascular tone: mapping the messenger maze. Clin Sci 82:597–605
11. Whitworth JA (1994) Studies on the mechanism of glucocorticoid hypertension in humans. Blood Pressure 3:24–32
12. Dallman MF, Strack AM, Akana SF, Bradbury MJ, Hanson ES, Scribner KA et al. (1993) Feast and famine: critical role of glucocorticoids with insulin in daily energy flow. Front Neuroendocrinol 14:303–347
13. Bouchard C, Despres J-P, Mauriege P. (1993) Genetic and nongenetic determinants of regional fat distribution. Endocr Rev 14:72–92
14. Sapolsky RM, Krey LC, McEwen BS. (1986) The neuroendocrinology of stress and ageing. The glucocorticoid cascade hypothesis. Endocr Rev 7:284–301
15. Addison T (1855) On the constitutional and local effects of disease of the supra-renal capsules. Reprinted by S. Highley, London
16. Bamberger CM, Schulte HM, Chrousos GP (1996) Molecular determinants of glucocorticoid receptor function and tissue sensitivity to glucocorticoids. Endocr Rev 17:245–261
17. Hammond GL, Smith CL, Paterson NAM, Sibbald WJ (1990) A role for corticosteroid-binding globulin in delivery of cortisol to activated neutrophils. J Clin Endocrinol Metab 71:34–39
18. Stephenson G, Krozowski Z, Funder JW. (1984) Extravascular CBG-like sites in rat kidney and mineralocorticoid receptor specificity. Am J Physiol 246:F227–F233
19. Mendel CM, Kuhn RW, Weisiger RA. (1991) Uptake of corticosterone by the perfused rat liver. Endocrinology 129:27–32
20. Kralli A, Bohen SP, Yamamoto KR (1995) LEM1, an ATP-binding-cassette transporter, selectively modulates the biological potency of steroid hormones. PNAS USA 92:4701–4705
21. Bamberger CM, Bamberger A-M, de Castro M, Chrousos GP. (1995) Glucocorticoid receptor β, a potential endogenous inhibitor of glucocorticoid action in humans. J Clin Invest 95:2435–2441
22. Funder JW, Feldman D, Edelman IS. (1973) Glucocorticoid receptors in rat kidney: the binding of tritiated-dexamethasone. Endocrinology 92:1005–1013
23. Burnstein KL, Cidlowski JA (1992) The downside of glucocorticoid receptor regulation. Mol Cell Endocrinol 83:C1–C8

24. Madan AP, DeFranco DB (1993) Bidirectional transport of glucocorticoid receptors across the nuclear envelope. PNAS USA 90:3588–3592
25. Pratt WB (1993) The role of heat shock proteins in regulating the function, folding, and trafficking of the glucocorticoid receptor. J Biol Chem 268:21455–21458
26. Sher ER, Leung DY, Surs W, Kam JC, Zeig G, Kamada AK et al. (1994) Steroid-resistant asthma. Cellular mechanisms contributing to inadequate response to glucocorticoid therapy. J Clin Invest 93:33–39
27. Vingerhoeds ACM, Thijssen JHH, Schwarz F. (1976) Spontaneous hypercortisolism without Cushing's syndrome. J Clin Endocrinol Metab 43:1128–1133
28. Brandon DD, Marwick AJ, Chrousos GP, Loriaux DL. (1989) Glucocorticoid resistance in humans and non-human primates. Cancer Res 49(Suppl):2203S–2213S
29. Iida S, Nakamura Y, Fujii H, Nishmura J, Tsugawa M, Gomi M et al. (1990) A patient with hypocortisolism and Cushing's syndrome-like manifestations: cortisol hyperreactive syndrome. J Clin Endocrinol Metab 70:729–737
30. Werder E, Zachmann M, Vollmin JA. (1974) Unusual steroid excretion in a child with low renin hypertension. Res Steroids 6:385–389
31. Ulick S, Levine LS, Gunczler P, Zanconato G, Ramirez LC, Rauh W et al. (1979) A syndrome of apparent mineralocorticoid excess associated with defects in the peripheral metabolism of cortisol. J Clin Endocrinol Metab 49:757–764
32. Stewart PM, Corrie JET, Shackleton CHL, Edwards CRW (1988) Syndrome of apparent mineralocorticoid excess: a defect in the cortisol-cortisone shuttle. J Clin Invest 82:340–349
33. Shackleton CHL, Stewart PM (1990) The hypertension of apparent mineralocorticoid excess syndrome. In: Biglieri EG, Melby JC (eds) Endocrine hypertension. Raven Press, New York, pp 155–173
34. Oberfield SE, Levine LS, Carey RM, Greig F, Ulick S, New MI (1983) Metabolic and blood pressure responses to hydrocortisone in the syndrome of apparent mineralocorticoid excess. J Clin Endocrinol Metab 56:332–339
35. Gustafsson J-A, Carlstedt-Duke J, Poellinger L, Okret S, Wikstrom A-C, Bronnegard M et al. (1987) Biochemistry, molecular biology, and physiology of the glucocorticoid receptor. Endocr Rev 8:185–234
36. Feldman D, Funder JW, Edelman IS (1973) Evidence for a new class of corticosterone receptors in the rat kidney. Endocrinology 92:1429–1441
37. Lombes M, Farman N, Oblin ME, Baulieu EE, Bonvalet JP, Erlanger BF et al. (1990) Immunohistochemical localization of renal mineralocorticoid receptor using an anti-idiotypic antibody that is an internal image of aldosterone. PNAS USA 87:1086–1088
38. Funder JW (1993) Mineralocorticoids, glucocorticoids, receptors and response elements. Science 259:1132–1133
39. Sheppard K, Funder JW (1987) Mineralocorticoid specificity of renal type 1 receptors: in vivo binding studies. Am J Physiol 252:E224–E229
40. Krozowski ZS, Funder JW (1983) Renal mineralocorticoid receptors and hippocampal corticosterone-binding species have identical intrinsic steroid specificity. PNAS USA 80:6056–6060
41. Reevers F (1948) Behandeling van uleus ventriculi in uleus duodeni met succus liquiritiae. Ned Tijdschr Geneesk 92:2968–2971
42. Conn JW, Rovner DR, Cohen EL (1968) Licorice-induced pseudoaldosteronism. JAMA 205:495–496
43. Epstein MT, Espiner EA, Donald RA, Hughes H (1977) Liquorice toxicity and the renin-angiotensin-aldosterone axis in man. BMJ I:209–210
44. Hoefnagels WHL, Kloppenborg PWC. (1983) Antimineralocorticoid effects of dexamethasone in subjects treated with glycyrrhetinic acid. J Hypertens 1 (Suppl 2):313–315

45. Borst JGG, Ten Holt SP, DeVries LA (1953) Synergistic action of liquorice and cortisone in Addison's and Simmond's disease. Lancet I:657–663
46. Stewart PM, Valentino R, Wallace AM, Burt D, Shackleton CHL, Edwards CRW (1987) Mineralocorticoid activity of liquorice: 11β-hydroxysteroid dehydrogenase deficiency comes of age. Lancet II:821–824
47. MacKenzie MA, Hoefnagels WHL, Jansen RWMM, Benraad TJ, Kloppenborg PWC (1990) The influence of glycyrrhetinic acid on plasma cortisol and cortisone in healthy young volunteers. J Clin Endocrinol Metab 70:1637–1643
48. Stewart PM, Wallace AM, Atherden SM, Shearing CH, Edwards CRW (1990) Mineralocorticoid activity of carbenoxolone: contrasting effects of carbenoxolone and liquorice on 11β-hydroxysteroid dehydrogenase activity in man. Clin Sci 78:49–54
49. Monder C, Stewart PM, Lakshmi V, Valentino R, Burt D, Edwards CRW. (1989) Licorice inhibits corticosteroid 11β-dehydrogenase of rat kidney and liver: in vivo and in vitro studies. Endocrinology 125:1046–1053
50. Edwards CRW, Stewart PM, Burt D, Brett L, McIntyre MA, Sutanto WS et al. (1988) Localisation of 11β-hydroxysteroid dehydrogenase tissue-specific protector of the mineralocorticoid receptor. Lancet II:986–989
51. Funder JW, Pearce PT, Smith R, Smith AI (1988) Mineralocorticoid action: target tissue specificity is enzyme, not receptor, mediated. Science 242:583–585
52. Wilson RC, Krozowski ZS, Li K, Obeyesekere VR, Razzaghy-Azar M, Harbison MD et al. (1995) A mutation in the HSD11B2 gene in a family with apparent mineralocorticoid excess. J Clin Endocrinol Metab 80:2263–2266
53. Stewart PM, Krozowski ZS, Gupta A, Milford DV, Howie AJ, Sheppard MC et al. (1996) Hypertension in the syndrome of apparent mineralocorticoid excess due to mutation of the 11β-hydroxysteroid dehydrogenase type 2 gene. Lancet 347:88–91
54. Mune T, Rogerson FM, Nikkila H, Agarwal AK, White PC (1995) Human hypertension caused by mutations in the kidney isozyme of 11β-hydroxysteroid dehydrogenase. Nature Genet 10:394–399
55. Wilson RC, Harbison MD, Krozowski ZS, Funder JW, Shackleton CHL, Hanauske-Abel HM et al. (1995) Several homozygous mutations in the gene for 11β-hydroxysteroid dehydrogenase type 2 in patients with apparent mineralocorticoid excess. J Clin Endocrinol Metab 80:3145–3150
56. Walker BR, Campbell JC, Fraser R, Stewart PM, Edwards CRW (1992) Mineralocorticoid excess and inhibition of 11β-hydroxysteroid dehydrogenase in patients with ectopic ACTH syndrome. Clin Endocrinol 27:483–492
57. Ulick S, Wang JZ, Blumenfeld JD, Pickering TG (1992) Cortisol inactivation overload: a mechanism of mineralocorticoid hypertension in the ectopic adrenocorticotropin syndrome. J Clin Endocrinol Metab 74:963–967
58. Stewart PM, Walker BR, Holder G, O'Halloran D, Shackleton CHL (1995) 11β-Hydroxysteroid dehydrogeanse activity in Cushing's syndrome: explaining the mineralocorticoid excess state of the ectopic ACTH syndrome. J Clin Endocrinol Metab 80:3617–3620
59. Walker BR, Stewart PM, Shackleton CHL, Padfield PL, Edwards CRW (1993) Deficient inactivation of cortisol by 11β-hydroxysteroid dehydrogenase in essential hypertension. Clin Endocrinol 39:221–227
60. Soro A, Ingram MC, Tonolo G, Glorioso N, Fraser R (1995) Evidence of coexisting changes in 11β-hydroxysteroid dehydrogenase and 5β-reductase activity in patients with untreated essential hypertension. Hypertension 25:67–70
61. Stewart PM, Sheppard MC (1992) Novel aspects of hormone action: intracellular ligand supply and its control by a series of tissue-specific enzymes. Mol Cell Endocrinol 83:C13–18

62. Monder C, White PC (1993) 11β-Hydroxysteroid dehydrogenase. Vitam Horm 47:187–271
63. Murphy BEP (1981) Ontogeny of cortisol-cortisone interconversion in human tissues; a role for cortisone in human fetal development. J Steroid Biochem 14:811–817
64. Murphy BEP (1978) Cortisol production and inactivation by the human lung during gestation and infancy. J Clin Endocrinol Metab 47:243–248
65. Lakshmi V, Monder C (1988) Purification and characterization of the corticosteroid 11β-dehydrogenase component of the rat liver 11β-hydroxysteroid dehydrogenase complex. Endocrinology 123:2390–2398
66. Monder C, Lakshmi V (1989) Evidence for kinetically distinct forms of corticosteroid 11β-dehydrogenase in rat liver microsomes. J Steroid Biochem 32:77–83
67. Lakshmi V, Monder C. (1985) Evidence for independent 11-oxidase and 11-reductase activities for 11β-hydroxysteroid dehydrogenase: enzyme latency, phase transitions and lipid requirement. Endocrinology 116:552–560
68. Agarwal AK, Monder C, Eckstein B, White PC (1989) Cloning and expression of rat cDNA encoding corticosteroid 11β-dehydrogenase. J Biol Chem 264:18939–18943
69. Agarwal AK, Tusie-Luna M-T, Monder C, White PC (1990) Expression of 11β-hydroxysteroid dehydrogenase using recombinant vaccinia virus. Mol Endocrinol 4:1827–1832
70. Low SC, Chapman KE, Edwards CRW, Seckl JR (1994) "Liver-type" 11β-hydroxysteroid dehydrogenase cDNA encodes reductase but not dehydrogenase activity in intact mammalian COS-7 cells. J Mol Endocrinol 13:167–174
71. Jamieson PM, Chapman KE, Edwards CRW, Seckl JR. (1995) 11β-Hydroxy steroid dehydrogenase is an exclusive 11β-reductase in primary cultures of rat hepatocytes: effect of physicochemical and hormonal manipulations. Endocrinology 136:4754–4761
72. Rajan V, Edwards CRW, Seckl JR (1996) 11β-Hydroxysteroid dehydrogenase in cultured hippocampal cells reactivates inert 11-dehydrocorticosterone, potentiating neurotoxicity. J Neurosci 16:65–70
73. Bush IE (1969) 11β-hydroxysteroid dehydrogenase: contrast between studies in vivo and studies in vitro. Adv Biosci 3:23–39
74. Hubbard WC, Bickel C, Schleimer RP (1994) Simultaneous quantitation of endogenous levels of cortisone and cortisol in human nasal and bronchoalveolar lavage fluids and plasma via gas chromatography-negative ion chemical ionization spectrometry. Ann Biochem 221:109–117
75. Albiston AL, Obeyesekere VR, Smith RE, Krozowski ZS (1994) Cloning and tissue distribution of the human 11β-hydroxysteroid dehydrogenase type 2 enzyme. Mol Cell Endocrinol 105:R11–R17
76. Agarwal AK, Mune T, Monder C, White PC (1994) NAD^+-dependent isoform of 11β-hydroxysteroid dehydrogenase. Cloning and characterisation of cDNA from sheep kidney. J Biol Chem 269:25959–25962
77. Brown RW, Chapman KE, Koteletsev Y, Yau JL, Lindsay RS, Brett LP et al. (1996) Cloning and production of anitisera to human placental 11β-hydroxysteroid dehydrogenase type 2. Biochem J 313:1007–1017
78. Naray-Fejes-Toth A, Fejes-Toth G (1995) Expression cloning of the aldosterone target cell-specific 11β-hydroxysteroid dehydrogenase from rabbit collecting duct cells. Endocrinology 136:2579–2586
79. Phillipou G, Higgins BA (1985) A new defect in the peripheral conversion of cortisone to cortisol. J Steroid Biochem 22:435–436
80. Rodin A, Thakkar H, Taylor N, Clayton R (1994) Hyperandrogenism in polycystic ovary syndrome: evidence of dysregulation of 11beta-hydroxysteroid dehydrogenase. N Engl J Med 330:460–465

81. Nikkila H, Tannin GM, New MI, Taylor NF, Kalaitzoglou G, Monder C et al. (1993) Defects in the HSD11 gene encoding 11β-hydroxysteroid dehydrogenase are not found in patients with apparent mineralocorticoid excess or 11-oxoreductase deficiency. J Clin Endocrinol Metab 77:687–691
82. Ulick S, Chan CK, Rao KN, Edassery J, Mantero F (1989) A new form of the syndrome of apparent mineralocorticoid excess. J Steroid Biochem 32:209–212
83. Ulick S, Tedde R, Mantero F (1990) Pathogenesis of the type 2 variant of the syndrome of apparent mineralocorticoid excess. J Clin Endocrinol Metab 70:200–206
84. Mantero F, Tedde R, Scaroni C, Opocher G, Campus S (1985) Dexamethasone-suppressible low renin low aldosterone hypertension: report of 3 cases. In: Mantero F, Biglieri EG, Funder JW, Scoggins BA (eds) The adrenal gland and hypertension. Raven Press, New York, pp 337–352
85. Tedde R, Pala A, Melis A, Ulick S (1992) Evidence for cortisol as the mineralocorticoid in the syndrome of apparent mineralocorticoid excess. J Endocrinol Invest 15:471–474
86. Mantero F, Tedde R, Opocher G, Fulgheri PD, Arnaldi G, Ulick S (1994) Apparent mineralocorticoid excess type II. Steroids 59:80–83
87. Best R, Nelson SM, Walker BR (1997) Dexamethasone and 11-dehydrodexamethasone as tools to investigate the isozymes of 11β-hydroxysteroid dehydrogenase in vitro and in vivo. J Endocrinol 153:41–48
88. Stewart PM, Murry BA, Mason JI. (1994) Human kidney 11β-hydroxysteroid dehydrogenase is a high affinity nicotinamide adenine dinucleotide-dependent enzyme and differs from the cloned type 1 isoform. J Clin Endocrinol Metab 79:480–484
89. Benediktsson R, Noble J, Calder AA, Edwards CRW, Seckl JR (1995) 11β-Hydroxysteroid dehydrogenase activity in intact dually-perfused fresh human placenta predicts birth weight. J Endocrinol 144:P161
90. Benediktsson R, Lindsay RS, Noble J, Seckl JR, Edwards CRW (1993) Glucocorticoid exposure in utero: new model for adult hypertension. Lancet 341:339–341
91. Lindsay RS, Lindsay RM, Waddell BJ, Seckl JR (1996) Prenatal glucocorticoid exposure leads to offspring hyperglycaemia in the rat: studies with the 11β-hydroxysteroid dehydrogenase inhibitor carbenoxolone. Diabetologia 39:1299–1305
92. Lindsay RS, Lindsay RM, Edwards CRW, Seckl JR (1996) Inhibition of 11β-hydroxysteroid dehydrogenase in pregnant rats and the programming of blood pressure in the offspring. Hypertension 27:1200–1204
93. Stewart PM, Rogerson FM, Mason JI (1995) Type 2 11β-hydroxysteroid dehydrogenase messenger RNA and activity in human placenta and fetal membranes: its relationship to birth weight and putative role in fetal steroidogenesis. J Clin Endocrinol Metab 80:885–890
94. Barker DJP, Osmond C, Golding J, Kuh D, Wadsworth MEJ (1989) Growth in utero, blood pressure in childhood and adult life, and mortality from cardiovascular disease. BMJ 298:564–567
95. Phillips DIW, Barker DJP, Hales CN, Hirst S, Osmond C (1994) Thinness at birth and insulin resistance in adult life. Diabetologia 37:150–154
96. Dazord A, Saez J, Bertrand J (1972) Metabolic clearance rates and interconversion of cortisol and cortisone. J Clin Endocrinol Metab 35:24–34
97. Barker PM, Markiewicz M, Parker KA, Walters DV, Strang LB (1990) Synergistic action of triiodothyronine and hydrocortisone on epinephrine-induced reabsorption of fetal lung liquid. Pediatr Res 27:588–591
98. Monder C, Miroff Y, Marandici A, Hardy MP (1994) 11β-Hydroxysteroid dehydrogenase alleviates glucocorticoid-mediated inhibition of steroidogensis in rat Leydig cells. Endocrinology 134:1199–1204
99. Low SC, Chapman KE, Edwards CRW, Wells T, Robinson ICAF, Seckl JR. (1994) Sexual dimorphism of hepatic 11β-hydroxysteroid dehydrogenase in the rat: the role of growth hormone patterns. J Endocrinol 143:541–548

100. Jamieson PM, Chapman KE, Walker BR, Seckl JR. (1996) Hepatic 11β-hydroxysteroid dehydrogenase type 1: evidence for a functional reductase which amplifies glucocorticoid action in vivo. J Endocrinol 148(Suppl):P44
101. Kotelevtsev YV, Jamieson PM, Edwards CRW, Seckl JR, Mullins JJ (1996) Gene targeting of 11β-hydroxysteroid dehydrogenase type 1. Proceedings of the International Congress of Endocrinology, San Francisco, OR52-1
102. Walker BR, Edwards CRW (1994) Licorice-induced hypertension and syndromes of apparent mineralocorticoid excess. Endocrinol Metab Clin North Am 23:359-377
103. Walker BR, Connacher AA, Lindsay RM, Webb DJ, Edwards CRW (1994) Carbenoxolone increases hepatic insulin sensitivity in man: in vivo evidence that ligand metabolism modulates activation of glucocorticoid receptors. J Endocrinol 104(Suppl):OC37
104. Moisan M-P, Seckl JR, Edwards CRW (1990) 11β-Hydroxysteroid dehydrogenase bioactivity and messenger RNA expression in rat forebrain: localization in hypothalamus, hippocampus and cortex. Endocrinology 127:1450-1455
105. Moisan M, Seckl JR, Monder C, Agarwal AK, White PC, Edwards CRW (1990) 11β-Hydroxysteroid dehydrogenase mRNA expression, bioactivity and immunoreactivity in rat cerebellum. Neuroendocrinology 2:853-858
106. Lakshmi V, Sakai RR, McEwen BS, Monder C (1991) Regional distribution of 11β-hydroxysteroid dehydrogenase in rat brain. Endocrinology 128:1741-1748
107. Sakai RR, Lakshmi V, Monder C, McEwen BS. (1992) Immunocytochemical localization of 11beta-hydroxysteroid dehydrogenase in hippocampus and other brain regions of the rat. J Neuroendocrinol 4:101-106
108. Seckl JR, Kelly PAT, Sharkey J (1991) Glycyrrhetinic acid, an inhibitor of 11β-hydroxysteroid dehydrogenase, alters local cerebral glucose utilisation in vivo. J Steroid Biochem Mol Biol 39:777-779
109. Seckl JR, Dow RC, Low SC, Edwards CRW, Fink G (1993) The 11β-hydroxysteroid dehydrogenase inhibitor glycyrrhetinic acid affects corticosteroid feedback regulation of hypothalamic corticotrophin-releasing peptides in rats. J Endocrinol 136:471-477
110. Jellinck PH, Monder C, McEwen BS, Sakai RR (1993) Differential inhibition of 11β-hydroxysteroid dehydrogenase by carbenoxolone in rat brain regions and peripheral tissues. J Steroid Biochem Mol Biol 46:209-213
111. Walker BR, Yau JL, Brett LP, Seckl JR, Monder C, Williams BC et al. (1991) 11β-Hydroxysteroid dehydrogenase in vascular smooth muscle and heart: implications for cardiovascular responses to glucocorticoids. Endocrinology 129:3305-3312
112. Walker BR, Sang KS, Smith JC, Dockrell MEC, Williams BC, Edwards CRW (1993) Direct and indirect effects of carbenoxolone (CBX) on vascular responses to glucocorticoids and catecholamines. J Endocrinol 137 (Suppl):P16
113. Brem AS, Bina RB, King T, Morris DJ (1995) Bidirectional activity of 11β-hydroxysteroid dehydrogenase in vascular smooth muscle cells. Steroids 60:406-410
114. Takeda Y, Miyamori I, Yoneda T, Hatakeyama H, Iki K, Takeda R (1994) Decreased activity of 11β-hydroxysteroid dehydrogenase in mesenteric arteries of Dahl salt-sensitive rats. Life Sci 54:1343-1349
115. Takeda Y, Miyamori I, Yoneda T, Iki K, Hatakeyama H, Takeda R (1994) Gene expression of 11β-hydroxysteroid dehydrogenase in the mesenteric arteries of genetically hypertensive rats. Hypertension 23:577-580
116. Takeda Y, Yoneda T, Miyamori I, Gathiram P, Takeda R. (1993) 11β-Hydroxysteroid dehydrogenase activity in mesenteric arteries of Spontaneously Hypertensive Rats. Clin Exp Pharmacol Physiol 20:627-631
117. Takeda Y, Miyamori I, Yoneda T, Ito Y, Takeda R (1994) Expression of 11β-hydroxysteroid dehydrogenase mRNA in rat vascular smooth muscle cells. Life Sci 54:281-285

118. Walker BR, Sang KS, Williams BC, Edwards CRW (1994) Direct and indirect effects of carbenoxolone on responses to glucocorticoids and noradrenaline in rat aorta. J Hypertens 12:33–39
119. Teelucksingh S, Mackie ADR, Burt D, McIntyre MA, Brett L, Edwards CRW (1990) Potentiation of hydrocortisone activity in skin by glycyrrhetinic acid. Lancet 335:1060–1063
120. Walker BR, Connacher AA, Webb DJ, Edwards CRW (1992) Glucocorticoids and blood pressure: a role for the cortisol/cortisone shuttle in the control of vascular tone in man. Clin Sci 83:171–178
121. Lombes M, Oblin MF, Gasc JM, Baulieu FE, Farman N, Bonvalet JP (1992) Immunohistochemical and biochemical evidence for a cardiovascular mineralocorticoid receptor. Circ Res 71:503–510
122. Marks R, Barlow JW, Funder JW (1982) Steroid-induced vasoconstriction; glucocorticoid antagonist studies. J Clin Endocrinol Metab 54:1075–1077
123. Gaillard RC, Poffet D, Riondel AM, Saurat J (1985) RU486 inhibits peripheral effects of glucocorticoids in humans. J Clin Endocrinol Metab 61:1009–1011
124. Walker BR, Best R, Shackleton CHL, Padfield PL, Edwards CRW (1996) Increased vasoconstrictor sensitivity to glucocorticoids in essential hypertension. Hypertension 27:190–196
125. Slight S, Ganjam VK, Nonneman DJ, Weber KT (1993) Glucocorticoid metabolism in the cardiac interstitium: 11β-hydroxysteroid dehydrogenase activity in cardiac fibroblasts. J Lab Clin Med 122:180–187
126. Brilla CG, Weber KT (1992) Mineralocorticoid excess, dietary sodium, and myocardial fibrosis. J Lab Clin Med 120:893–901
127. Phillips DM, Lakshmi V, Monder C (1989) Corticosteroid 11β-dehydrogenase in rat testis. Endocrinology 125:209–216
128. Monder C, Sakai RR, Miroff Y, Blanchard DC, Blanchard RJ (1994) Reciprocal changes in plasma corticosterone and testosterone in stressed male rats maintained in a visible burrow system: evidence for a mediating role of testicular 11β-hydroxysteroid dehydrogenase. Endocrinology 134:1193–1198
129. Benediktsson R, Yau JLW, Low S, Brett LP, Cooke BE, Edwards CRW et al. (1992) 11β-Hydroxysteroid dehydrogenase in the rat ovary: high expression in the oocyte. J Endocrinol 135:53–58
130. Michael AE, Pester LA, Curtis P, Shaw RW, Edwards CRW, Cooke BA (1993) Direct inhibition of ovarian steroidogenesis by cortisol and the modulatory role of 11β-hydroxysteroid dehydrogenase. Clin Endocrinol 38:641–644
131. Michael AF, Gregory L, Walker SM, Antoniw JW, Shaw RW, Edwards CRW et al. (1993) Ovarian 11β-hydroxysteroid dehydrogenase: potential predictor of conception by in vitro fertilisation and embryo transfer. Lancet 342:711–712
132. Ten-Cate W-JF, Monder C, Marandici A, Rarey KE (1994) 11β-Hydroxysteroid dehydrogenase in the rat inner ear. Am J Physiol 266:E269–E273
133. Gaeggeler H-P, Edwards CRW, Rossier BC (1989) Steroid metabolism determines mineralocorticoid specificity in the toad bladder. Am J Physiol 257:F690–F695
134. Whitehouse BJ, Vinson GP (1967) Effect of blood on the metabolism of cortisol by the adrenal gland of the golden hamster. J Endocrinol 39:117–118
135. Bailey E, West HF (1969) The secretion, interconversion and catabolism of cortisol, cortisone and some of their metabolites in man. Acta Endocrinol (Copenh) 62:339–359
136. Williams AC, Walker BR, Edwards CRW, Burt D, Williams BC, Walker SW (1992) Differential effects of acute and chronic stimulation by ACTH on the secretion of cortisol, cortisone and 18-hydroxycortisol from bovine zona fasciculata cells in primary culture. J Endocrinol 132 (Suppl):111

137. Naray-Fejes-Toth A, Watlington CO, Fejes-Toth G (1991) 11β-hydroxysteroid dehydrogenase activity in the renal target cells of aldosterone. Endocrinology 129:17–21
138. Naray-Fejes-Toth A, Rusvai E, Fejes-Toth G (1994) Mineralocorticoid receptors and 11β-steroid dehydrogenase activity in renal principal and intercalated cells. Am J Physiol 266:F76–F80
139. Naray-Fejes-Toth A, Fejes-Toth G (1996) Intracellular localization of 11β-hydroxysteroid dehydrogenase type 2. Proceedings of the International Congress of Endocrinology, San Francisco, P1-236
140. Low SC, Moisan M-P, Edwards CRW, Seckl JR (1994) Glucocorticoids regulate 11β-hydroxysteroid dehydrogenase activity and gene expression in vivo in the rat. J Neuroendocrinol 6:285–290
141. Christy NP, Laragh JH (1961) Pathogenesis of hypokalemic alkalosis in Cushing's syndrome. N Engl J Med 265:1083–1088
142. Crane MG, Harris JJ (1966) Desoxycorticosterone secretion rates in hyperadrenocorticism. J Clin Endocrinol 26:1135–1143
143. Schambelan M, Slaton PE, Biglieri EG (1971) Mineralocorticoid production in hyperadrenocorticism. Am J Med 51:299–303
144. Hermus A, Hobma S, Pieters G, van de Calseyde J, Smals A, Kloppenborg P (1991) Are the hypokalaemia and hypertension in Cushing's disease caused by apparent mineralocorticoid excess. Horm Metab Res 23:572–573
145. Walker BR, Williams BC, Edwards CRW (1994) Regulation of 11β-hydroxysteroid dehydrogenase activity by the hypothalamic-pituitary-adrenal axis in the rat. J Endocrinol 141:467–472
146. Walker BR, Aggarwal I, Stewart PM, Padfield PL, Edwards CRW (1995) Endogenous inhibitors of 11β-hydroxysteroid dehydrogenase in hypertension. J Clin Endocrinol Metab 80:529–533
147. Morris DJ, Semafuko WEB, Latif SA, Vogel B, Grimes C, Sheff MF (1992) Detection of glycyrrhetinic acid-like factors (GALFs) in human urine. Hypertension 20:356–360
148. Semafuko WEB, Sheff MF, Grimes CA, Latif SA, Sadaniantz A, Levinson P et al. (1993) Inhibitors of 11β-hydroxysteroid dehydrogenase and 5β-steroid reductase in urine from patients with congestive heart failure. Ann Clin Lab Sci 23:456–461
149. Walker BR, Williamson PM, Brown MA, Honour JW, Edwards CRW, Whitworth JA (1995) 11β-Hydroxysteroid dehydrogenase and its inhibitors in hypertensive pregnancy. Hypertension 25:626–630
150. Walker BR, Murad P, Smith JC, Edwards CRW (1995) Isoform-specificity of endogenous inhibitors of 11β-hydroxysteroid dehydrogenase (11β-HSD) may explain their lack of influence on blood pressure. J Endocrinol 144(Suppl):P182
151. Takeda Y, Miyamori I, Iki K, Inaba S, Furukawa K, Hatakeyama H et al. (1996) Endogenous renal 11β-hydroxysteroid dehydrogenase inhibitory factors in patients with low-renin essential hypertension. Hypertension 27:197–201
152. Kalimi M, Shafagoj Y, Loria R, Padgett D, Regelson W (1994) Anti-glucocorticoid effects of dehydroepiandrosterone (DHEA). Mol Cell Biochem 131:99–104
153. Khan MH, Sullivan FM (1967) The pharmacology of carbenoxolone sodium. In: Robson JM, Sullivan FM (eds) Symposium on carbenoxolone sodium. Butterworth, London, pp 5–13
154. Baker ME, Fanestil DD (1991) Licorice, computer-based analyses of dehydrogenase sequences, and the regulation of steroid and prostaglandin action. Mol Cell Endocrinol 78:C99–C102
155. Walker BR, Connacher AA, Lindsay RM, Webb DJ, Edwards CRW (1995) Carbenoxolone increases hepatic insulin sensitivity in man: a novel role for 11-oxosteroid reductase in enhancing glucocorticoid receptor activation. J Clin Endocrinol Metab 80:3155–3159

156. Tannin GM, Agarwal AK, Monder C, White PC (1990) Cloning and sequencing of the human cDNA for corticosteroid 11-beta-dehydrogenase. Programme of the 73rd Meeting of the Endocrine Society 72:1049
157. Tannin GM, Agarwal AK, Monder C, New MI, White PC (1991) The human gene for 11β-hydroxysteroid dehydrogenase. J Biol Chem 266:16653–16658
158. Yang K, Smith CL, Dales D, Hammond GL, Challis JR (1992) Cloning of an ovine 11beta-hydroxysteroid dehydrogenase complementary deoxyribonucleic acid: tissue and temporal distribution of its messenger ribonucleic acid during fetal and neonatal development. Endocrinology 131:2120–2126
159. Rajan V, Chapman KE, Lyons V, Jamieson PM, Mullins JJ, Edwards CRW et al. (1995) Cloning, sequencing and tissue-distribution of mouse 11beta-hydroxysteroid dehydrogenase-l cDNA. J Steroid Biochem Mol Biol 52:141–147
160. Brown RW, Chapman KE, Koteletsev Y, Yau JL, Leckie CM, Murad P et al. (1995) Cloning of 11β-hydroxysteroid dehydrogenase type 2 from human placenta: characterisation and distribution. J Endocrinol 144(Suppl):OC35
161. Mercer WR, Krozowski ZS (1992) Localization of an 11β-hydroxysteroid dehydrogenase activity to the distal nephron. Evidence for the existence of two species of dehydrogenase in the rat kidney. Endocrinology 130:540–543
162. Walker BR, Campbell JC, Williams BC, Edwards CRW (1992) Tissue-specific distribution of the NAD^+-dependent isoform of 11β-hydroxysteroid dehydrogenase. Endocrinology 131:970–972
163. Rusvai E, Naray-Fejes-Toth A (1993) A new isoform of 11β-hydroxysteroid dehydrogenase in aldosterone target cells. J Biol Chem 268:10717–10720
164. Brown RW, Chapman KE, Edwards CRW, Seckl JR (1993) Human placental 11β-hydroxysteroid dehydrogenase: evidence for and partial purification of a distinct NAD-dependent isoform. Endocrinology 132:2614–2621
165. Lugg MA, Nicholas TE (1978) The effect of dexamethasone on the activity of 11β-hydroxysteroid dehydrogenase in the foetal rabbit lung during the final stage of gestation. J Pharm Pharmacol 30:587–589
166. Hammami MM, Siiteri PK (1991) Regulation of 11β-hydroxysteroid dehydrogenase activity in human skin fibroblasts: enzymatic modulation of glucocorticoid action. J Clin Endocrinol Metab 73:326–334
167. Smith RE, Funder JW (1991) Renal 11β-hydroxysteroid dehydrogenase activity: effects of age, sex and altered hormonal status. J Steroid Biochem Mol Biol 38:265–267
168. Zumoff B, Bradlow HL, Levin J, Fukushima DK (1983) Influence of thyroid function on the in vivo cortisol-cortisone equilibrium in man. J Steroid Biochem 18:437–440
169. Ichikawa Y, Yoshida K, Kawagoe M, Saito E, Abe Y, Arikawa K et al. (1977) Altered equilibrium between cortisol and cortisone in plasma in thyroid dysfunction and inflammatory diseases. Metabolism 26:989–997
170. Hellman L, Bradlow HL, Zumoff B, Gallagher TF (1961) The influence of thyroid hormone on hydrocortisone production and metabolism. J Clin Endocrinol Metab 21:1231–1247
171. Whorwood CB, Sheppard MC, Stewart PM (1993) Tissue specific effects of thyroid hormone on 11β-hydroxysteroid dehydrogenase gene expression. J Steroid Biochem Mol Biol 46:539–547
172. Koerner DR, Hellman L (1964) Effect of thyroxine administration on the 11β-hydroxysteroid dehydrogenases in rat liver and kidney. Endocrinology 75:592–601
173. Low SC, Assaad SN, Rajan V, Chapman KE, Edwards CRW, Seckl JR (1993) Regulation of 11β-hydroxysteroid dehydrogenase by sex steroids in vivo: further evidence for the existence of a second dehydrogenase in rat kidney. J Endocrinol 139:27–35
174. Souness GW, Latif SA, Laurenzo JL, Morris DJ (1995) 11alpha- and 11beta-hydroxyprogesterone, potent inhibitors of 11beta-hydroxysteroid dehydrogenase (isoforms 1

and 2), confer marked mineralocorticoid activity on corticosterone in the ADX rat. Endocrinology 136:1809–1812
175. Latif SA, Hartman LR, Souness GW, Morris DJ (1994) Possible endogenous regulators of steroid inactivating enzymes and glucocorticoid-induced Na+ retention. Steroids 59:352–356
176. Buhler H, Perschel FH, Hierholzer K (1991) Inhibition of rat renal 11β-hydroxysteroid dehydrogenase by steroidal compounds and triterpenoids; structure/function relationship. Biochim Biophys Act 1075:206–212
177. Perschel FH, Buhler H, Hierholzer K (1991) Bile acids and their amidates inhibit 11β-hydroxysteroid dehydrogenase obtained from rat kidney. Pflugers Arch 418:538–543
178. Homma M, Oka K, Niitsuma T, Itoh H (1994) A novel 11β-hydroxysteroid dehydrogenase inhibitor contained in Saiboku-To, a herbal remedy for steroid-dependent bronchial asthma. J Pharm Pharmacol 46:305–309
179. Song D, Lorenzo B, Reidenberg MM (1992) Inhibition of 11β-hydroxysteroid dehydrogenase by gossypol and bioflavonoids. J Lab Clin Med 120:792–797
180. Raven PW, Checkley SA, Taylor NF (1995) Extra-adrenal effects of metyrapone include inhibition of the 11-oxoreductase activity of 11beta-hydroxysteroid dehydrogenase: a model for 11-HSD I deficiency. Clin Endocrinol 43:637–644

Chapter 5

Microenvironmental control of glucocorticoid functions in immune regulation

Jon D. Hennebold and Raymond A. Daynes

Introduction

Lipophilic hormones, including the steroids, thyroid hormones, retinoids, 1,25 dihydroxyvitamin D_3, members of the prostaglandin J_2 series, and leukotriene B_4 represent a diverse group of molecules that can regulate gene expression through a common molecular mechanism. This diverse set of ligands traverse the plasma membrane and bind to ligand-specific intracellular receptors present within the cytosol, perinuclear space, or the nucleus of responsive cell types. Once activated by ligand binding, these intracellular receptors, which are all members of the steroid/nuclear receptor superfamily bind to specific sites on DNA and function as enhancers or repressors of gene transcription [1,2].

A large number of studies have demonstrated that several distinct species of lipophilic hormones have the capacity to influence particular aspects of the mammalian immune system. These substances facilitate the modulation of numerous lymphoid cell activities including proliferation, differentiation, and the capacity to produce bioactive molecules. For example, specific retinoid metabolites are critical for the ability of T lymphocytes to enter and progress through the cell cycle [3]. The lipophilic hormone 1,25 dihydroxyvitamin D_3 can inhibit the ability of activated lymphocytes to proliferate and produce cytokines that are involved in inflammatory processes [4,5]. Estrogen has been reported to function as a potent inhibitor of B-cell development [6]. These represent a few examples of the varied immunomodulatory effects possessed by the lipophilic hormones.

Perhaps the most studied and the best characterized class of lipophilic hormones are the glucocorticoids (GCS). The endogenous GCS, including cortisol and corticosterone, represent steroid hormones produced by the adrenal glands that are well established as possessing potent anti-inflammatory activities [7–10]. From research initially conducted in the 1950s, it was established that GCS are capable of inhibiting the pathologic inflammatory processes associated with arthritis [11]. The initial promise of GCS as safe and effective therapeutic agents for chronic inflammatory diseases, however, was soon brought into question as serious side-effects came to light following the sustained use of GCS at pharmacological levels. When used clinically, chronic GCS treatments are known to significantly suppress immune function and increase host susceptibility to a number of infectious diseases [12,13]. Chronic GCS administration also causes a variety of other undesirable side-effects [14].

Over the past 40 years studies have been conducted that have focused on the mechanisms by which GCS modulate lymphoid cell physiology. From these studies, it has become apparent that GCS are not carrying out their diverse effects through a single mechanism. Rather, these steroid hormones seem to affect a wide range of activities of multiple cellular constituents of the immune system. GCS are now appreciated to be directly involved in the regulation of the tissue distribution, survival, proliferative capacity, and effector functions of granulocytes, macrophages, B-cells and T-cells [15–17].

Regulation of the immune system by glucocorticoids

GCS influence several molecular and cellular processes that are important in inflammation and innate immunity. The ability of macrophages to phagocytose and digest antigen particles, as well as to present antigen, can all be regulated by GCS [18,19]. The expression of macrophage class II major histocompatibility complex (MHC) molecules is reduced under conditions of elevated GCS levels (20). Additionally, the ability of macrophages to limit the growth of some species of intracellular bacteria is reduced by elevations in GCS [12]. This occursparthy through the ability of GCS to regulate negatively the expression of enzymes responsible for the production of the bacteriostatic compound nitric oxide [21]. Neutrophil physiology is also profoundly affected by conditions which elevate GCS levels. High GCS levels depress neutrophil production of oxygen radicals such as superoxide, bioactive arachidonate metabolites, and also inhibit the ability of these cells to release their granule contents [22–24]. Treatment of neutrophils with supraphysiological levels of GCS also inhibits their capacity to adhere to platelet-activating factor (PAF) or thrombin-treated endothelial cells [25]. Similar effects by GCS have also been reported to occur on mast cells [26].

GCS are able to influence both the development and effector functions of the adaptive immune response through their direct actions on lymphocytes. It has been well documented over the past 30 years that GCS can regulate lymphocyte development, proliferation, and survival [15,17]. In the 1960s Dougherty and coworkers [27] initially demonstrated that lymphocytes cultured in the presence of pharmacological levels of GCS undergo distinct morphological changes that include cytoplasmic removal, plasma membrane blebbing, and chromatin condensation. It was later demonstrated that the lympholytic actions of GCS are mediated through the induction of programmed cell death or apoptosis [28,29]. The ability of GCS to induce apoptosis also extends to thymocytes [30] where studies of thymocyte development and apoptosis has led to the conclusion that natural GCS are playing important physiological roles in the maturation of T-cells within the thymus [31,32]. Additionally, GCS play an important role in the development of B-cells. It has been reported that modest elevations in circulating GCS levels in vivo can significantly reduce the numbers of B-cell precursors that develop in the bone marrow [33].

GCS inhibit the ability of lymphocytes to divide following cellular activation [15]. The mechanism by which lymphocyte proliferative capacity is diminished by GCS is mediated through an inhibition of the production of autocrine-acting substances that promote progression through the cell cycle [17]. One such factor whose production is inhibited by GCS is the cytokine IL-2. This cytokine is critical for lymphocyte proliferation via an autocrine-acting pathway [34].

The list of cytokines whose synthesis can be regulated by GCS has been extended considerably over the past several years (Table 5.1). It is now well established that elevations in GCS can inhibit the production of cytokines that are of critical importance for the generation of an inflammatory response. These include the cytokines IL-1, IL-6, and TNFα [20,35,36]. With regard to T-cells, GCS appear to regulate differentially their capacity to produce cytokines. It is now well accepted that naive T-cells have the capacity to differentiate down various developmental pathways (type 1 versus type 2), evaluated by their preferential production of certain patterns of cytokines with activation. These distinctions differentially facilitate subsequent development of immune effector pathways. The differentiation of T-cells that produce predominantly the type 1 cytokines IL-2 and IFNγ are important for promoting strong inflammatory and cell-mediated immune responses. Elevations in GCS potently inhibits T-cell production of those substances [37]. In contrast to the type 1 cytokines, elevations in GCS levels actually increase T-cell production of the type 2 cytokines IL-4, IL-10, and IL-13 [38,39]. These molecules are important for B-cell development, immunoglobulin class switching, and actually serve to inhibit inflammatory processes including the production of many of the inflammatory cytokine species [34,40–42].

Table 5.1. The effects of glucocorticoids (GCS) on lymphoid cell cytokine production

Cell target	Cytokine	Effect of GCS[a]
Lymphocytes	IL-2	–
	IFNγ	–
	IL-4	+[b]
	IL-5	+[b]
	IL-10	+
	IL-13	+
	TGFβ	+
Macrophages	IL-1	–
	IL-6	–
	IL-8	–
	IL-10	+
	IL-12	–
	TNFα	–
	IFNα[d]	–
	RANTES[e]	–
	MIF	+[c]

[a] + Indicates increased cytokine production; – indicates decreased cytokine production.
[b] GCS have been reported to inhibit the production of these cytokines [Byron K.A. et al. (1992) Immunology 77 : 624–626; Rolfe F.G. et al. (1992) Immunology 77 : 494–499].
[c] GCS increase the production of macrophage migration inhibitory factor (MIF) at low GCS concentrations (10^{-12} to 10^{-9} M) but inhibit MIF synthesis at high concentrations (>10^{-9} M) [140].
[d] [Rossel S. et al. (1985) J Infect Dis 150 : 815–821.]
[e] [Marfaing-Koka A. et al. (1996) Int Immunol 8 : 1587–1594.]

Additionally, GCS enhance the production of lipocortin-1 and bioactive TGFβ, cytokines that also possess potent anti-inflammatory activities [43–45].

Glucocorticoid mechanism of action

The ability of GCS to regulate lymphoid cell function is dependent upon their ability to function as ligands for specific intracellular receptors [46,47]. Following ligand binding, the GCS receptors gain the capacity to regulate gene transcription in either a positive or negative manner. The type 1 and type 2 GCS receptors belong to the steroid hormone receptor gene superfamily [2]. The type 1 GCS receptor, also known as the mineralocorticoid receptor, binds to both GCS and aldosterone with high affinity [48]. The type 2 receptor evolved with specificity for the naturally occurring GCS cortisol and corticosterone [49]. Most, if not all, of the effects of GCS on immune function are mediated as a consequence of activation of the type 2 GCS receptor.

Type 2 GCS receptor activation is dependent upon the activities of three domains contained within the protein. These domains comprise the ligand binding domain, the DNA binding domain, and the transcriptional activation domain. It is the ligand binding domain that is responsible for controlling the ability of the receptor to regulate gene transcription. In the absence of ligand, heat shock proteins (hsp90, hsp70, and hsp56) associate with the ligand binding domain [50,51]. The interaction of the hsp and the ligand binding domain maintains the GCS receptor in a state that has high affinity for GCS ligand but is incapable of translocating to the nucleus [47]. Upon binding of the receptor with GCS, however, the hsp are released from the receptor which can now translocate to the nucleus and become competent to regulate gene transcription in either a positive or negative manner. Positive or negative regulation of gene expression by GCS are mediated through distinct molecular mechanisms.

GCS-dependent enhancement of gene transcription is facilitated through the ability of the activated receptor complex to move from the cytoplasm to the promoter region of target genes in the nucleus (Fig. 5.1A). Following its translocation to the nucleus, the GCS receptor complex binds to a specific DNA sequence in the 5′ promoter region of corticosteroid responsive genes. This promoter sequence is

Fig. 5.1. Glucocorticoids (GCS) positively and negatively regulate target gene expression. Ligand-bound GCS receptors bind directly to specific DNA sequences (glucocorticoid response element, GRE) in the 5′ promoter regions of target genes and enhance their rate of transcription (**A**). The ability of ligand-bound GCS receptors negatively to regulate gene transcription occurs through multiple mechanisms (**B**). Following activation with ligand, GCS receptors induce the expression of molecules that can directly inhibit the production of inflammatory substances. Alternatively, activated GCS receptors bind to other transcription factors (e.g. c-*jun*, NF-κB) and prevent them from interacting with specific response elements in target genes, thereby inhibiting gene transcription.

termed the GCS response element (GRE) [47]. Receptor binding to the GRE is mediated through the interaction of two zinc fingers in the DNA binding domain with a specific DNA sequence [52]. The GRE itself is palindromic and is composed of two half sites. Each of the half sites is sufficient to allow the binding of

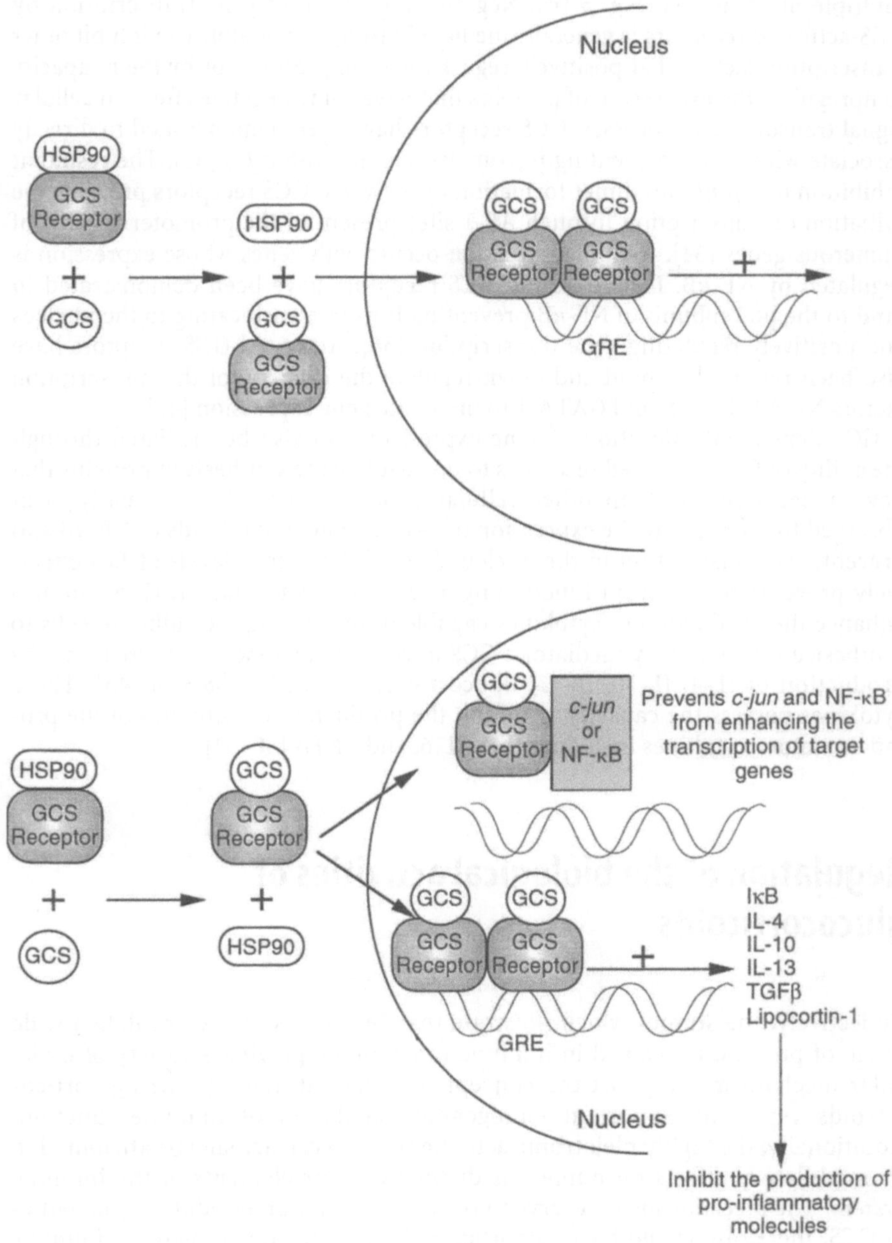

one activated GCS receptor, thereby facilitating the formation of GCS receptor homodimers on the GRE [52]. Following dimerization, the GCS receptors then stabilize the transcription complexes at the tumour-associated transplantation antigen (TATA) box through their transactivation domain [46]. The stabilization of the transcription complex increases or initiates target gene expression.

Negative regulation of the rate of gene transcription can be facilitated through multiple mechanisms (Fig. 5.1B). Negative regulation of gene transcription by GCS-activated receptors is generally mediated through their ability to inhibit other transcription factors that positively regulate gene expression, or by their capacity to upregulate the expression of proteins that have some negative effect on cellular signal transduction processes. GCS receptors have been demonstrated to directly associate with c-*jun*, preventing it from associating with c-*fos* [53]. The resultant inhibition of c-*jun*/c-*fos* dimer formation by activated GCS receptors prevents the initiation of transcription through AP-1 sites present in the promoter regions of numerous genes [54]. A similar situation occurs with genes whose expression is regulated by NF-κB. Ligand-bound GCS receptors have been demonstrated to bind to the p65 subunit of NF-κB preventing it from translocating to the nucleus and positively regulating gene transcription [36]. Activated GCS receptors have also been reported to bind and down-regulate the capacity of the transcription factors NF-AT, OTF-1, and GATA-1 to modulate gene expression [47].

GCS-dependent inhibition of gene expression can also be mediated through the ability of GCS-activated receptors to up-regulate the synthesis of proteins that have a negative effect on other cellular processes. GCS have recently been observed to up-regulate the expression of 1κB, a protein that binds to NF-κB and prevents its translocation to the nucleus [55,56]. Enhanced levels of IκB effectively prevents NF-κB from functioning as a transcription factor. GCS can also enhance the production of cytokines capable of preventing the ability of cells to synthesize inflammatory mediators. GCS have been documented to increase the production of IL-4, IL-10, IL-13, lipocortin-1, and TGFβ [38,39,44,45]. These cytokines possess the capacity to inhibit the production or activities of the pro-inflammatory cytokines including IL-1, IL-6, and TNFα [40–42].

Regulation of the biological activities of glucocorticoids

Collectively, the studies which illustrate that GCS can serve to regulate a wide range of processes involved in immune function, employing a variety of molecular mechanisms, support the concept that the naturally occurring corticosteroids represent important endogenous regulators of immune function. Additionally, the highly pleiotropic activities of GCS can actually be attributed to its modulatory effects on numerous distinct cellular elements of the immune system. The mechanisms that serve to regulate the immunomodulatory activities of GCS, therefore, would have important effects on the outcome of an immune response.

At least three distinct mechanisms act in a coordinated manner to regulate the many GCS-dependent effects on the immune system. GCS activities can be controlled in a systemic fashion by molecular processes that regulate adrenal corticosteroid biosynthesis. Regulation of GCS actions can also occur at the level of the cell through a modulation of cellular sensitivity to particular concentrations of GCS. This is either accomplished by processes that regulate the number and activity of GCS receptors or through the actions of counterregulatory molecules which control cell responsiveness to hormone action like macrophage migration inhibitory factor (MIF) [57]. These processes are important for determining GCS sensitivity at the cellular level. Finally, the concentration of GCS within specific tissues of the body are dramatically regulated via the actions of enzymes that control the ratio of biologically active GCS and biologically inactive GCS metabolites.

GCS are produced almost exclusively by cells of the adrenal cortex [7,58]. The rate of adrenal GCS production is under tight regulation by factors that are produced by the hypothalamus and pituitary. Following appropriate stimulation, the hypothalamus produces corticotrophin-releasing hormone (CRH) and arginine vasopressin (AVP), peptides that are transported to the pituitary via the hypothalamic-pituitary portal system. CRH and AVP, then in turn, stimulate the release of adrenocorticotrophin hormone (ACTH) from the pituitary into the systemic circulation. ACTH interaction with the cells present in the adrenal cortex induces the synthesis of GCS [7]. In humans, the majority of the GCS that is produced and released is in the form of cortisol, while in rodents corticosterone is the predominant circulating GCS. Hypothalamic and pituitary production of CRH and ACTH, respectively can be negatively regulated by GCS, providing a feedback loop [58]. Under non-stress conditions, GCS down-regulate the level of CRH and ACTH production thereby providing a mechanism for tightly controlling the concentration of circulating GCS. Under stress conditions, however, signals from the hippocampus induce increased release of CRH which causes a rapid rise in circulating GCS levels [16].

Many of the inflammatory cytokines possess the direct capacity to regulate the level of adrenal GCS production and, therefore, plasma GCS levels. It has been demonstrated that the cytokines IL-1α, IL-1β, IL-6, IL-12 and TNFα can each cause increases in adrenal GCS biosynthesis [59]. These cytokines function to induce the production of GCS directly by acting on cells of the adrenal gland or by inducing hypothalamic and pituitary production of CRH and ACTH, respectively [59].

Responsiveness to GCS can be very efficiently regulated at the receptor level. It has been well established that type 2 GCS receptors are ubiquitously expressed in most mammalian tissues [60,61]. The level of receptor expression, however, can vary significantly from cell type to cell type even in the same organ system. In the brain, for example, GCS receptor numbers vary greatly within the different cellular constituents of the hippocampus [62]. The differential expression of GCS receptors in various cell types is probably due to the presence of at least three distinct upstream promoter regions of the GCS receptor gene [63]. The most active of these promoters is only utilized in T-cells and may be responsible for the high level of GCS receptor expression within this cell type [63].

GCS receptor activity in cells can be effectively modulated via post-transcriptional mechanisms. Proliferating cells exhibit cell cycle dependence in the ability of GCS to modulate gene expression [64]. Cells are more sensitive to the actions of GCS while in the G1/S phase but become resistant during their progression through the G2/M period of the cell cycle [65]. This differential sensitivity to GCS during different phases of the cell cycle has been correlated with the level of GCS receptor phosphorylation [65]. During the G2/M phase of the cell cycle, when cells are more resistant to the effects of GCS, the GCS receptor becomes hyperphosphorylated [66]. Experimental evidence suggests that in the hyperphosphorylated state, the GCS receptor becomes tightly associated with hsp [66].

The ability of GCS to regulate cellular processes within different organ systems is absolutely dependent upon the physical association of the GCS with the GCS receptor. Different tissues have the capacity to limit this physical association by regulating the intracellular concentration of corticosteroids, regardless of extracellular hormone levels, via the actions of enzymes that efficiently convert biologically active GCS to inactive compounds [27]. The level of expression of GCS metabolizing enzymes within tissues such as lymphoid organs can play a role in regulating the degree to which distinct organ systems are influenced by circulating GCS.

11β-Hydroxysteroid dehydrogenase

The tissue levels of biologically active GCS within distinct tissues are regulated primarily through the actions of the microsomal enzyme 11β-hydroxysteroid dehydrogenase (11β-HSD). 11β-HSD can interconvert biologically active 11-hydroxy GCS (cortisol, corticosterone) to biologically inactive 11-keto metabolites (cortisone, 11-dehydrocorticosterone, respectively; (Fig. 5.2) [67,68]. It was first demonstrated in 1953 that a large amount of 11β-HSD activity is present in the liver [69]. Since this initial description, 11β-HSD enzyme activity has been observed in wide variety of other tissues and organ systems. Tissues known to possess high levels of 11β-HSD activity include the liver, kidney, lung, brain, skin, intestine, placenta, adipose tissue and spleen [70].

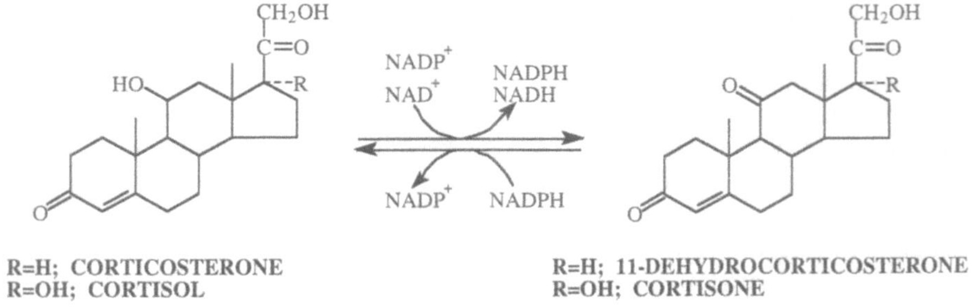

Fig. 5.2. 11β-Hydroxysteroid dehydrogenase (11β-HSD) interconverts the biologically active 11-hydroxy GCS (cortisol, corticosterone) with their biologically inactive 11-keto metabolites (cortisone, 11-dehydrocorticosterone). These oxidation/reduction reactions require the presence of nicotinamide adenine dinucleotide co-factors.

The studies describing the gross tissue distribution of 11β-HSD enzyme activity were completed in the 1960s and 1970s. Subsequently, there have been numerous studies demonstrating that 11β-HSD plays an important role in regulating the concentration of biologically active GCS within the numerous tissues where this enzyme is expressed. The importance of 11β-HSD in controlling GCS/GCS receptor interaction has been best studies in the kidney. The distal convoluted tubule epithelium regulates Na^+ and K^+ absorption through the expression of a Na^+/K^+ ATPase [71]. The expression of this transporter is significantly elevated through the association of the mineralocorticoid aldosterone with the type 2 GCS receptor [71]. The increased expression of the transporter leads to an increased rate of Na^+ absorption and K^+ excretion and subsequently an elevation in blood pressure [72]. It was originally hypothesized that the ability of only aldosterone to regulate the expression of Na^+/K^+ ATPase was due to its specific binding with the type 2 or mineralocorticoid receptor. Following the cloning of the mineralocorticoid receptor, however, it was determined that both GCS and aldosterone could bind to recombinantly expressed mineralocorticoid receptors with equally high affinity [48]. This occurred despite the observations that the mineralocorticoid receptor of the kidney distal convoluted tubule is activated selectively by aldosterone in vivo [73]. A selective binding of aldosterone to the mineralocorticoid receptor seems highly unlikely, since circulating levels of GCS are at a 1000-fold higher concentration then aldosterone. The aldosterone selectivity was originally proposed to be due to the preferential binding of GCS to corticosteroid-binding globulins (CBG) compared with aldosterone. Investigators hypothesized that CBG bound to GCS and thus prevented their entry into mineralocorticoid receptor-expressing cells in the kidney. This mechanism turned out not to be responsible for the aldosterone selectivity since aldosterone still preferentially binds to the mineralocorticoid receptor expressed in the distal convoluted tubules in the kidneys of neonatal animals. Neonates are relatively deficient in CBG [74].

The mechanism of how aldosterone selectivity is maintained came from an understanding of the molecular basis for the disease "apparent mineralocorticoid excess" (AME), a rare hypertensive disease of children. It had been known for some time that AME associates with normal plasma levels of aldosterone and an increased ratio of 11-hydroxy corticosteroids to 11-keto GCS derivatives in the urine [75,76]. These observations led to the realization that the kidney utilizes 11β-HSD to rapidly convert biologically active GCS, forms that have the capacity to bind to the mineralocorticoid receptor, to biologically inert 11-keto derivatives that bind negligibly to the mineralocorticoid receptor [77]. Further evidence for 11β-HSD playing a role in mineralocorticoid selectivity came from studies demonstrating that treatment of rats with inhibitors of 11β-HSD activity, glycyrrhetinic acid (GA) or carbenoxolone, could abolish the specificity of aldosterone for the mineralocorticoid receptor in vivo [78]. In vivo autoradiographic studies in the rat showed that renal binding of ^{3}H-corticosterone to the mineralocorticoid receptor in the distal convoluted tubule was normally very low [73]. When the animals were treated with GA, however, ^{3}H-corticosterone binding was elevated, becoming indistinguishable from the binding of ^{3}H-aldosterone. This mechanism of end organ catabolism has been conserved over 300 million years of evolution with amphibian, avian and mammalian mineralocorticoid receptor specificity within numerous tissues of many species being maintained by the catabolic actions of 11β-HSD [79].

Based on recent studies, it has been proposed that 11β-HSD can also serve to regulate the association of GCS and the GCS receptor. For example the 11β-HSD activity in the testes of a number of mammalian species appears to be necessary for reproduction. Chronic elevations in GCS reduce the level of testicular testosterone biosynthesis. GCS suppression of testosterone biosynthesis is mediated by the suppression of side-chain cleavage enzyme and 17α-hydroxylase, both of which are critically important for the Leydig cells of the testes to produce testosterone [80,81]. Monder and coworkers demonstrated by immunohistochemical means that the 11β-HSD within the tests localizes to the Leydig cells [82,83]. The expression of this enzyme in the testes is sufficient to lower tissue concentrations of GCS to a level that allows testosterone biosynthesis to occur.

Further support for 11β-HSD playing a role in regulating GCS/GCS receptor interactions in vivo has been provided by additional studies. In the skin, for example, the topical administration of GCS results in a dose-dependent vasoconstriction and skin blanching. This phenomenon represents a GCS-specific event [84]. Cortisol alone exerts only a weak effect in this particular assay system. Topical administration of GA with cortisol, however, results in a much more effective vasoconstrictor response [84]. Furthermore, the expression of 11β-HSD in the skin is localized to the same cellular elements that are known to possess GCS receptors [84,85]. GCS also serve to up-regulate the expression of the α1-subunit of Na^+/K^+ ATPase in epithelial cells via its activation of the GCS receptor [71]. Induction of the α1-subunit expression is greatly increased in vivo by the co-administration of both corticosterone and the 11β-HSD inhibitor carbenoxolone [86]. These studies demonstrate that 11β-HSD can modulate the ability of GCS to physically associate with its receptor through the regulation of microenvironmental concentrations of biologically active corticosteroids.

Molecular characteristics of 11β-HSD

The original observation that enzymatically active 11β-HSD is present in a wide range of tissues was made over the 45 years ago. Over the past several years, however, the purification and cloning of genes encoding proteins that express 11β-HSD activity has occurred. 11β-HSD was first purified from detergent-solubilized rat liver microsomes through the use of a $NADP^+$ affinity chromatography matrix [87]. The purified enzyme was observed to be a 34 kDa, microsome-associated protein. Polyclonal antiserum was generated from this liver-enriched enzyme preparation and used to analyse different tissues for expression of the enzyme by Western blot and immunohistochemical techniques [88]. In rat tissues, including brain, kidney, testes, lung, stomach and intestine, a detectable immunoreactivity was observed [88].

Shortly after the purification of the enzyme from liver, the gene encoding the 11β-HSD in rat liver was cloned and sequenced [89]. Since then, the genes encoding the liver enzyme have been cloned and sequenced for a number of different species including human, sheep, squirrel monkey and mouse [90–93]. From the deduced nucleotide sequence it was determined that the liver 11β-HSD enzyme is a member of the family of enzymes termed the short-chain alcohol dehydrogenases (SCAD). A high degree of homology between 11β-HSD and other SCAD members is present in the regions comprising the adenine dinucleotide cofactor binding site and the active site [79]. With the 11β-HSD cDNA being used as a probe, Northern blot analysis demonstrated that rat liver and most other tissues that a single species of mRNA was present (~1.7 kb) [89]. When the gene responsible for encoding the rat liver 11β-HSD was transfected into chinese hamster ovary (CHO) cells it was observed that the expressed enzyme was capable of functioning as both a dehydrogenase and a reductase [89]. This is supportive of previous findings that 11β-HSD functions primarily as a reductase in the liver [94]. The rat liver enzyme possesses a K_m of 17 μM for cortisol and 2 μM for corticosterone [89].

Several experiments have suggested that additional isoforms of 11β-HSD exist. Western blot analysis of rat tissues using the polyclonal 11β-HSD antiserum revealed multiple immunoreactive species in different tissues [88]. In the liver there appears to be only a single 34 kDa species. In other tissues, including the kidney, there is a far greater heterogeneity of immunoreactivity. The kidney possesses additional 40 kDa and 68 kDa species [88]. From these observations, it was proposed that there may be antigenically related proteins in other tissues.

Using the polyclonal antiserum generated against the purified liver 11β-HSD it was demonstrated that proximal tubules of the kidney stain intensely by immunohistochemical analysis [95]. These results came as a surprise since it was thought that the 11β-HSD enzyme would have to localize to the distal convoluted tubules since this is where the mineralocorticoid receptors are localized. Subsequently, it was observed that the distal convoluted tubule epithelium contains 11β-HSD activity that has enzymatic characteristics distinct from the liver enzyme [96–98]. It was determined that distal tubule epithelium 11β-HSD activity had a very low K_m (<100 nM), utilized NAD^+ almost exclusively as a cofactor, and functioned exclusively as a dehydrogenase.

Additional experimental support for the existence of a second 11β-HSD isoform came from the analysis of rat liver and kidney sex-dependent expression of 11β-HSD mRNA and activity. From this study, it was observed that enzyme activity was different in the livers of males and females [99]. 11β-HSD activity was higher in the liver of male rats when compared with the liver activity of female rats. mRNA levels and enzymatic activity for liver 11β-HSD in ovarectomized females were observed to be similar to levels normally present in males. Male animals that had been gonadectomized and given oestrogen supplementation exhibited low levels of liver activity and mRNA that was similar to liver activity and mRNA levels of normal females. In the kidneys of oestrogen-treated and gonadectomized male rats, however, there was an observed enhancement in overall 11β-HSD activity despite the complete absence of 11β-HSD mRNA corresponding to the cloned liver 11β-HSD. These results demonstrated that within the kidney a second 11β-HSD isoform must be present.

In a first step towards characterizing the gene encoding a new 11β-HSD isoform that is responsible for maintaining mineralocorticoid receptor specificity, mRNA from dissected cortical collecting ducts of rats was isolated, size fractionated, and injected into *Xenopus* oocytes [100]. It was observed that the mRNA corresponding to a size of approximately 2 kb encoded significant levels of 11β-HSD activity. This 11β-HSD activity exhibited similar characteristics to the previously described enzyme present in the distal convoluted tubule epithelium [96]. Shortly thereafter, this 11β-HSD isoform was cloned from a human kidney cDNA library via an expression cloning strategy in CHO cells [101]. This second isoform was termed 11β-HSD2 whereas the liver isoform was termed 11β-HSD1. Analysis of the enzymatic properties of the 11β-HSD2 isoform resulted in the observation that this enzyme possessed characteristics that were quite similar to those reported for the enzyme present in isolated distal convoluted tubule epithelium and placenta [101]. The 11β-HSD2 enzyme exhibits a low K_m for substrate (4 nM for corticosterone, 47 nM for cortisol), utilizes NAD^+ almost exclusively as a cofactor, and functions only as a dehydrogenase [101,102]. More recently, the 11β-HSD2 genes have been cloned for a number of different species including sheep, rabbit, and mouse [101,103,104]. All of these genes encode a protein of approximately 44 kDa that possess similar enzymatic properties.

Upon comparing the sequences between the gene for 11β-HSD1, 11β-HSD2 and other members of the SCAD family, it was determined that the 11β-HSD2 enzyme is a member of the SCAD family of dehydrogenases. It was somewhat of surprise, however, to find that the human 11β-HSD2 sequence has low homology to 11β-HSD1 (14%) while possessing significant homology with the enzyme 17β-hydroxysteroid dehydrogenase 2 (17β-HSD2) [101]. The similarities that exist between the 11β-HSD1 and 11β-HSD2 enzymes lie within a region that is responsible for cofactor-binding and the active site of the enzymes [101].

By Northern blot analysis it has been observed that kidney, small intestine and colon posses high levels of mRNA for the 11β-HSD2 enzyme [101]. The expression of 11β-HSD2 closely parallels the tissue-restricted expression of the mineralocorticoid receptor [102,104,105]. From these studies, it has been proposed that the 11β-HSD2 enzyme is required for the maintenance of mineralocorticoid specificity for aldosterone in vivo. This turns out to be the case since mutations in the human 11β-HSD2 gene have been identified in individuals with AME that result in the expression of a non-functional enzyme [106,107]. These mutations are due to incorrect exon splicing which in turn produces mRNA that has lower stability and point mutations [106,107].

A high level of expression of the 11β-HSD2 isoform has also been detected in most early fetal tissues and fetal-derived placental constituents [101,107–109]. By in situ analysis using an 11β-HSD2 probe, it was demonstrated that most fetal tissues contain high levels of 11β-HSD2 mRNA [110]. This high level of message expression correlates with an extremely high level of 11β-hydroxy dehydrogenase activity in both fetus-derived placental tissues and fetal tissues themselves [111,112]. The high rate of GCS catabolism via the actions of 11β-HSD2 is thought to protect the fetus from the teratogenic and growth-retarding effects of GCS on the fetus [111,112]. In experiments utilizing rats, the idea that 11β-HSD2 activity protects the fetus from the deleterious effects of endogenous GCS has gained support. Chronic administration of dexamethasone, a weakly metabolizable synthetic GCS, to pregnant female rats resulted in offspring that had significantly lower birthweight and an increased risk of becoming and staying hypertensive throughout their lifespan [113]. Pregnant rats treated with the 11β-HSD inhibitor carbenoxolone also produced offspring that had reduced birthweight and elevations in basal blood pressure [114]. A similar situation has been reported to occur in humans where decreased placental 11β-HSD activity correlates with low birthweights babies [107]. These studies support the hypothesis that fetal and placental GCS catabolism plays an important role in proper fetal development by protecting the fetus from negative influences by GCS.

At the present time it is unclear whether there are more than just two 11β-HSD isoforms. Five different isoforms of 17β-HSD that differ in their tissue distribution and physical parameters have been cloned and sequenced to date [115]. The enzyme 3β-hydroxysteroid dehydrogenase/Δ4-Δ5 isomerase (3β-HSD), another steroid dehydrogenase, is also composed of multiple isoforms [116]. There have been four distinct 3β-HSD enzymes that have been cloned and characterized in humans, mice, and rats [116]. The apparent gene duplication events that have given rise to a number of steroid dehydrogenase enzymes may have also occurred with the 11β-HSD genes.

Several observations have led to the proposal that additional 11β-HSD isoforms might exist beyond those that have already been identified [70,117]. The polyclonal antiserum generated against $NADP^+$-agarose purified liver 11β-HSD1 cross-reacts with a 44 kDa protein in both the kidney and the colon [88]. It is assumed that this immunoreactive species probably represents the recently characterized 11β-HSD2 enzyme. It was also noted that homogenates of rat testes possessed a 47 kDa cross-reactive immunoreactive component [88,117]. At the present time it is unclear whether this protein possesses 11β-HSD activity. Support for another 11β-HSD enzyme also comes from the observation that rat epididymis contains significant enzymatic activity but exhibits no 34 or 44 kDa immunoreactive proteins by Western blot analysis [118]. Since this is not an aldosterone responsive tissue, it is doubtful that the enzymatic activity is due to the presence of the 11β-HSD2 enzyme within the epididymis.

Direct experimental support for the existence of a third 11β-HSD enzyme has come from studies analysing the characteristics of the 11β-HSD enzyme activity in the human choriocarcinoma cell line JEG-3. Analysis of 11β-HSD activity within JEG-3 revealed an enzyme with characteristics that are distinct from both the type 1 and type 2 11β-HSD isoforms [119]. JEG-3 11β-HSD possesses a K_m (~250 nM) that is intermediate between what has been reported for either 11β-HSD1 or 11β-HSD2. Additionally, it was demonstrated that JEG-3 11β-HSD could utilize $NADP^+$ or NAD^+ equivalently as a cofactor, another fact that distinguishes it from the 11β-HSD1 or 11β-HSD2 isoforms.

Characteristics and function of lymphoid organ 11β-HSD

Following the first description of 11β-HSD activity within the liver, it was soon discovered that lymphoid tissues also possess the capacity to metabolize biologically active 11-hydroxy GCS to biologically inactive 11-keto metabolites.

Dougherty and colleagues demonstrated in 1964 that lymphoid organs possess the capacity to convert cortisol to cortisone. In these studies, Dougherty and co-workers also determined that high concentrations of GCS can act directly to induce cell death within lymphocytes [27,120]. From these observations, it was hypothesized that the metabolic action of 11β-HSD in lymphoid organs were important for protecting lymphocytes from the cytolytic and anti-proliferative action of GCS (27, 120).

Recent studies have supported the findings of Dougherty and colleagues that lymphoid organs possess significant levels of 11β-HSD activity [121,122]. It is now appreciated that 11β-HSD activity is not equally expressed in all lymphoid organs. Microsomal preparations of lymphoid organs that drain peripheral tissues (axillary, brachial, inguinal) were found to possess the greatest amount of 11β-HSD activity, whereas lymphoid organs that receive their drainage from mucosal surfaces (Peyer's patches) were observed to contain significantly lower levels of 11β-HSD activity [122]. Interestingly, the amount of 11β-HSD activity in a lymphoid organ directly correlated with the ability of resident T-cells to produce the type 1 cytokines IL-2 and IFNγ following activation. Peripheral lymphoid organs possess the highest amount of 11β-HSD activity and T-cells resident within these tissues produce good quantities of IL-2 and IFNγ following activation [122,123]. T-cell production of the type 1 cytokines is highly sensitive to the inhibitory effects of GCS [37]. In contrast, mucosal lymphoid organs contain low levels of 11β-HSD activity and harbor T-cells that have a reduced potential to produce the type 1 cytokines while simultaneously producing greater amounts of the type 2 cytokines following stimulation [122,123]. These results suggest that the level of lymphoid organ 11β-HSD might be controlling the amount of biologically active GCS available for the regulation of lymphoid cell function.

Dougherty and Berliner demonstrated nearly 40 years ago that the 11β-HSD activity within lymphoid organs is primarily localized to the stromal, non-recirculating cellular elements of the lymphoid organ [120]. These conclusions came from studies in which animals were injected with a large enough dose of GCS to induce lymphocyte death in vivo. Following this treatment it was observed that lymphoid organs posses a two-fold higher level of 11β-HSD activity on a per gram basis then lymphoid organs from animals that did not receive the GCS injections [120]. This increase in enzymatic activity occurred even though there was a dramatic loss in lymphocyte numbers within the lymphoid organ. Their studies suggested that the remaining cellular constituents, which includes the stromal cells, must possess the majority of the lymphoid organ 11β-HSD. By separating stromal cells from the lymphoid cells (B-cells, T-cells and macrophages) it has subsequently been confirmed that the stromal cells possess nearly all of the lymphoid organ 11β-HSD activity [122]. It is the immobile cellular elements of the lymphoid organ, therefore, that possess a majority of the enzymatic capacity to regulate the amount of biologically active GCS that are present within this tissue.

The differences in the reactions catalysed by the recently cloned and characterized 11β-HSD1 and 11β-HSD2 enzymes raises an important question with regard to the role of the 11β-HSD in lymphoid organs. Is lymphoid organ 11β-HSD activity due to the expression of the 11β-HSD1 or 11β-HSD2 isoform? The differential expression of these two isoforms would be an important factor in determining if lymphoid organ GCS concentrations are lower or higher than what is present in the circulation since 11β-HSD1 functions predominantly as a reductase and 11β-HSD2 functions exclusively as a dehydrogenase.

Results from various studies of the 11β-HSD activity present in lymphoid organs such as the spleen actually indicate that there is a 11β-HSD activity that is distinct from the characterized 11β-HSD1 and 11β-HSD2 isoforms. Two independent studies have demonstrated that the K_m for the major species of 11β-HSD within the spleen is between 160 and 250 nM (122,124). These values for the spleen 11β-HSD is distinct from the K_m reported for both 11β-HSD1 and 11β-HSD2 enzymes [89,104]. A unique 11β-HSD enzyme in the spleen was further supported by experiments demonstrating that spleen 11β-HSD can utilize both NAD^+ and $NADP^+$ equivalently as a co-factor (122). It has been determined that 11β-HSD1 uses $NADP^+$ preferentially as a co-factor whereas 11β-HSD2 exclusively requires NAD^+ for maximal activity [89,101,102,125]. These results are very similar to reported characteristics of an apparently novel 11β-HSD enzyme within JEG-3 human choriocarcinoma cells [119]. Furthermore, it was demonstrated that murine spleen fragments possessed greater 11β-dehydrogenase activity than 11-reductase activity [122], suggesting that the lymphoid organ enzyme prefers to function predominantly as an inactivator of biologically active GCS. These results are very different from the net reaction catalysed by either the 11β-HSD1 or 11β-HSD2 isoforms. Additionally, very little 11β-HSD1 protein was observed in lymphoid organs by Western blot analysis utilizing polyclonal antiserum prepared from $NADP^+$ purified liver 11β-HSD1 enzyme [122]. Analysis of lymphoid organ mRNA levels for both 11β-HSD1 and 11β-HSD2 isoforms revealed that 11β-HSD1 message was barely detectable and mRNA for 11β-HSD2 was completely absent (Fig. 5.3). Collectively, these studies support the idea that another 11β-HSD isoform exists within lymphoid tissues.

The experiments demonstrating that lymphoid organs possess an 11β-HSD that is functioning as a dehydrogenase in vivo has been further supported by studies that have utilized chemical inhibitors of 11β-HSD. When animals are treated with the 11β-HSD inhibitor GA, there is a decrease in the production of the type 1 cytokines (IL-2 and IFNγ) by activated lymphocytes derived from the spleen, along with an increase in the production of the type 2 cytokines IL-4 and IL-10 [122,126]. The observed changes in immune cell function following GA administration are similar to effects mediate by GCS in vitro or in vivo [38,39].

The observed changes in the cytokine patterns produced by splenic lymphocytes removed from animals treated with GA (to inhibit lymphoid organ 11β-HSD activity) correlated with changes in immune effector function that occur when GCS concentrations are therapeutically elevated in vivo. For example, the ability of animals to develop contact hypersensitivity responses was markedly inhibited in sensitized animals that had been injected intraperitoneally with GA prior to challenge with the sensitizer [122]. The depression in the contact hypersensitivity response following GA treatment occurred without any elevation in plasma corticosterone levels. Contact hypersensitivity responses are known to be inhibited by elevations in systemic or local GCS concentrations [127]. Additionally, the inhibition of lymphoid 11β-HSD activity also reduced the

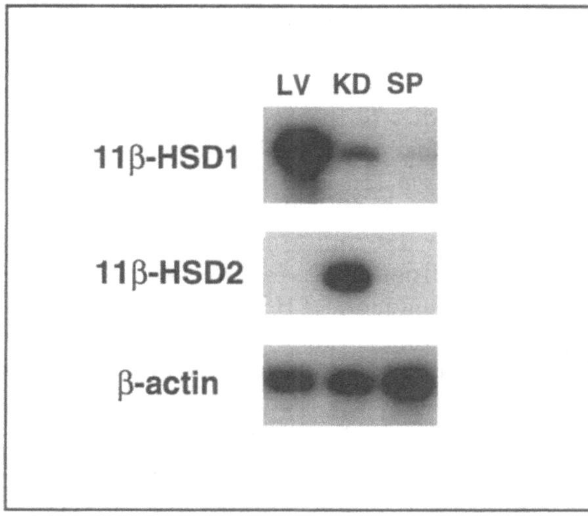

Fig. 5.3. Lymphoid organs from C57BL/6 mice possess no mRNA for 11β-HSD2 and low levels of mRNA for 11β-HSD1. RT-PCR was used to analyse mRNA levels in liver (LV), kidney (KD), and spleen (SP) for 11β-HSD1, 11β-HSD2, and β-actin as an internal control.

ability of the innate immune system to effectively handle the growth of the intracellular pathogen *Listeria monocytogenes* [126]. From these studies, it was observed that inhibition of 11β-HSD activity lowered the ability of splenocytes to produce IFNγ following a *Listeria* challenge [126]. It is known that IFNγ production is critical for limiting the cellular spread of *Listeria*, probably through its ability to induce the production of the bacteriostatic actions of NO [128]. In addition to inhibiting the ability of the host to produce the necessary IFNγ response, injection of mice with GA prior to *Listeria* challenge also enhanced host production of IL-10. Increased IL-10 levels have been shown to directly increase susceptibility to challenge with *Listeria* [129]. Furthermore, the spleens of mice given GA prior to *Listeria* infection were found to harbour significantly more bacteria than the spleens from mice with normal 11β-HSD activity (126). The increased susceptibility to infection following GA treatment was similar in effect to the administration of small doses of non-metabolizable GCS such as dexamethasone to normal mice followed by infection.

The finding that inhibition of lymphoid organ 11β-HSD activity in vivo mimicked the conditions associated with an elevated GCS level, supports the hypothesis that this enzyme functions in vivo as a dehydrogenase. If the lymphoid organ enzyme was functioning primarily as a reductase, then an inhibition of its activity would result in an opposite effect from what was observed with regard to the cytokine patterns produced by activated lymphocytes and the immune effector responses that are elicited. Depressions in reductase activity, achieved following the administration of GA, would lower intra-organ levels of biologically active GCS by inhibiting the rate of back-conversion of biologically inactive 11-keto GCS to their biologically active 11-hydroxy forms. In this situation, immune effector responses would appear as though there was less of a GCS influence. The predicted results from inhibiting reductase activity in lymphoid organs are in contrast to the observed effects that an 11β-HSD inhibitor actually has on immune effector function.

Changes in 11β-HSD activity: physiological and pathological conditions

The observed alterations in cytokine production and immune effector function that are induced by the administration of an 11β-HSD inhibitor to normal animals indicates that this enzyme plays an important role in reducing GCS-mediated activities within lymphoid organs. These results also suggest that any situation where lymphoid organ 11β-HSD activity is naturally modulated could have significant consequences on immune function and possibly in the development of pathological conditions.

Stress, trauma, chronic infection and the natural consequences of aging each associates with a dysregulation in a variety of immunological activities including the production of specific species of macrophage and lymphocyte cytokines [130–133]. In many of these conditions, it has been reported that the activation-induced T-cell production of the type 1 cytokine IL-2 is depressed whereas inducible expression of the type 2 cytokines IL-4 and IL-10 is actually augmented. These effects are consistent with the biological activities associated with elevated GCS concentrations. In fact, in many clinical situations where a marked alteration in immune function occurs, adrenal production of GCS has been reported to be elevated [134–136]. The immunological changes that occur following stress or trauma are also similar to what is observed after the in vivo inhibition of 11β-HSD in normal animals by pharmacological means [122]. It is conceivable, therefore, that the changes in immunological function which take place in these conditions may be mediated by a decrease in tissue GCS catabolism, and/or an elevation in steroid biosynthesis by the adrenal gland.

The analysis of 11β-HSD activity in spleen fragments from mature adult (1 to 2 months of age) and aged mice (20 to 24 months of age), has recently determined that aging associates with a depression in 11β-HSD dehydrogenase activity (Fig. 5.4A). The reduced capacity of lymphoid organs to catabolize GCS to biologically inactive compounds was observed to correlate with an age-associated elevation in the expression of the 11β-HSD1 isoform (Figure 4B). This isoform functions predominantly as a reductase in vivo, actually working to back-convert biologically inactive 11-keto GCS to biologically active 11-hydroxy GCS. A similar reduction in 11-hydroxy GCS catabolism occurs in the spleens of experimental animals that have received a thermal injury (J.D.H., unpublished observations). It is presently unclear, however, if the depressions in spleen 11β-HSD are also associated with an elevation in the expression of the 11β-HSD1 reductase following thermal injury. An increased stromal cell expression of 11β-HSD1 during the processes of aging or trauma could create higher GCS concentrations in lymphoid organs which then would influence resident T-cells and inhibit their production of IL-2 while increasing their capacity to produce the type 2 cytokines IL-4 and IL-10 in response to stimulation.

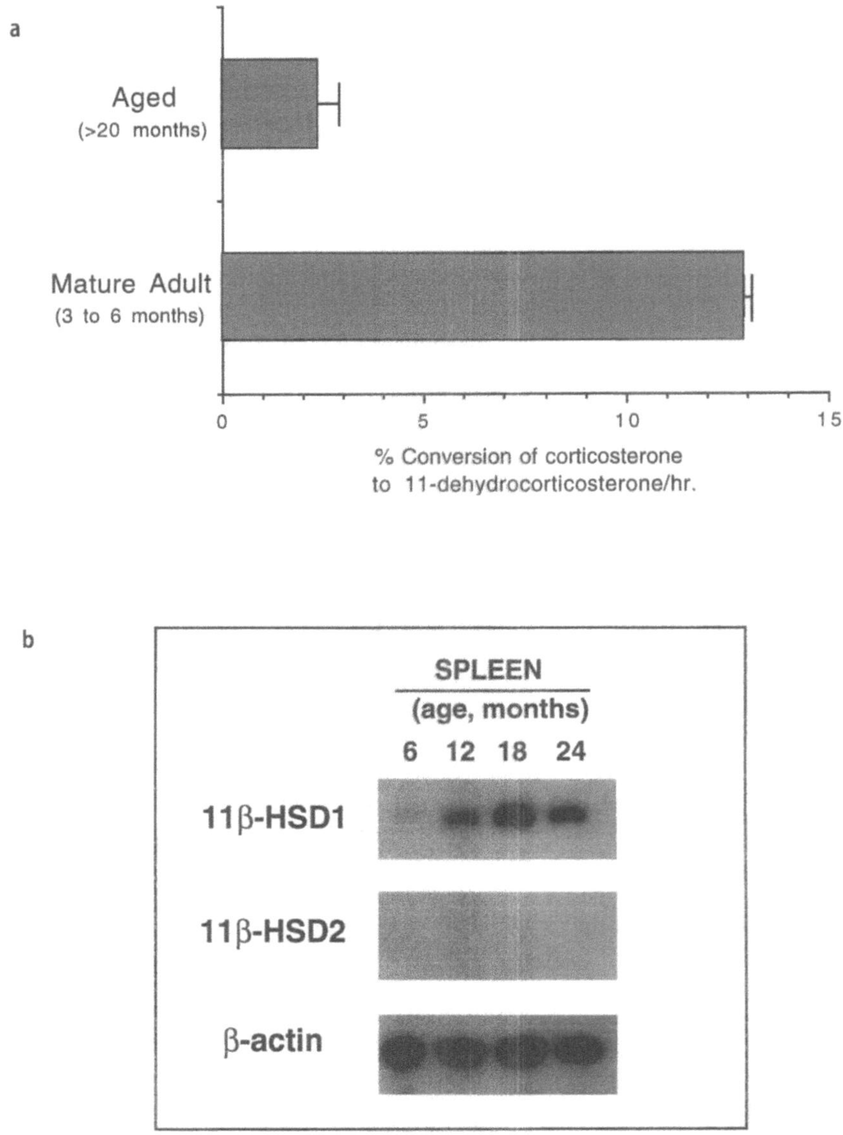

Fig. 5.4. There is an age-associated depression in 11β-dehydrogenase activity that associates with an elevation of the 11β-HSD1 oxidoreductase isoform. The ability of spleen fragments from aged C57BL/6 mice to metabolize radiolabelled corticosterone is lower than spleen fragments from mature adult C57BL/6 mice (**a**). This depression in spleen 11β-HSD activity associates with an elevation in the expression of the 11β-HSD1 isoform as measured by RT-PCR (**b**). The 11-β-HSD1 isoform functions predominantly as a reductase in vivo, back converting biologically inactive 11-keto GCS to their biologically active 11-hydroxy metabolites.

Additional physiologic mechanisms involved in the regulation of GCS activities: the role of macrophage migration inhibitor factor

Specific cellular elements of the innate immune system appear to become refractory to the effects of GCS under certain clinical conditions. It has been observed that aging, trauma, and chronic microbial infection represent three such conditions where both the production of the inflammatory cytokines and the plasma levels of GCS are coordinately elevated. It has been demonstrated that despite an elevation in adrenal GCS output and a reduced rate of lymphoid organ GCS catabolism in aged animals [137,138], there is a concomitant augmentation in the production of certain pro-inflammatory cytokines (e.g. TNFα, IL-6) [130,139]. This presents something of a paradox since the production of these inflammatory cytokines is appreciated to be effectively inhibited by relatively low levels of GCS under normal conditions. Collectively, these findings suggest that some of the cell types responsible for the production of the inflammatory cytokines may acquire a reduced sensitivity to the inhibitory actions of the elevated plasma and tissue levels of GCS that generally accompany these conditions.

The results of recent studies have demonstrated that a 12.5 kDa protein, known historically as macrophage migration inhibitory factor (MIF), can effectively limit cellular responsiveness to GCS. Lipopolysaccharide (LPS)-stimulated macrophages, pretreated with low doses of MIF (1–10 ng/ml), were found to become refractory to GCS-mediated inhibition of IL-1, IL-6, IL-8 and TNFα production [140]. Such effects could translate into very significant consequences with regards to endotoxin shock since the administration of recombinant MIF to mice greatly lowered the amount of LPS required to achieve lethality in an animal model system [141]. Conversely, the administration of neutralizing anti-MIF antibodies actually served to protect the treated mice from a lethal dose of endotoxin [141]. In delayed hypersensitivity reactions, the administration of neutralizing anti-MIF antibodies was found to reduce the swelling response induced by foreign antigen challenge [142]. Collectively, these results suggest that MIF appears to be functioning in vivo to limit the ability of endogenously produced GCS from exerting their inhibitory effects on the production of pro-inflammatory cytokines by macrophages. The inducible expression of MIF, therefore, would represent an additional means by which cellular GCS sensitivity or action could be functionally regulated in vivo.

MIF is produced by a variety of distinct cell types including T-cells, macrophages, keratinocytes, the lens cells of the eye, and the pituitary gland [141, 143–145]. The gene for MIF in humans and mice has recently been cloned and sequenced [146,147]. Analysis of the 5′ promoter region of MIF has revealed the presence of several consensus sequences for a variety of transcription factors including NF-κB [147]. The presence of a NF-κB binding site in the 5′ promoter region of MIF suggests that the expression of this cytokine might also be up regulated by inflammatory stimuli that induce the activation and nuclear translocation of NF-κB. Based on the presence of a NF-κB site it appears that MIF is up-regulated in parallel with other inflammatory cytokines.

The production of MIF can be induced by a variety of inflammatory substances including LPS and the cytokines TNFα and IFNγ [144]. It appears, therefore, that under conditions of inflammation, the inflammatory cytokines may be inducing the local production of MIF which, in turn, would lower responsiveness to GCS of cells at the inflammatory site. It has been demonstrated that GCS themselves can also induce the production of MIF [140]. The ability of GCS to up-regulate the secretion of MIF, however, occurs in a very narrow dose–response range (10^{-9} to 10^{-12} M). At GCS concentrations above 10^{-9} M, the production of MIF is actually inhibited. Therefore, in lymphoid organs or tissues where macrophages and infiltrating T lymphocytes are present, the tissue concentration of GCS, plus the presence of inflammatory cytokines, both represent key determinants in regulating the sensitivity of MIF responsive cells to GCS via their ability to regulate the function of this GCS counter-regulatory cytokine.

In addition to being potent immunomodulators, GCS are also involved in the control of glucose and lipid metabolism, developmental processes, neuronal function and cellular proliferation [16]. Therefore, if the expression of MIF somehow becomes dysregulated such that it is abnormally or constitutively expressed, the tissues responsive to this cytokine would become refractory to the important regulatory effects of GCS. As previously stated, aging, trauma and chronic infection represent conditions in which there are coordinate increases in tissue concentrations of GCS and the spontaneous expression of several inflammatory cytokines [131,137–139]. In these clinical situations it is conceivable that the cellular production of MIF is also elevated. In support of this hypothesis, we have recently demonstrated that the expression of MIF mRNA is significantly increased in the spleens of aged mice compared with the spleens of mature adult animals (Fig. 5.5).

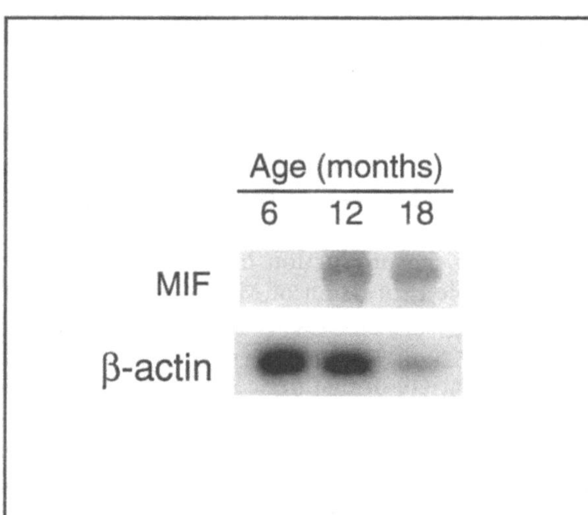

Fig. 5.5. The expression of macrophage migration inhibitory factor (MIF) mRNA increases in an age-associated fashion. RT-PCR was used to analyse the level of MIF mRNA expression in the spleens of 6-, 12-, or 18-month-old C57BL/6 mice.

The elevated levels of MIF mRNA were accompanied by significant elevations in the expression of other cytokines including TNFα, IL-6, IL-12 and the enzyme cyclooxygenase 2 (COX2) which is responsible for converting arachidonic acid to prostaglandins [Spencer et al., submitted for publication]. The results from these studies demonstrate that there is an age-associated constitutive expression of genes, such as MIF, that are normally only induced following an inflammatory insult.

It has been demonstrated that supplementation of aged animals with dehydroepiandrosterone (DHEA) or its sulfated derivative dehydroepiandrosterone sulfate (DHEAS) restores the age-associated constitutive expression of the pro-inflammatory cytokines (e.g. TNFα, IL-6, IL-12, MIF) and COX2 back to very low or undetectable levels [129, Spencer et al., submitted for publication]. DHEAS represents an endogenous steroid that is found in the circulation of mature adults of most mammalian species [148].

The production and circulating levels of DHEAS decline dramatically during the course of aging. During the seventh decade of life, humans possess 10% or less of the circulating DHEAS levels that were present between the second or third decade of life [149]. The age-associated loss in adrenal DHEAS production has been proposed to be an important contributor to the physiological changes that occur with advancing age, including the appearance of the spontaneous production of the inflammatory cytokines (e.g. TNFα, IL-12, IL-6) [131,150,151].

The mechanism by which DHEAS abolishes constitutive cytokine production in the lymphoid cells of aged animals is not fully understood at the present time. From the pharmacology literature, however, it has been demonstrated that DHEAS is an inducer of the expression of peroxisomal enzymes and peroxisome biogenesis [152]. DHEAS regulates peroxisome activities through its ability to activate a member of the steroid hormone receptor superfamily termed peroxisome proliferator activated receptor α (PPARα) [153]. In support of the hypothesis that DHEAS is restoring dysregulated cytokine production in aged animals through a mechanism that involves the activation of PPARα, and as a consequence the up-regulation of PPARα-controlled gene products, it was observed that Wy 14 643, another known activator of PPARα, can also reverse the age-associated expression of inflammatory cytokines [Spencer et al., submitted for publication]. From these studies, it was concluded that the up-regulation of some peroxisomal proteins must be important for keeping gene products involved in inflammatory responses from being constitutively expressed.

Peroxisomes are single membrane organelles containing numerous enzymes that are critical for a wide range of cellular activities. These include enzymes that are involved in the β-oxidation of long-chain fatty acids, plasmalogen biosynthesis, and conversion of H_2O_2 to water and oxygen through the actions of catalase [154]. Nearly all of the catalase present within a cell is located within the peroxisome and it is this enzymatic activity that is thought to be important in lowering the accumulation of intracellular reactive oxygen intermediates (ROIs). It has been observed that the important anti-oxidant activities of catalase are depressed in cells from aged animals when compared with the same cell types removed from normal adult animals [155].

Theories exist which attribute many of the physiological changes that are associated with aging to abnormal cellular generation of ROIs [156,157]. ROIs induce

damage to many cellular constituents including DNA, proteins and lipids [156, 158–160]. In addition to randomly modifying the structure of basic cellular molecules, ROIs also are involved in regulating the ability of certain cellular proteins to perform their specific molecular function. For example, it has been demonstrated that ROIs can serve to directly activate the transcription factor NF-κB

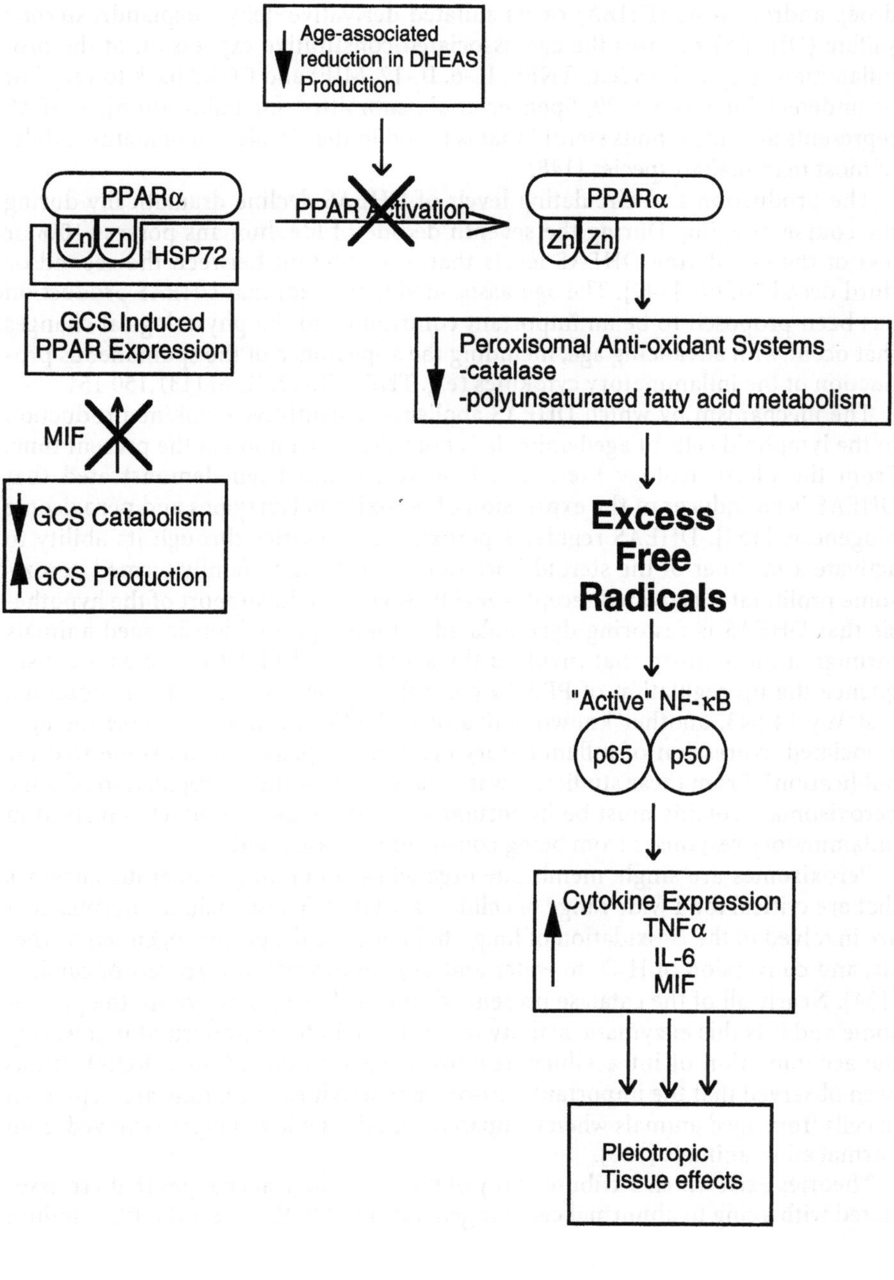

[161]. Normally NF-κB is kept within the cytoplasm by binding to its inhibitor IκB. Following an inflammatory stimuli or an increase in intracellular ROIs, IκB is released and rapidly degraded [162]. NF-κB is left free to translocate to the nucleus and positively regulate the expression of numerous pro-inflammatory gene products [162].

There is a loss in the expression of the anti-oxidant enzyme catalase with aging [155]. This change might be linked to the age-dependent depression of circulating DHEAS levels. A depression in catalase activity would result in a decreased removal of intracellular ROIs, which then would inappropriately activate NF-κB. The expression of pro-inflammatory cytokines caused by NF-κB activation would be further enhanced by an increase in MIF through the decrease in the capacity of GCS effectively to inhibit further secretion of the inflammatory cytokines caused by this counterregulatory molecule.

A MIF-dependent reduction in GCS activities might further reduce peroxisome catalase expression through a secondary pathway. This hypothesis is based on the recent experimental evidence showing that GCS regulate the expression of PPARα [163-165]. In fact, the expression of PPARα fluctuates in parallel with the circadian changes in circulating GCS concentrations [164]. It is plausible, therefore, that when there is an elevation in the expression of MIF (e.g. in the spleens of aged mice) the ability of GCS to increase the expression of PPARα might be inhibited. A reduction in PPARα expression would reduce the ability of endogenous PPARα activators such as DHEAS and leukotriene B_4 to promote the induction of an adequate expression of catalase and other anti-oxidant enzymes [153,166]. The lowered level of natural antioxidants would allow the activation of NF-κB to be more easily facilitated and result in the continued production of MIF and other inflammatory cytokines. Pathologies or situations that directly induce the expression of MIF could result in the inability of GCS to directly inhibit inflammatory cytokine synthesis as well as promote the production of NF-κB-driven inflammatory cytokines through the depression of PPARα expression [Fig. 5.6].

Fig. 5.6 The age-associated increase in MIF may lead to the inability of GCS to up-regulate an important anti-inflammatory molecular process. The loss in the ability of GCS to induce the expression of peroxisome proliferator activated receptorα (PPARα) via an age-associated increase in MIF synthesis would lead to a lowered cellular response to the endogenous PPARα activator dehydroeplandrosterone sulfate (DHEAS). This in turn, would lead to an elevation in the activation of NF-κB through a reduction in anti-oxidant defenses. Increased NF-κB activity would then allow the expression of inflammatory cytokines, including MIF. The MIF-dependent reduction in PPARα expression occurs despite the fact that circulating and tissue levels of GCS are elevated due to a depression in GCS catabolism and increased adrenal GCS biosynthesis. Filled arrows indicate either a positive or negative effect on the production or action of specific components.

Summary

GCS represent a class of steroid hormones that circulate in large quantities in all mammalian species and possess very potent biological and immunomodulatory properties. The activities of GCS can be controlled grossly by regulating the level of GCS biosynthesis by the adrenal gland. Historically and presently, the measurement of biologically active GCS concentrations in the circulation have been used as the only indicator of GCS bioactivity. This may be inappropriate, however, since many biochemical mechanisms exist that serve to regulate GCS activities at the cellular and tissue-specific levels. At the cellular level, the effectiveness of GCS actions can be limited by the regulation of GCS receptor affinities and numbers. Additionally, the presence of MIF, through a yet to be identified mechanism, has the potential to inhibit the ability of cytokine-responsive cells to exhibit susceptibility to GCS influences. Microenvironmental regulation of tissue-specific concentrations of GCS also occurs through the expression of the enzyme 11β-HSD. Through the ability of this enzyme to interconvert biologically active GCS with inactive metabolites, 11β-HSD can effectively control GCS levels in distinct tissues. There exist numerous processes which collectively control the ability of GCS to affect cellular and immunological activities in vivo. Fluctuations in any of these mechanisms, therefore, could lead to pathological consequences.

References

1. Chin WW (1989) Hormonal regulation of gene expression. In: DeGroot LJ (ed) Endocrinology. Saunders, Philadelphia, p 6
2. Manglesdorf DJ, Thummel C, Beato M, Herrlich P, Schütz G, Umesono K, Blumberg B, Kastner P, Mark M, Chambon P, Evans, RM (1995) The nuclear receptor superfamily: the second decade. Cell 83:835
3. Garbe A, Buck J, Hämmerling U (1992) Retinoids are important cofactors in T cell activation. J Exp Med 176:109
4. Rigby WFC, Noelle RJ, Krause K, Fanger MW (1985) The effects of 1,25-dihydroxyvitamin D_3 on human T lymphocyte activation and proliferation: a cell cycle analysis. J Immunol 135:2279
5. Rigby WFC Denome S, Fanger MW (1987) Regulation of lymphokine production and human T lymphocyte activation by 1,25-dihydroxyvitamin D_3. Specific inhibition at the level of messanger RNA. J Clin Invest 79:1659
6. Smithson G, Medina K, Ponting I, Kincade PW (1995) Estrogen suppresses stromal cell-dependent lymphopoiesis in culture. J Immunol 155:3409
7. Axelrod J, Reisine TD (1984) Stress hormones: their interaction and regulation. Science 244:452
8. Fraser R (1992) Biosynthesis of adrenocortical steroids. In: James VHT (ed) The adrenal gland. Raven Press, New York, p 117
9. Reichlin S (1993) Nueroendocrine-immune interactions. New Engl J Med 329:1246
10. Schleimer RP (1993) An overview of glucocorticoid anti-inflammatory actions. Eur J Clin Pharmacol 45:S3

11. Hench PS (1952) The reversibility of certain rheumatic and nonrheumatic conditions by the use of cortisone or of the pituitary adrenocorticotropic hormone. Ann Intern Med 36:1
12. Kass EH, Finland M (1953) Adrenocortical hormones in infection and immunity. Ann Rev Microbiol 7:361
13. Sheagren JN (1989) Glucocorticoid action: infectious diseases. In: Schleimer RP, Clamen HN, Oronsky A (eds) Anti-inflammatory steroid action. Academic Press, San Diego, p525
14. Axelrod L (1989) Side effects of glucocorticoid therapy. In: Schleimer RP, Clamen HN, Oronsky A (eds) Anti-inflammatory steroid action. Academic Press, San Diego, p 377
15. Cupps TR, Fauci AS (1982) Corticosteroid-mediated immunoregulation in man. Immunol Rev 65:133
16. Munck A, Guyre PM, Holbrook NJ (1984) Physiological functions of glucocorticoids in stress and their relation to pharmacological actions. Endocr Rev 5:25
17. Cupps TR (1989) Effects of glucocorticoids on lymphocyte function. In: Schleimer RP, Clamen, HN, Oronsky A, (eds) Anti-inflammatory steroid action. Academic Press, San Diego, p132
18. Lurie MB, Zappadosi P, Dannenberg AMJ, Swartz IB (1951) Constitutional factors in resistance to infection: the effect of cortisone on the pathogenesis of tuberculosis. Science 113:234
19. Vernon-Roberts B (1969) The effects of steroid hormones on macrophage activity. In: Bourne GH, Danielli JF (eds) International Review of Cytology Academic Press, New York, p131
20. Snyder DS, Unanue ER (1982) Corticosteroids inhibit murine macrophage Ia expression and interleukin 1 production. J Immunol 129:1803
21. Rosa MD, Radomski M, Carnuccio R Moncada S (1990) Glucocorticoids inhibit the induction of nitric oxide synthase in macrophages. Biochem Biophys Res Comm 172:1246
22. Smith RJ, Iden SS (1980) Pharmacological modulation of chemotactic factor-elicited release of granule-associated enzymes from human neutrophils. Effects of prostaglandins, non-steroid anti-inflammatory agents and corticosteroids. Biochem Pharmacol 29:2389
23. Nelson DH, Wennhold AR, Murray DK (1981) Corticosteroid-induced simultaneous changes in leukocyte phospholipids and superoxide anion production. J Steroid Biochem 14:321
24. Kurihara A, Ojima F, Tsurufuji S (1984) Chemotactic factor production by rat polymorphonuclear leukocytes: stimulation with opsonized zymosan particles and inhibition by dexamethasone. Biochem Biophys Res Comm 119:720
25. Watanabe M, Yagi M, Omata M, Hirasawa N, Mue S, Tsurufuji S, Ohuchi K (1991) Stimulation of neutrophil adherence to vascular endothelial cells by histamine and thrombin and its inhibition by PAF antagonists and dexamethasone. Br J Pharmacol 102:239
26. Schleimer RP (1989) The effects of glucocorticoids on mast cells and basophils. In: Schleimer RP, Clamen HN, Oronsky A (eds) Anti-inflammatory steroid action. Academic Press, San Diego, p 226
27. Dougherty TF, Berliner DL, Berliner ML (1961) Corticosteroid-tissue interactions. Metab Clin Exp 10:966
28. Wyllie AH (1980) Glucocorticoid-induced thymocyte apoptosis is associated with endogenous endonuclease activation. Nature 284:555
29. Fernandez-Ruiz E, Rebollo A, Nieto MA, Sanz E, Somoza C, Ramirez F, Lopez-Rivas A, Silva A (1989) IL-2 protects T cell hybrids from the cytolytic effect of glucocorticoids. Synergistic effect of IL-2 and dexamethasone in the induction of high-affinity IL-2 receptors. J Immunol 143:4146

30. Cohen JJ (1992) Glucocorticoid-induced apoptosis in the thymus. Semin Immunol 4:363
31. Zacharchuk CM, Mercep M, Chakraborti P, Simons SSJ, Ashwell JD (1990) Programmed T lymphocyte death: cell activation- and steroid-induced pathways are mutually antagonistic. J Immunol 145:4037
32. King LB, Vacchio MS, Dixon K, Hunziker R, Marguiles DH, Ashwell JD (1995) A targeted glucocorticoid receptor antisense transgene increases thymocyte apoptosis and alters thymocyte development. Immunity 3:647
33. Garvy BA, King LE, Telford WG, Morford LA, Fraker PJ (1993) Chronic elevation of plasma corticosterone causes reductions in the number of cycling cells of the B lineage in murine bone marrow and induces apoptosis. Immunology 80:587
34. Paul WE, Seder RA (1994) Lymphocyte responses and cytokines. Cell 76:241
35. Beutler B, Krochin N, Milsark IW, Luedke C, Cerami A (1986) Control of cachectin (tumor necrosis factor) synthesis: mechanism of endotoxin resistance. Science 232:977
36. Ray A, Prefontaine KE (1994) Physical association and functional antagonism between the p65 subunit of transcription factor NF-kB and the glucocorticoid receptor. PNAS USA 91:752
37. Arya SK, Wong-Staal F, Gallo RC (1984) Dexamethasone-mediated inhibition of human T cell growth factor and gamma-interferon messenger RNA. J Immunol 133:273
38. Daynes RA, Araneo BA (1989) Contrasting effects of glucocorticoids on the capacity of T cells to produce the growth factors interleukine 2 and interleukin 4. Eur J Immunol 19:2319
39. Ramírez F, Fowell DJ, Puklavec M, Simmonds S, Mason D (1996) Glucocorticoids promote a Th2 cytokine response by $CD4^+$ T cells in vitro J Immunol 156:2406
40. de Waal Malefyt RJ, Abrams J, Bennet B, Figdor CG, de Vries JE (1991) Interleukin-10 (IL-10) inhibits cytokine synthesis by human monocytes: an autoregulatory role of IL-10 produced by monocytes. J Exp Med 174:1209
41. de waal Malefyt R, Figdor CG, Huijbens R, Mohan-Peterson S, Bennett B, Culpepper J, Dang W, Zurawski G, de Vries JE (1993) Effects of IL-13 on phenotype, cytokine production, and cytotoxic function of human monocytes: comparison with IL-4 and modulation by IFNγ of IL-10 J Immunol 151:6370
42. te Velde AA, Huijbens RJF, Heije K, de Vries JE, Figdor CG (1990) Interleukin-4 (IL-4) inhibits secretion of IL-1β, tumor necrosis factor α, and IL-6 by human monocytes. Blood 76:1392
43. AyanlarBatuman O, Ferrero AP, Diaz A, Jimenez SA (1991) Regulation of transforming growth factor-b1 gene expression by glucocorticoids in normal human T lymphocytes. J Clin Invest 88:1574
44. Errasfa M, Russo-Marie F (1989) A purified lipocortin shares the anti-inflammatory effect of glucocorticosteroids in vivo in mice. Br J Pharmacol 97:1051
45. Oursler MJ, Riggs BL, Spelslaerg TC (1993) Glucocorticoid-induced activation of latent transforming growth factor-β by normal human osteoblast-like cells. Endocrinology 133:2187
46. Wright APH, Zilliacus J, McEwan IJ, Dahlman-Wright K, Almof T, Carlstedt-Duke J, Gustafsson J-A (1993) Structure and function of the glucocorticoid receptor. J Steroid Biochem Mol Biol 47:11
47. Simons SSJ (1994) Function/activity of specific amino acids in glucocorticoid receptors. Vitam Horm 49:49
48. Arriza JL, Weinberger C, Cerelli G, Glaser TM, Handelin BL, Housman DE, Evans RM (1987) Cloning of human mineralocorticoid receptor complementary DNA: structural and functional kinship with the glucocorticoid receptor. Science 237:268

49. Miesfeld RL, Rusconi S, Godowski PJ, Maler BA, Okret S, Wikstrom A-C, Gustafsson A, Yamamoto KR (1986) Genetic complementation of a glucocorticoid receptor deficiency by expression of a cloned receptor cDNA. Cell 46:389
50. Pratt WB (1993) The role of heat shock proteins in regulating the function, folding, and trafficking of the glucocorticoid receptor. J Biol Chem 268:21455
51. Czar MJ, Lyons RH, Welsh MJ, Renoir JM, Pratt WB (1995) Evidence that the FK506-binding immunophilin heat shock protein 56 is required for trafficking of the glucocorticoid receptor from the cytoplasm to the nucleus. Mol Endocrinol 9:1549
52. Luisi BF Xu WX, Otwinowski Z, Freedman LP, Yamamoto KR, Sigler PB (1991) Crystallographic analysis of the interaction of the glucocorticoid receptor with DNA. Nature 352:497
53. Yang-Yen H-S, Chambard J-C, Sun Y-L, Smeal T, Schmidt TJ, Drouin J, Karin M (1990) Transcriptional interference between c-Jun and the glucocorticoid receptor: mutual inhibition of DNA binding due to direct protein-protein interaction. Cell 62:1205
54. Jonat C, Rahmsdorf HJ, Park K-K, Cato ACB, Gebel S, Ponta H, Herrlich P (1990) Antitumor promotion and antiinflammation: down-modulation of AP-1 (Fos/Jun) activity by glucocorticoid hormone. Cell 62:1189
55. Auphan N, DiDonato JA, Rosette C, Helmberg A, Karin M (1995) Immunosuppression by glucocorticoids: inhibition of NF-$\kappa\beta$ activity through induction of I$\kappa\beta$ synthesis. Science 270:286
56. Scheinman RI, Cogswell PC, Lofquist AK, Baldwin ASJ (1995) Role of transcriptional activation of IκBα in mediation of Immunosuppression by glucocorticoids. Science 270:283
57. Calandra T, Bucala R (1996) Macrophage migration inhibitory factor: a counter-regulator of glucocorticoid action and critical mediator of septic shock. J Inflam 47:39
58. Keller-Wood, ME Dallman, MF (1984) Corticosteroid inhibition of ACTH secretion. Endocr Rev 5:1
59. Mandrup-Poulsen T, Nerup J, Reimers JI, Pciot F, Andersen HU, Karlsen A, Bjerre U, Bergholdt R (1995) Cytokines and the endocrine system. I. The immunoendocrine network. Eur. J. Endocrinol 133:660
60. Munck A, Leung K (1977) Glucocorticoid receptors and mechanism of action. In: Pasqualini JR (ed) Receptors and mechanism of action of steroid hormones. Marcel Dekker, New York, p 311
61. Ballard PL, Baxter JD, Higgins SJ, Rousseau GG, Tomkins G M (1974) General presence of glucocorticoid receptors in mammalian tissues. Endocrinology 94:998
62. Herman JP, Baxter JD, Higgins SJ, Rousseau, GG, Tomkins GM (1989) Localization and regulation of glucocorticoid and mineralocorticoid receptor messenger RNAs in the hippocampal formation of the rat. Mol. Endocrinol 3:1886
63. Cole TJ, Blendy JA, Schmid W, Strahle U, Schutz G (1993) Expression of the mouse glucocorticoid receptor and its role during development. J Steroid Biochem Mol Biol 47:49
64. Bodwell JE, Hu L-M, Hu J.-M, Orti E, Munck A (1993) Glucocorticoid receptors: ATP-dependent cycling and hormone-dependent hyperphosphorylation. J Steroid Biochem Mol Biol 47:31
65. Hu J-M, Bodwell JE, Munck A (1994) Cell cycle-dependent glucocorticoid receptor phosphorylation and activity. Mol Endocrinol 8:1709
66. Hu L-M, Bodwell J, Hu J-M, Orti E, Munck A (1994) Glucocorticoid receptors in ATP-depleted cells. Dephosphorylation, loss of hormone binding, hsp90 dissociation, and ATP-dependent cycling. J Biol Chem 269:6571
67. Monder C (1991) Corticosteroids, receptors, and the organ-specific functions of 11β-hydroxysteroid dehydrogenase. FASEB J. 5:3047

68. Seckl JR (1993) 11β-hydroxysteroid dehydrogenase isoforms and their implications for blood pressure regulation. Eur J Clin Invest 23:589
69. Amelung D, Huebner HJ, Roka L, Meyerheim G (1953) Conversion of cortisone to compound F. J Clin Endocrinol Metab 13:1125
70. Monder C, White PC (1993) 11β-Hydroxysteroid dehydrogenase. Vitam Horm 47:187
71. Whorwood CB, Ricketetts ML, Stewart PM (1994) Regulation of sodium-potassium adenosine triphosphate subunit gene expression by corticosteroids and 11β-hydroxysteroid dehydrogenase activity. Endocrinology 135:901
72. Katz AI (1982) Na-K-ATPase: its role in tubular sodium and potassium transport. Am J Physiol 242:F207
73. Edwards CRW, Stewart PM (1991) The cortisol-cortisone shuttle and the apparent specificity of glucocorticoid and mineralocorticoid receptors. J Steroid Biochem Mol Biol 39:859
74. Sheppard K, Funder JW (1987) Mineralocorticoid specificity of renal Type I receptors: in vivo binding studies. Am J Physiol 252:E224
75. Ulick S, Levine LS, Gunczler P, Zanconato G, Ramirez G, Rauh W, Rosler A, Bradlow HL, New MI (1979) A syndrome of apparent mineralocorticoid excess associated with defects in the peripheral metabolism of cortisol. J Clin Endocrinol Metab 49:757
76. Stewart PM, Corrie JET, Shackleton CHL, Edwards CRW (1988) Syndrome of apparent mineralocorticoid excess. A defect in the cortisol cortisone shuttle. J Clin Invest 82:340
77. Funder JW, Pearce PT, Smith R, Smith AI (1988) Mineralocorticoid action: target-tissue specificity is enzyme, not receptor, mediated. Science 242:583
78. Monder C, Stewart PM, Lakshmi V, Valentino R, Burt D, Edwards CRW (1989) Licorice inhibits corticosteroid 11β-dehydrogenase of rat kidney and liver: In vivo and in vitro studies. Endocrinology 125:1046
79. Krozowski Z (1992) 11β-hydroxysteroid dehydrogenase and the short-chain alcohol dehydrogenase (SCAD) superfamily. Mol Cell Endocrinol 84:C25
80. Hales DB, Payne AH (1989) Glucocorticoid-mediated repression of $P450_{SCC}$ mRNA and de novo synthesis in cultured Leydig cells. Endocrinology 124:2099
81. Payne AH, Sha L (1991) Multiple mechanisms for regulation of 3β-hydroxysteroid dehydrogenase/$\Delta^5 \rightarrow \Delta^4$-isomerase, 17$\alpha$-hydroxylase/$C_{17-20}$ lyase cytochrome P450, and cholesterol side-chain cleavage cytochrome P450 messenger ribonucleic acid levels in primary cultures of mouse Leydig cells. Endocrinology 129:1429
82. Phillips DM, Lakshmi V, Monder C (1989) Corticosteroid 11β-dehydrogenase in rat testis. Endocrinology 125:209
83. Monder C, Hardy MP, Blanchard RJ, Blanchard DC (1994) Comparative aspects of 11β-hydroxysteroid dehydrogenase. Testicular 11β-hydroxysteroid dehydrogenase: development of a model for the mediation of Leydig cell function by corticosteroids. Steroids 59:69
84. Teelucksingh S, Mackie ADR, Burt D, McIntyre MA, Brett L, Edwards CRW (1990) Potentiation of hydrocortisone activity in skin by glycyrrhetinic acid. Lancet 335:1060
85. Karstila T, Rechardt L, Honkaniemi J, Gustafsson J-A, Wikstroms A-C, Karppinen A, Pelto-Huikko M (1994) Immunocytochemical localization of glucocorticoid receptor in rat skin. Histochemistry 102:305
86. Fuller PJ, Verity K (1990) Colonic sodium potassium adenosine triphosphate subunit gene expression: ontogeny and regulation by adrenocortical steroids. Endocrinology 127:32
87. Lakshmi V, Monder C (1988) Purification and characterization of the corticosteroid 11β-dehydrogenase component of the rat liver 11β-hydroxysteroid dehydrogenase complex. Endocrinology 123:2390
88. Monder C, Lakshmi V (1990) Corticosteroid 11β-dehydrogenase of rat tissues: Immunological studies. Endocrinology 126:2435

89. Agarwal AK, Monder C, Eckstein B, White PC (1989) Cloning and expression of rat cDNA encoding corticosteroid 11β-dehydrogenase. J Biol Chem 264:18939
90. Tannin GM, Agarwal AK, Monder C, New MI, White PC (1991) The human gene for 11β-hydroxysteroid dehydrogenase. J Biol Chem 266:16653
91. Yang K, Smith CL, Dales D, Hammond GL, Challis JR (1992) Cloning of an ovine 11 beta-hydroxysteroid dehydrogenase complementary deoxyribonucleic acid: tissue and temporal distribution of its messanger ribonucleic acid during fetal and neonatal development. Endocrinology 131:2120
92. Moore CCD, Mellon SH, Murai J, Siiteri PK, Miller WL (1993) Structure and function of the hepatic form of 11β-hydroxysteroid dehydrogenase in the squirrel monkey, an animal model of glucocorticoid resistance. Endocrinology 133:368
93. Rajan V, Chapman KE, Lyons V, Jamieson P, Mullins JJ, Edwards C RW, Seckl JR (1995) Cloning, sequencing and tissue-distribution of mouse 11β-hydroxysteroid dehydrogenase-1 cDNA. J Steroid Biochem Mol Biol 52:141
94. Jamieson PM, Chapman KE, Edwards CRW, Seckl JR (1995) 11β-hydroxysteroid dehydrogenase is an exclusive 11β-reductase in primary cultures of rat hepatocytes: effect of physiochemical and hormonal manipulations. Endocrinology 136:4754
95. Rundle SE, Funder JW, Lakshmi V, Monder C (1989) The intrarenal localization of mineralocorticoid receptors and 11β-dehydrogenase: immunocytochemical studies. Endocrinology 125:1700
96. Náray-Fejes-Tóth A, Watlington CO, Fejes-Tóth G (1991) 11β-hydroxysteroid dehydrogenase activity in the renal target cells of aldosterone. Endocrinology 129:17
97. Rusvai E, Náray-Fejes-Tóth A (1993) A new isoform of 11β-hydroxysteroid dehydrogenase in aldosterone target cells. J Biol Chem 268:10717
98. Stewart PM, Murry BA, Mason JI (1994) Human kidney 11β-hydroxysteroid dehydrogenase is a high affinity nicotinamide adenine dinucleotide-dependent enzyme and differs from the cloned type 1 isoform. J Clin Endocrinol Metab 79:480
99. Low SC, Assaad SN, Rajan V, Chapman KE, Edwards CRW, Seckl JR (1993) Regulation of 11β-hydroxysteroid dehydrogenase by sex steroids in vivo: further evidence for the existence of a second dehydrogenase in rat kidney. J Endocrinol 139:27
100. Náray-Fejes-Tóth A, Rusvaia E, Denault DL, Germain DLS, Fejes-Toth G (1993) Expression and characterization of a new species of 11β-hydroxysteroid dehydrogenase in *Xenopus* oocytes: Am J Physiol 265:F896
101. Albiston AL, Obeyesekere VR, Smith RE, Krozowski ZS (1994) Cloning and tissue distribution of the human 11β-hydroxysteroid dehydrogenase type 2 enzyme. Mol Cell Endocrinol 105:R11
102. Agarwal AK, Rogerson FM, Mune T, White PC (1995) Analysis of the human gene encoding the kidney isozyme of 11β-hydroxysteroid dehydrogenase. J Steroid Biochem Mol Biol 55:473
103. Náray-Fejes-Tóth, A, Fejes-Tóth G (1995) Expression cloning of the aldosterone target cell-specific 11β-hydroxysteroid dehydrogenase from rabbit collecting duct cells. Endocrinology 136:2579
104. Cole TJ (1995) Cloning of the mouse 11β-hydroxysteroid dehydrogenase type 2 gene: tissue specific expression and localization in distal convoluted tubules and collecting ducts of the kidney. Endocrinology 136:4693
105. Whorwood CR, Ricketts ML, Stewart PM (1994) Epithelial cell localization of type 2 11β-hydroxysteroid dehydrogenase in rat and human colon. Endocrinology 135:2533
106. Obeyesekere VR, Ferrari P, Andrews RK, Wilson RC, New MI, Funder JW, Krozowski ZS (1995) The R337C mutation generates a high Km 11β-hydroxysteroid dehydrogenase Type II enzyme in a family with apparent mineralocorticoid excess. J Clin Endocrinol Metab 80:3381
107. Stewart PM, Rogerson FM, Mason JI (1995) Type 2 11β-hydroxysteroid dehydrogenase messanger ribonucleic acid and activity in human placenta and fetal

membranes: its relationship to birth weight and putative role in fetal adrenal steroidogenesis. J Clin Endocrinol Metab 80:885
108. Krozowski Z, Maguire JA, Stein-Oakley AN, Dowling J, Smith RE, Andrews RK (1995) Immunohistochemical localization of the 11β-hydroxysteroid dehydrogenase Type II in human kidney and placenta. J Clin Endocrinol Metab 80:2203
109. Langlois DA, Matthews SG, Yu M, Yang K (1995) Differential expression of 11β-hydroxysteroid dehydrogenase 1 and 2 in the developing ovine fetal liver and kidney. J Endocrinol 147:405
110. Brown RW, Diaz R, Robson A, Kotelevstev YV, Mullins JJ, Kaufman MH, Seckl JR (1996) The ontogeny of 11β-hydroxysteroid dehydrogenase Type 2 and mineralocorticoid receptor gene expression reveal intricate control of glucocorticoid action in development. Endocrinology 137:794
111. Murphy BEP (1979) Cortisol and cortisone in human fetal development. J *Steroid Biochem* 11:509
112. Murphy BEP (1981) Ontogeny of cortisol-cortisone interconversion in human tissues: a role for cortisone in human fetal development. J Steroid Biochem 14:811
113. Benediktsson R, Lindsay RS, Noble J, Seckl JR, Edwards CR W (1993) Glucocorticoid exposure in utero: new model for adult hypertension. Lancet 341:339
114. Lindsay RS, Noble JM, Edwards CRW, Seckl JR (1994) Maternal carbenoxolone treatment reduces birth weight in the rat. J Endocrinol 140 (Suppl):18
115. Andersson S (1995) 17β-hydroxysteroid dehydrogenase: isozymes and mutations. J Endocrinol 146:197
116. Labrie F, Simard J, Luu-The V, Belanger A, Pelletier G (1992) Structure, function and tissue-specific gene expression of 3β-hydroxysteroid dehydrogenase/5-ene-4-ene isomerase enzymes in classical and peripheral intracrine steroidogenic tissues. J Steroid Biochem Mol Biol 43:805
117. Walker BR, Moisan, MP (1992) Multiple isoforms of the cortisol-cortisone shuttle. J Endocrinol 133:1
118. Whorwood CB, Franklyn JA, Sheppard MC, Stewart PM (1992) Tissue localization of 11β-hydroxysteroid dehydrogenase and its relationship to the glucocorticoid receptor. J Steroid Biochem Mol Biol 41:21
119. Gomez-Sanchez EP, Cox D, Foecking M, Ganjam V, Gomez-Sanchez CE (1996) 11β-hydroxysteroid dehydrogenase of the choriocarcinoma cell line JEG-3 and their inhibition by glycyrrhetinic acid and other natural substances. *Steroids* 61:110
120. Dougherty TF, Berliner ML, Berliner DL (1960) 11β-Hydroxy dehydrogenase system activity in thymi of mice following prolonged cortisol treatment. *Endocrinology* 66:550
121. Marandici A, Monder C (1993) Inhibition by glycyrrhetinic acid of rat tissue 11β-hydroxysteroid dehydrogenase in vivo. Steroids 58:153
122. Hennebold JD, Ryu SY, Mu H-H, Galbraith A, Daynes RA (1996) 11β-Hydroxysteroid dehydrogenase modulation of glucocorticoid activities in lymphoid organs. Am J Physiol 270:R1296
123. Daynes RA, Araneo BA, Dowell TA, Huang K, Dudley D (1990) Regulation of murine lymphokine production in vivo. III. The lymphoid tissue microenvironment exerts regulatory influences over T helper cell function. J Exp Med 171:979
124. Deckx R, De Moor P (1966) Study of the 11β-hydroxysteroid dehydrogenase in vitro. I. Biochemical characterization in spleen homogenate. Pflugers Arch 289:59
125. Agarwal AK, Mune T, Monder C, White PC (1994) NAD^+ dependent isoform of 11β-hydroxysteroid dehydrogenase. J Biol Chem 269:25959.
126. Hennebold JD, Mu H-H, Poynter ME, Chen X-P, Daynes RA (1997) Active catabolism of glucocorticoids by 11β-hydroxysteroid dehydrogenase in vivo is a necessary requirement for natural resistance to infection with *Listeria monocytogenes*. Int Immunol 9:105

127. Brattsand R, Thalen A, Roemke K, Kallstrom L, Gruvstad E (1982) Development of new glucocorticoids with a very high ratio between topical and systemic activities. Eur J Resp Dis 63:62
128. Boockvar KS, Granger DL, Poston RM, Maybodi M, Washington M K, Hibbs Jr, JB, Kurlander RL (1994) Nitric oxide produced during murine Listeriosis is protective. Infect Immun 62:1089
129. Kelly JP, Bancroft GJ (1996) Administration of interleukin-10 abolishes innate resistance to *Listeria monocytogenes*. Eur J Immunol 26:356
130. Daynes RA, Araneo BA, (1992) Prevention and reversal of some age-associated changes in immunologic responses by supplemental dehydroepiandrosterone sulfate therapy. Aging Immunol Infect Dis 3:135
131. Daynes RA, Araneo BA, Ershler WB, Maloney C, Li G-Z, Ryu S-Y (1993) Altered regulation of IL-6 production with normal aging. Possible linkage to the age-associated decline in dehydroepiandrosterone and its sulfated derivative. J Immunol 150:5219
132. Clerici M, Shearer GM (1994) The Th1-Th2 hypothesis of HIV infection: new insights. Immunol Today 15:575
133. Araneo BA, Shelby J, Li G-Z, Ku W, Daynes RA (1993) Administration of dehydroepiandrosterone to burned mice preserves normal immunologic competence. Arch Surgery 128:318
134. Vaugham VM, Becker RA, Allen JP, Goodwin CV, Pruitt BA, Mason AD (1982) Cortisol and corticotropin in burned patients. J Trauma 22:263
135. Montanini VM, Simoni M, Chiossi G, Baraghini GF, Verlardo A, Barald E, Marrama P (1988) Age-related changes in plasma dehydroepiandrosterone sulphate, cortisol, testosterone and free testosterone circadian rhythms in adult men. Horm Res 29:1
136. Clerici M, Bevilacqua M, Vago T, Villa ML, Shearer GM, Norbiato G (1994) An immunoendocrinological hypothesis of HIV infection. Lancet 373:1552
137. Dodt C, Theine KJ, Uthgenannt D, Born J, Fehm HL (1994) Basal secretory activity of the hypothalamo-pituitary-adrenocortical axis is enhanced in healthy elderly. An assessment during undisturbed night-time sleep. Eur J Endocrinol 131:443
138. Yau JLW, Olsson T, Morris RGM, Meaney MJ, Seckl JR (1995) Glucocorticoids, hippocampal corticosteroid receptor gene expression and antidepressant treatment: relationship with spatial learning in young and aged rats. Neuroscience 66:571
139. Chorinchath BB, Kong L-Y, Mao L, McCallum RE (1996) Age-associated differences in TNF-α and nitric oxide production in endotoxic mice. J Immunol 156:1525
140. Calandra T, Bernhagen J, Metz CN, Spiegel LA, Bacher M, Donnelly T, Cerami A, Bucala R (1995) MIF as a glucocorticoid-induced modulator of cytokine production. Nature 377:68
141. Bernhagen J, Calandra T, Mitchell RA, Martin SB, Tracey KJ, Voelter W, Manogue KR, Cerami A, Bucala R (1993) MIF is a pituitarty-derived cytokine that potentiates lethal endotoxaemia. Nature 365:756
142. Bernhagen J, Bacher M, Calandra T, Metz CN, Doty SB, Donnelly T, Bucala R (1996) An essential role for macrophage migration inhibitory factor in the tuberculine delayed-type hypersensitivity reaction. J Exp Med 183:277
143. Bernhagen J, Mitchell RA, Calandra T, Voelter W, Cerami A, Bucala R (1994) Purification, bioactivity, and secondary structure analysis of mouse and human macrophage migration inhibitory factor (MIF). Biochemistry 33:14144
144. Calandra T, Bernhagen J, Mitchell RA, Bucala R (1994) The macrophage is an important and previously unrecognized source of macrophage migration inhibitory factor. J Exp Med 179:1895

145. Shimuzu T, Ohkawara A, Nishihira J, Sakamoto W (1996) Identification of macrophage migration inhibitory factor (MIF) in human skin and its immunohistochemical localization. FEBS Lett 381:199
146. Weiser WY, Temple PA, Witek-Giannotti JS, Remold HG, Clark S C, David JR (1989) Molecular cloning of a cDNA encoding a human macrophage inhibitory factor. Proc Natl Sci Counc Repub China [B] 86:7522
147. Mitchell R, Bacher M, Bernhagen J, Pushkarskaya T, Seldin MF, Bucala R (1995) Cloning and characterization of the gene for mouse macrophage migration inhibitory factor (MIF). J Immunol 154:3863
148. Meikle WA, Daynes RA, Araneo BA (1991) Adrenal androgen secretion and biologic effects. In: Nelson DH (ed) Endocrinology and metabolism clinics of north america. New aspects of adrenal cortical disease. Saunders, Philadelphia, p 381
149. Orentreich N, Brind JL, Rizer RL, Vogelman JH (1984) Age changes and sex differences in serum dehydroepiandrosterone sulfate concentrations throughout adulthood. J Clin Endocrinol Metab 59:551
150. Spencer NFL, Norton SD, Harrison LL, Li G-L, Daynes RA (1995) Dysregulation of IL-10 production with aging: possible linkage to the age-associated decline in DHEA and its sulfated derivative. Exp Gerontol 31:393
151. Hennebold JD, Poynter ME, Daynes RA (1995) DHEA and immune function: activities and mechanism of action. Semin Reprod Endocrinol 13:257
152. Frenkel RA, Slaughter CA, Orth K, Moomaw CR, Hicks SH, Snyder JM, Bennett M, Prough RA, Putnam RS, Milewich L (1990) Peroxisome proliferation and induction of peroxisomal enzymes in mouse and rat liver by dehydroepiandrosterone feeding. J Steroid Biochem 35:333
153. Peters JM, Zhou YC, Ram PA, Lee SST, Gonzalez FJ, Waxman DJ (1996) Peroxisome proliferator-activated receptor α required for gene induction by dehydroepiandrosterone-3β-sulfate. Mol Pharmacol 50:67–74
154. van der Bosch H, Schutgens RBH, Wanders RJA, Tager JM (1992) Biochemistry of peroxisomes. Annu Rev Biochem 61:157
155. Beier K, Völkl A, Fahimi HD (1993) The impact of aging on enzyme proteins of rat liver peroxisomes: quantitative analysis by immunoblotting and immunoelectron microscopy. Virchows Arch [B] 63:139
156. Bunker VW (1992) Free radicals, antioxidants and ageing. Med Lab Sci 42:299
157. Harman D (1956) Aging: A theory based on free radical and radiation chemistry. J Gerontol 11:298
158. Greenwald RA, Moy WW (1980) Effect of oxygen-free radicals on hyaluronic acid. Arthritis Rheum 23:455
159. Wolf SP, Dean RT (1986) Fragmentation of proteins by free radicals and its effect on their susceptibility to enzymatic hydrolysis. Biochem J 234:399
160. Stadtman ER (1992) Protein oxidation and aging. Science 257:1220
161. Schreck R, Rieber P, Baeuerle PA (1991) Reactive oxygen intermediates as apparently widely used messengers in the activation of the NF-kappaB transcription factor and HIV-1. EMBO J 10:2247
162. Kopp EB, Ghosh S (1995) NF-κB and Rel proteins in innate immunity. Adv Immunol 58:1
163. Lemberger T, Staels B, Saladin R, Desvergne B, Auwerx J, Wahli W (1994) Regulation of the peroxisome proliferator-activated receptor α gene by glucocorticoids. J Biol Chem 269:24527
164. Lemberger T, Saladin R, Vazquez M, Assimacopoulos F, Staels B, Desvergne B, Wahli W, Auwerx J (1996) Expression of the peroxisome proliferator-activated receptor α gene is stimulated by stress and follows a diurnal rhythm. J Biol Chem 271:1764
165. Steineger HH, Sorensen HN, Tugwood JD, Skrede S, Spydevold O, Gautvik KM (1994) Dexamethasone and insulin demonstrate marked and opposite regulation of the

steady-state mRNA level of the peroxisomal proliferator-activated receptor (PPAR) in hepatic cells: hormonal modulation of fatty-acid-induced transcription. Eur J Biochem 225:967

166. Devchand PR, Keller H, Peters JM, Vazquez M, Gonzalez FJ, Wahli W (1996) The PPARα-leukotriene B_4 pathway to inflammation control. Nature 384:39

Chapter 6

Immunological consequences of inhibiting dehydroepiandrosterone (DHEA) sulfatase in vivo

Roly Foulkes, Stevan Shaw and Amanda Suitters

Introduction

Background

The host response to combat invasion of infectious agents needs to be rapid, flexible, effective and above all coordinated. Recognition and elimination of antigen is usually a very efficient process, requiring the integrated activity of a variety of cell types. It is the recruitment and activation of these cells that leads to an appropriate immune response. Indeed an inappropriate immune response can often lead to pathology such as that seen in autoimmune disease and allergy.

Central to the initiation and the coordination of the immune response is the helper T-cell (T_H cell). By secreting the appropriate amount and variety of cytokines, the T_H cell instructs other cell types to elicit the required effect. T_H cells can be divided into T_H0, T_H1 and T_H2 cells characterized by their cytokine secretory profile. For cell-mediated immunity (CMI), which leads to the activation of CD8+ T-cells, macrophages and secretion of cytotoxic antibodies (e.g. IgG2a in mice) by B-cells, the T_H cell needs to secrete a T_H1 profile of cytokines (i.e. IL-2, IFNγ). In contrast, a humoral response manifested by isotype switching by B-cells to secrete IgG1 in mice and IgE is characterized by a T_H2 secretory profile (i.e. IL-4, IL-5, IL-10, IL-13). These specific cytokine secretory patterns can be easily demonstrated using T-cell clones and primary T-cells studied in vitro and ex vivo [1], but what is less clear is how these responses are coordinated in vivo. Since all primary T-cells are capable of secreting the full spectrum of T-cell cytokines there is clearly a need for close control over these responses. Factors such as antigen density, presence of accessory molecules, type of antigen presenting cell and crucially the local microenvironment in which the immune response is taking place all contribute to this regulatory process. Indeed some immune responses are anatomically compartmentalized, for example, cell-mediated immune responses mainly occur in peripheral lymph nodes and in the spleen, whereas humoral responses are localized often to mucosal lymphoid sites such as Peyer's patches and the gut-associated lymphoid tissue. What is still unclear is how these differing responses, occurring in anatomically distinct regions, are so closely regulated.

It has become appreciated that the microenvironment within these secondary lymphoid areas can be greatly affected by factors from outside. In particular, the hypothalamic-pituitary-adrenocortical (HPA) axis can influence the immune system, especially adrenal steroids and steroid-like molecules [2]. Steroids function, in part, by influencing cytokine gene transcription following binding to intracellular receptors followed by translocation of the steroid/receptor complex to the nucleus where they can interact with specific elements such as the glucocorticoid response element. Glucocorticoid response elements are found abundantly in diverse genes encoding cytokines as well as a variety of other mediators for cellular regulatory elements of inflammation and immune responses.

Effects of adrenal steroids on the immune system

It has long been recognized that glucocorticoids produced by the adrenal cortex can affect immune function especially if they are given exogenously. Dexamethasone, for example, can inhibit IL-2 and IFNγ but augment IL-4 release from activated T-cells whilst suppressing TNFα and IL-1 release from inflammatory cells [3,4]. Glucocorticoids are released following activation of the HPA axis into the circulation and can access all lymphoid areas to cause effects. More recently it has become appreciated that in addition to these immunosuppressive steroids, other counter-regulatory factors including other steroids also circulate. Dehydroepiandrosterone (DHEA) is an example of an endogenous immunostimulatory steroid which has marked effects on the immune system (see below). Indeed it is just such a balance between opposing influences on the immune response that may regulate discrete patterns of T_H cell cytokine release in anatomically distinct sites [5].

DHEA, an immunostimulatory steroid

DHEA is an adrenal androgen in man and other species including primates, pigs and dogs, although in rodents it is likely to be of extra-adrenal origin, since rodent adrenals lack the 17-hydroxylase enzyme [6]. In all species DHEA is formed from its common precursors, cholesterol and pregnenolone [7], and displays circadian rythmicity, at least in man [8,9]. DHEA has intrinsic biological activity but is rapidly sulfated in the adrenal cortex and other tissues to DHEA sulfate (DHEAS) [10] by a specific sulfotransferase. This sulfate may be active in its own right in the brain (see Rook et al., Chapter 9) and perhaps also as a peroxinome proliferator in other tissues (see Hennebold and Daynes, Chapter 5). However, DHEAS lacks biological activity in the systems we have studied and therefore it must be hydrolysed to DHEA by a steroid sulfatase enzyme to elicit a biological effect. This enzyme is a microsomal enzyme [11,12] also known as aryl sulfatase C. It can hydrolyse a variety of sulphated steroids and is found in variety of tissues including lymphoid tissue [13]. Amongst its postulated biological roles,

DHEA can inhibit glucose-6-phosphate dehydrogenase [14], act as a potential precursor for other steroid hormones [6,7,15] and reverse the insulin resistance associated with glucocorticoid treatment in genetically obese animals [16–18]. What remains unclear, however, is whether any of these proposed functions of DHEA are of physiological importance given that it circulates in as DHEAS in millimolar concentrations [9,19–21].

Interest in the effects of DHEA on immune and inflammatory reactions was stimulated by the work of Daynes et al [22], who showed that splenocytes removed from mice treated with exogenous DHEA or DHEAS showed an increased ability to secrete IL-2 but not IL-4 in response to mitogen or antigen (T helper (T_H1)-type cytokine profile). These observations contrasted with the effects seen with glucocorticoid administration which elicited an enhanced IL-4 but reduced IL-2 secretory pattern from splenocytes (T_H2-type cytokine profile) [23]. If DHEA and glucocorticoid were administered concomitantly, the DHEA effect predominated. The authors concluded that, unlike glucocorticoids which are immunosuppressive, DHEA is an immunostimulant. The fact that exogenous DHEAS, which lacks biological activity in the systems we are studying, can elicit biological effects indicates that steroid sulfatase may be an important enzyme in these responses. These and similar observations have led to the proposal that ongoing immune responses, in terms of the T-cell cytokine profile (T_H1 or T_H2-like), can be influenced by the opposing effects of DHEA (or related steroids) and glucocorticoids [22–28]. Indeed, the amount of steroid sulfatase correlates anatomically with those lymphoid areas which normally give rise to a high IL-2 T-cell cytokine profile (spleen, peripheral lymph nodes) and not those that give a low IL-2/high IL-4 profile (Peyer's patches) [24]. However, all data published so far have described the interaction between exogenously administered steroids on immune responses and have not investigated the role of endogenous steroids on these responses. We postulate that the key regulatory steroid is DHEA which maintains the immune system in a state of readiness (i.e. T_H1-primed) by preventing the natural immunosuppressive/anti-inflammatory effects of endogenous glucocorticoids. If DHEA availability is regulated by steroid sulfatase then inhibition of this enzyme should result in a decreased immune responsiveness by allowing endogenous glucocorticoid effects to predominate.

The aims of our studies were twofold: first, to ascertain whether the immunostimulatory effects seen with exogenous DHEA/S administration on cellular responses translates into in vivo physiological effects, and second, to investigate our postulate that inhibition of steroid sulfatase activity may be anti-inflammatory by allowing endogenous glucocorticoid effects to predominate. We have addressed these in a mouse contact sensitization model (a form of delayed type hypersensitivity, DTH) to show that exogenous DHEA and DHEAS do indeed enhance the inflammatory response to antigen. DTH is generally regarded as being a T_H1-type immune reaction. We have used a specific inhibitor of steroid sulfatase (oestrone-3-sulfamate, CT2251) in this model to determine the consequences of limiting the conversion of inactive DHEAS to active DHEA. Oestrone-3-sulfamate (CT2251) has been shown to be a potent ($K_I = 0.67\ \mu M$), irreversible, active site-directed inhibitor of steroid sulphatase, a microsomal enzyme [11,12]. In addition we have studied the role of steroid sulfatase on inflammatory responses in carrageenan pleurisy and in collagen-induced arthritis in mice The data indicate that steroid sulfatase is a physiologically important enzyme in these responses and that DHEA (or related steroids) are the likely effector hormones.

Experimental studies

In vivo characterization of steroid sulfatase inhibitor activity

Because of the irreversible binding nature of oestrone-3-sulfamate (CT2251), its likely duration of action was estimated by treating mice with the drug and measuring liver steroid sulfatase activity at various time points thereafter. In addition, animals treated with varying doses of CT2251 during a contact sensitization experiment were killed and liver steroid sulfatase activity measured both pre-challenge (day 4) and 24 h post-challenge (see below for details of model).

Liver supernatants (containing the microsomal fraction) were assayed for DHEAS sulfatase activity using an adapted form of the method of Purohit et al., [12].

Treatment of mice with varying doses of CT2251 on day 0 and measurement of liver steroid sulfatase activity on subsequent days revealed that the number of days with activity below the normal range increased with increasing dose of the inhibitor. Examination of liver steroid sulfatase activities in animals treated with CT2251, during the contact sensitization model revealed that levels of liver steroid sulfatase activity below the normal range correlated with increasing dose of inhibitor both pre- and post-challenge. Doses of CT2251 above 0.01 mg/kg in pre-challenge livers and 0.1 mg/kg in post-challenge livers resulted in enzyme activity below the normal range ($P<0.05$ ANOVA)

Effect of steroids and CT2251 on contact sensitization

The contact sensitization model allows the effect of agents that may either augment or inhibit the ear swelling response to be studied. On day 0 the mice ($n = 10$ per group) were painted on their shaven flank with 50 μl of 2.5% oxazalone in 4 : 1 acetone:olive oil or vehicle only (4 : 1 acetone:olive oil). On day 5 the animals were challenged on their right dorsal ear surface with 25 μl of 0.25% oxazalone unless otherwise stated. Prior to ear challenge, measurements of right and left ear thickness were taken using an engineer's micrometer. Further ear measurements were taken 24 h post-challenge, which we have shown to be the time of maximal ear swelling in previous experiments. Controls included animals which were challenged only. Results were expressed as change in ear thickness from the prior measurement. The changes in ear thickness from individual mice were further analysed to give percentage inhibition from vehicle or positive control group using the following formula:

$$\%\ \text{inhibition} = \frac{\begin{array}{c}\text{Change in ear thickness}\\ \text{(individual mice)}\end{array} - \begin{array}{c}\text{Challenge only}\\ \text{(mean of group)}\end{array}}{\begin{array}{c}\text{Vehicle control}\\ \text{(mean of group)}\end{array} - \begin{array}{c}\text{Challenge only}\\ \text{(mean)}\end{array}}$$

In a pilot study we determined that the challenge dose of oxazalone was critical in determining the ear swelling; 2.5% giving the maximal response and 0.25% (the dose used in these experiments) giving approximately 50% maximal response.

Steroids (DHEA, DHEAS), dexamethasone (dexamethasone), corticosterone or CT2251 were administered subcutaneously (s.c.) at doses stated later in the text, in olive oil or 20% dimethylsulfoxide (DMSO)/80% olive oil generally on day 0 and 4.

DHEA and DHEAS can augment a contact sensitization response

(Fig. 6.1)

DHEA and DHEAS at 5 mg/kg augmented the contact sensitization response with an increase in ear thickness of 39.6% and 45.9% respectively when given on day 0, 4 and 5. In contrast, dexamethasone at 5 mg/kg effectively inhibited the response by 78.8%. In subsequent experiments similar results were seen when the steroid hormones were given on day 0 and 4 (data not shown). These data confirm the immunostimulatory effects of DHEA and DHEAS in vivo.

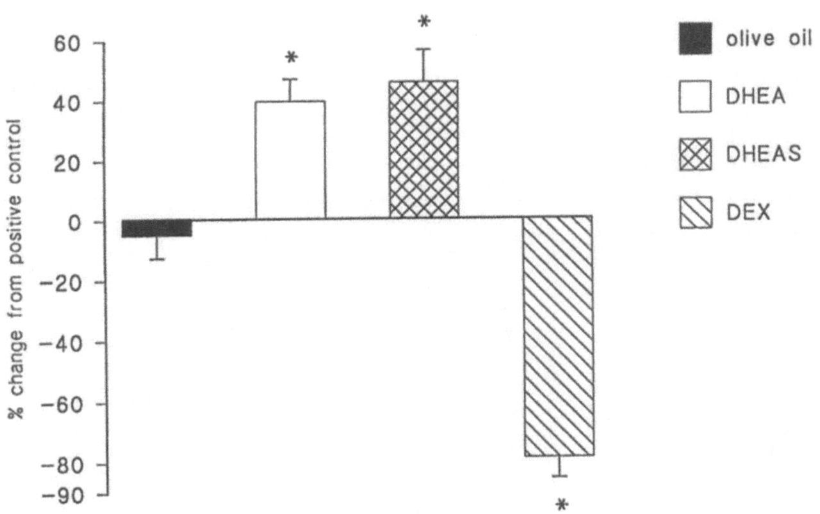

Fig. 6.1. Effect of DHEA, DHEAS, dexamethasone (DEX) and vehicle in contact sensitization. Animals were sensitized with 2.5% oxazalone on day 0, and challenged with 0.25% oxazalone on day 5. Steroids were given on day 0, 4 and 5 s.c. at 5 mg/kg. Ear measurements were taken at 24 h post-challenge, and expressed as percentage change from positive control. Mean ± SEM, $n = 7$ or 8 per group. *$P < 0.001$ from positive control (ANOVA).

Reversal of corticosterone inhibition with increasing doses of DHEAS (Fig. 6.2)

One prerequesite for the hypothesis that immunosuppressive and immunostimulatory steroids can counter-regulate each other is that the effects of glucocorticoids can be overcome by DHEAS. Corticosterone alone inhibited the response by 42.7%, while DHEAS alone given on day 0 and 4 augmented the contact sensitization response. Increasing doses of DHEAS from 0.0005 to 50 mg/kg reversed the inhibitory effect of corticosterone in a dose-dependent manner, with doses of DHEAS at 0.5, 5 and 50 mg/kg completely reversing the effect of corticosterone.

Inhibition of steroid sulphatase abrogates a contact sensitization response

(Table 6.1) Next we tested whether steroid sulfatase is an important enzyme in an in vivo model of cell-mediated immunity by treating animals with the enzyme inhibitor. CT2251 given s.c. on day 0 and 4 in olive oil at 0.3 and 0.1 mg/kg inhibited the contact sensitization response by 42.9% and 40.9% respectively. Doses below 0.1 mg/kg had no suppressive effect on the contact sensitization response. Hence, steroid sulfatase is a key enzyme in determining whether these mice could mount an effective immune response to oxazalone.

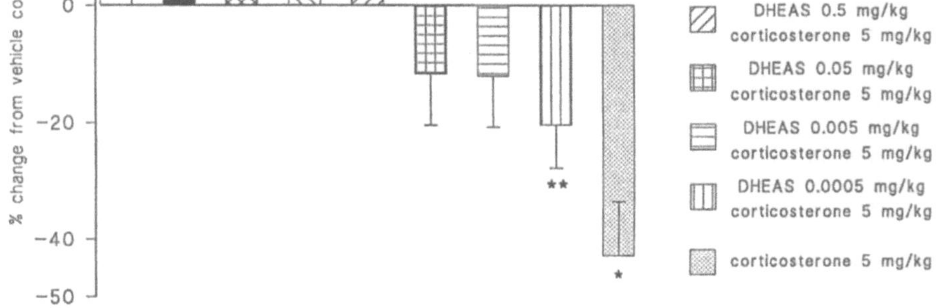

Fig. 6.2. Reversal of corticosterone inhibition with DHEAS. Animals were sensitized with 2.5% oxazalone on day 0, and challenged with 0.25% oxazalone on day 5. Steroids were given on day 0 and 4 s.c. Ear measurements were taken at 24 h post-challenge, and expressed as percentage change from vehicle control. Mean ± SEM, $n = 10$ per group. *$P < 0.05$ from vehicle control, **$P < 0.05$ from DHEAS 50 mg/kg, †$P < 0.01$ and ††$P < 0.001$ from corticosterone (ANOVA).

Table 6.1. Dose response to CT2251. Animals were sensitized with 2.5% oxazalone on day 0, and challenged with 0.25% oxazalone on day 5. CT2251 was given on day 0 and 4 s.c. Ear measurements were taken at 24 h post-challenge, and expressed as percentage change from vehicle control. Change in ear thickness from prior measurement in vehicle control = 0.136 mm and challenge only control = 0.0075 mm. Mean ± SEM (in brackets), n = 7–14 per group. *P<0.05 from vehicle control

	CT2251					
	0.3 mg/kg	0.1 mg/kg	0.03 mg/kg	0.01 mg/kg	0.003 mg/kg	0.001 mg/kg
Change in ear thickness from prior measurement (mm)	0.081* (0.045)	0.083* (0.006)	0.128 (0.007)	0.131 (0.006)	0.135 (0.009)	0.128 (0.011)
% Inhibition from vehicle control	42.93* (14.43)	40.94* (5.37)	5.84 (5.79)	3.89 (5.45)	0.825 (7.36)	5.75 (8.87)

Effect of irreversible sulfatase inhibitor (CT2251) alone and on DHEA and DHEAS-augmented responses (Fig. 6.3)

CT2251 given on day 0 and 4 in olive oil at 10 and 0.1 mg/kg inhibited the contact sensitization response by 61.6 and 38.6% respectively again confirming a critical role for steroid sulfatase in this response. In contrast, DHEA and DHEAS given on day 0 and 4 in olive oil at 5 mg/kg augmented the contact sensitization response by 43.1% and 45%, respectively. When given in combination with DHEA (5 mg/kg), CT2251 had no effect on the DHEA-augmented response. However, when given in combination with DHEAS (5 mg/kg), CT2251 reversed the DHEAS-augmented contact sensitization response causing an inhibition of 48.9% and 35% at 10 and 0.1 mg/kg respectively. These observations confirm that DHEAS needs to be converted to the active DHEA to express this biological activity in vivo. Dexamethasone (5 mg/kg) alone inhibited the contact sensitization response by 82.5%.

Reversal of CT2251 inhibition by DHEAS (Fig. 6.4)

In order to investigate whether the effect of CT2251 could be overcome by increasing concentrations of exogenous DHEAS, CT2251 at 0.1 mg/kg and DHEAS at 5, 15 and 50 mg/kg were given in olive oil on day 0 and 4. The inhibitory effect of 0.1 mg/kg CT2251 (51.3%) could be not be reversed by 5 mg/kg DHEAS, was partially reversed by 15 mg/kg DHEAS and completely reversed by 50 mg/kg DHEAS, with an augmentation of 35.8% seen. DHEAS alone at 5, 15 and 50 mg/kg augmented the contact sensitization response by 19.7%, 20.2% and 76.3%, respectively.

Effect of sulfatase inhibition on the early and late phases of the contact sensitization response (Fig. 6.5)

Since the contact sensitization model and DTH models in general are dependent on the activation of both the immune and inflammatory processes, CT2251 could

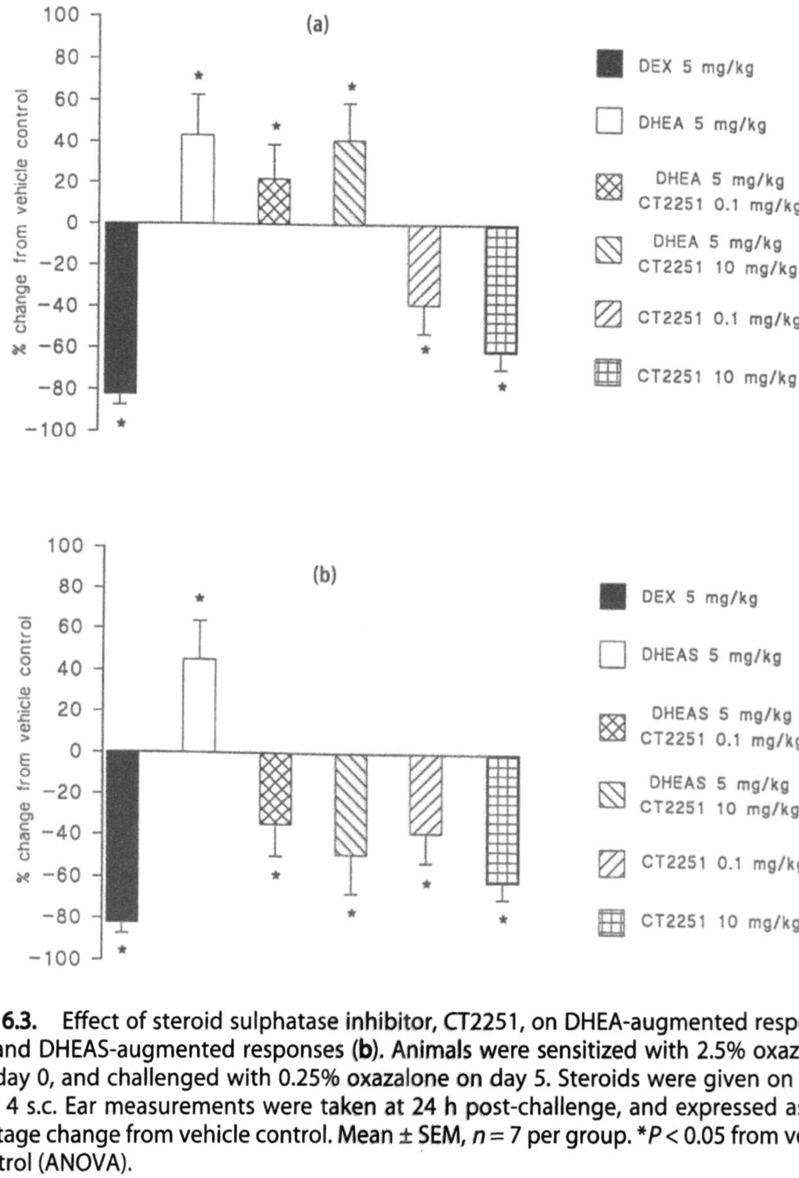

Fig. 6.3. Effect of steroid sulphatase inhibitor, CT2251, on DHEA-augmented responses (**a**) and DHEAS-augmented responses (**b**). Animals were sensitized with 2.5% oxazalone on day 0, and challenged with 0.25% oxazalone on day 5. Steroids were given on day 0 and 4 s.c. Ear measurements were taken at 24 h post-challenge, and expressed as percentage change from vehicle control. Mean ± SEM, $n = 7$ per group. *$P < 0.05$ from vehicle control (ANOVA).

have affected either or both of these. Therefore animals were treated with CT2251 during the early (immune system priming) phase and during the later post-challenge (inflammatory) phase. CT2251 was given at 0.1 mg/kg, a dose which will inhibit steroid sulfatase for up to 2 days (data not shown) on day 0, day 4 or days 0 and 4. In all cases the ear swelling response was inhibited by CT2251 indicating the importance of steroid sulfatase in both the early and late phases of this response.

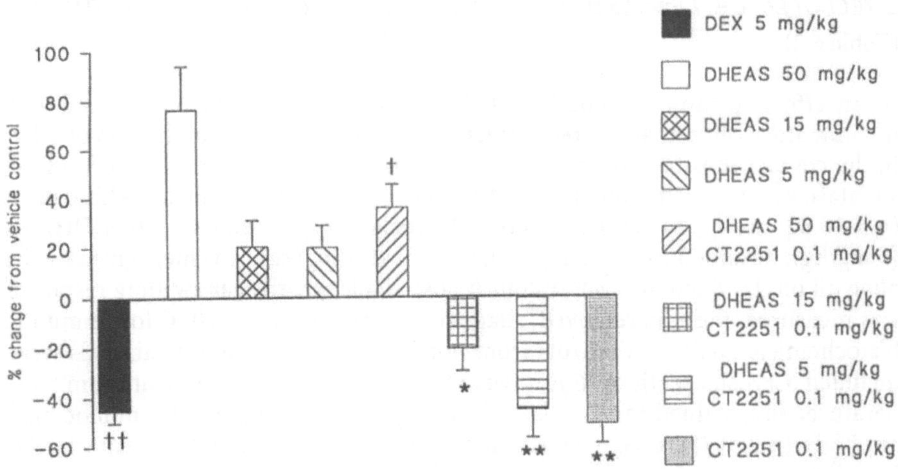

Fig. 6.4. Reversal of CT2251 inhibition by DHEAS. Animals were sensitized with 2.5% oxazalone on day 0, and challenged with 0.25% oxazalone on day 5. CT2251 and DHEAS was given on day 0 and 4 s.c. Ear measurements were taken at 24 h post-challenge, and expressed as percentage change from vehicle control. Mean ± SEM, $n = 7–14$ per group. *$P < 0.05$ from DHEAS 50 mg/kg, **$P < 0.01$ from DHEAS 50 mg/kg, †$P < 0.05$ from DHEAS 5 mg/kg + CT2251 0.1 mg/kg and CT2251 0.1 mg/kg, **$P < 0.01$ from DHEAS 50 mg/kg and DHEAS 50 mg/kg + CT2251 0.1 mg/kg (ANOVA).

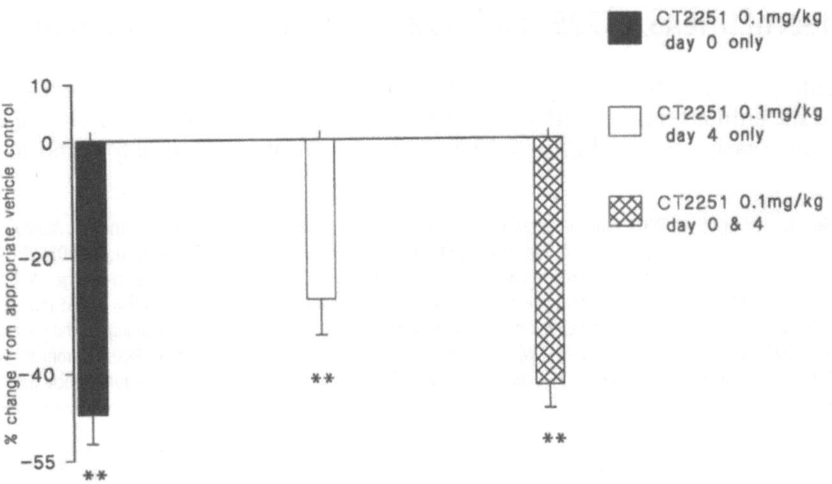

Fig. 6.5. Effect of CT2251 on contact sensitization given at different time points. Animals were sensitized with 2.5% oxazalone on day 0, and challenged with 0.25% oxazalone on day 5. CT2251 was given at 0.1 mg/kg on day 0 only, day 4 only or on days 0 and 4 s.c. Ear measurements were taken at 24 h post-challenge and expressed as percentage change from vehicle control. Mean ± SEM, $n = 19$ or 20 per group. **$P < 0.05$ from vehicle control.

Effect of DHEA, DHEAS and CT2251 on the cellular infiltrate in the ear (Table 6.2)

In an effort to monitor whether DHEA/S was important in both T-cell and inflammatory cell activation/recruitment, the cellular infiltrate in the ears of mice in the contact sensitization model was measured. In addition the role of steroid sulfatase was assessed. Contact sensitization was performed in male BALB/c mice (n = 10 per group) as described above. Animals were treated with either DHEAS (5 mg/kg), dexamethasone (5 mg/kg), CT2251 (0.1 mg/kg) or vehicle given s.c. in olive oil on day 0 and 4. Twenty-four h post-challenge, the ear swelling response was measured, the ears removed, snap-frozen and stored at −70 °C for immunohistochemical analysis. Controls from normal and challenge only animals were included. Cryostat sections (6 μm) were fixed in acetone for 10 min at room temperature, and, stained for T-cells and macrophages using anti-CD3 monoclonal antibody (mAb) (Serotec, Oxford) and anti-Mac 1 mAb (Serotec) respectively. Positive staining was visualized using the alkaline-phosphatase anti-alkaline phosphatase (APAAP) technique. Numbers of positive cells were counted blind in 10 random fields per tissue section and expressed as numbers per mm^2 of skin.

DHEAS increased the numbers of both immune (CD3 positive T-cells) and inflammatory cells (Mac-1 positive macrophages) by 63.8 and 107.0% respectively compared with vehicle control (Table 6.2). Dexamethasone effectively inhibited the influx of both lymphocytes (66.8%,), and, macrophages (90.1%,). In a separate experiment, CT2251 decreased the number of T-cells by 65.3% and macrophages by 80.3%, when compared with vehicle control (Table 6.2).

Effect of DHEAS, CT2251 and dexamethasone on oedema in the ear

Contact sensitization was performed in male BALB/c mice as described above, and the animals treated with either DHEAS (5 mg/kg) and/or CT2251 (10 mg/kg), dexamethasone (5 mg/kg) or vehicle given s.c. in olive oil on day 0 and 4. Ear

Table 6.2. Effect of (A), DHEAS, dexamethasone (DEX), and (B) CT2251 on cellular infiltration in the ear. Animals were sensitized with 2.5% oxazalone on day 0, and challenged with 0.25% oxazalone on day 5. DHEAS (5 mg/kg), DEX (5 mg/kg) and CT2251 (0.1 mg/kg) were given on day 0 and 4 s.c. Ear measurements were taken at 24 h post-challenge. (A) Change in ear thickness from prior measurement in vehicle control = 0.107 mm and challenge only control = 0.0065 mm, and (B) vehicle control = 0.221 mm and challenge only control = 0.008 mm. Ears were removed for staining for the presence of T- cells (CD3+) and macrophages (Mac-1+). Mean ± SEM (in brackets), (A) n = 9 or 10 per group. *P<0.001 from all groups, **P<0.001 from all groups except challenge only and normal. (B) n = 19 or 20 per group, *P<0.001 from vehicle control

	A					B	
	Challenge only	Vehicle control	DHEAS	DEX	Normal	Vehicle control	CT2251
T-cells per mm skin	0.51 (0.169)	23.81 (3.344)	39.00* (2.455)	7.91** (2.681)	1.44 (0.329)	36.2 (3.83)	12.56* (2.02)
Macrophages per mm² skin	0	93.82 (13.39)	194.23* (26.98)	9.31** (2.88)	0.70 (0.16)	131.67 (16.39)	25.99* (7.54)

measurements were taken prior to, and 24 h post-challenge. At 24 h, ears were excised, and weight measurements made prior to and after drying in an oven overnight (120 °C), to estimate tissue water content. Results were expressed as change from prior weight. Controls were included from normal and challenge only animals.

The percentage water content in a normal mouse ear was 55.4, which increased to 61.5% in animals which had been challenged on day 5 only. In animals which had been both sensitized and challenged, and treated with vehicle, the water content in the ear was further increased to 69.7%. There was no effect of treatment with DHEAS with or without CT2251 or dexamethasone on water content within the ear.

Effect of DHEA/S, dexamethasone and CT2251 in carrageenan-induced pleurisy in the rat

As stated above, the contact sensitization model is dependent on the activation of both the immune and inflammatory processes and CT2251 can affect both the early and late phases of the contact sensitization response. To confirm the role of steroid sulfatase on inflammatory responses we studied the effects of CT2251 in a carrageenan pleurisy model whose pathology is dependent solely on inflammatory components. Wistar rats (n=6) were given intrapleural injections of 0.5% carrageenan then killed 6 h later and the pleural cavity lavaged with 1 ml citrated saline. The volume of fluid exudate and exudate cell concentration were measured. The data indicate that dexamethasone can completely inhibit fluid leakage and markedly reduce cellular influx (by 68%, $P<0.01$, ANOVA). Treatment with CT2251 (10 mg/kg) at –24 h and –1 h reduced both exudate volume (by 43%, $P<0.05$, ANOVA) and cell infiltrate (by 28%, $P<0.05$, ANOVA) indicating that inhibition of steroid sulfatase can reduce an inflammatory-based response.

Steroid sulfatase is an important enzyme in collagen-induced arthritis in mice

Since steroid sulfatase clearly is an important enzyme in the immune and inflammatory responses associated with contact sensitization and carrageenan pleurisy, we were interested to see if steroid sulfatase was a key enzyme in a more chronic disease model. Collagen-induced arthritis is an immune-mediated form of progressive joint swelling and erosion. CT2251 was studied in this model by treating from the time of sensitization for 6 weeks and also by treating only following the onset of disease symptoms. Male DBA/1 mice were sensitized with type II collagen at the base of the tail in Freund's complete adjuvant on day 0 and again (in Freund's incomplete adjuvant) on day 18. Animals were treated with either CT2251 (10 mg/kg p.o. once a week) or olive oil or saline from day –1 (n=10). In the saline- and olive oil-treated groups disease incidence was 60% and

40% respectively with severe disease seen in all affected animals. In contrast, none of the CT2251-treated animals developed disease (P<0.05. Fisher's exact test).

If CT2251 treatment was witheld until arthritic symptoms developed, then this agent prevented any further progression of disease such that 35 days post-disease onset the area under curve (AUC) for disease severity was 31.6 ± 21 in the CT2251 group compared with 144 ± 20 in the vehicle group (P<0.001, Student's t-test). Furthermore there was no histological evidence of joint destruction in the CT2251-treated animals. These data indicate the central role that steroid sulfatase can play in a variety of immune and inflammatory processes and that inhibition of this enzyme can markedly improve disease outcome in this model.

Metabolites of DHEA may be the active form of the steroid

(Fig. 6.6)

DHEA itself or DHEAS can be further metabolized by a variety of enzymes. Indeed 17-β hydroxysteroid dehydrogenase can metabolize these steroids to androstenediol (AED) and androstenediol sulfate (AEDS) and there may be even subsequent metabolism. We tested whether AED and AEDS had activity in the contact sensitization model and further whether steroid sulfatase was important in their regulation. Animals given AED or AEDS on days 0 and 4 showed an

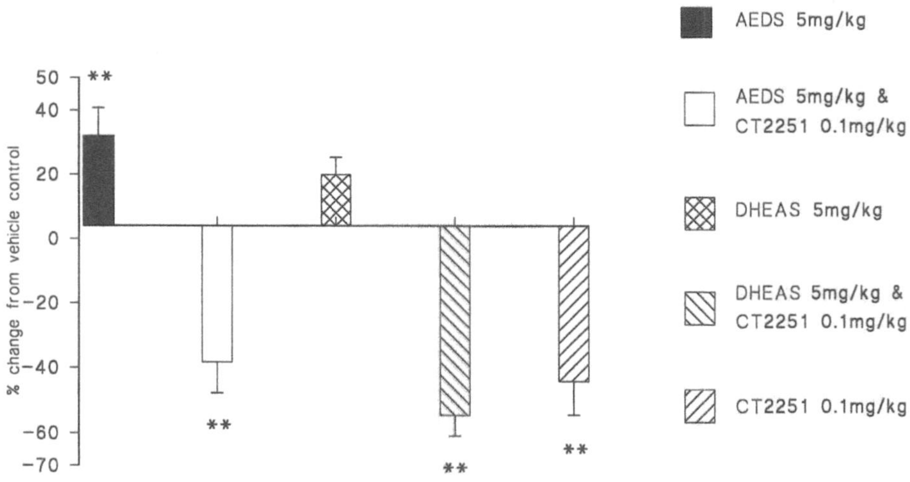

Fig. 6.6. Effect of androstenediol sulfate (AEDS), DHEAS and CT2251 on contact sensitization. Animals were sensitized with 2.5% oxazalone on day 0, and challenged with 0.25% oxazalone on day 5. DHEAS and AEDS were given at 5 mg/kg and CT2251 at 0.1 mg/kg on days 0 and 4 s.c. Ear measurements were taken at 24 h post-challenge, and expressed as percentage change from vehicle control. Mean ± SEM, n = 10 per group. **P < 0.05 from vehicle control.

augmented ear swelling response to oxazalone (+51 ± 7% and +20 ± 7% at 0.05 mg/kg, respectively). CT2251 at 0.1 mg/kg given on days 0 and 4 prevented the ear swelling response in both DHEAS- and AEDS-treated animals (Fig. 6.6). Hence metabolites of DHEA/S do have activity in vivo and steroid sulfatase is necessary for that activity.

Discussion

We have attempted to investigate the role of putative immunostimulatory steroids in the coordination of cell-mediated-type immune responses. It is well accepted that glucocorticoid hormones are endogenous immunosuppressive regulators and, if acting alone, would therefore maintain the immune system in an basal immunosuppressed state. However, the immune system needs to mount an appropriate response which may be cell-mediated in nature and one hypothesis is that immunostimulatory steroids normally function to override the effects of glucocorticoids. This would then maintain immune homeostasis in a "primed" condition allowing T_H1 responses normally to occur in the presence of antigen. DHEA has been shown to be immunostimulatory [22–25] but its regulation would need to occur at a local level. One of the enzymes thought to be central to this process is steroid sulfatase which can regulate the availability of DHEA (or related steroids) from the blood-borne but inactive DHEAS. We have used an inhibitor of this enzyme to characterize its role in a range of immune and inflammatory models.

We initially undertook to further the observations made by Daynes et al. [22–25] showing that DHEA can augment an immune response and that this is dependent on the action of steroid sulfatase. We investigated the effect of DHEA and DHEAS, and, the role played by steroid sulfatase in vivo on a T-cell-dependent model of immune function and inflammatory response. We have reproducibly demonstrated that both DHEA and DHEAS augment the contact sensitization-induced ear swelling response in mice, whereas glucocorticoids, such as dexamethasone and corticosterone, inhibit the response. These results are consistent with the observations of Daynes et al [22–25], who presented evidence from in vitro and ex vivo data that glucocorticoid down-regulated IL-2, and increased IL-4 production (i.e. up-regulating a T_H2-type response). In contrast, DHEA acts by increasing the production of IL-2 from T cells, and increasing proliferation of these cells (i.e. up-regulating a T_H1-type response). In addition, they suggested that DHEA exerts its effects directly on the T-cells, since DHEA exposure of antigen presenting cells was without effect [22] and DHEA receptors have been found in T-cells [29]. However, more recent data suggests the presence of a DHEA receptor in monocytes [30]. Nevertheless Daynes et al. [22,24] concluded that DHEA acts as an immunostimulant, whereas glucocorticoid acts as an immunosuppressant during immune responses. Moreover, these and other authors suggested that immune responses may be influnced by opposing effects of these two types of steroid in a counter-regulatory fashion [5,18,22,24].

Although the exact mechanism of action of DHEA remains unclear, the consistency of their in vitro data with our results, suggests that the effect of DHEA in vivo may be similar, by potentiating the production of T_H1-type cytokines and increasing proliferation of T-cells. In contact sensitization, a T-cell-dependent T_H1 immune reaction, these effects would certainly lead to a positive effect on the ear swelling end point of the response. Moreover, Suzuki and colleagues [31] have shown that in patients with systemic lupus erythematosus (SLE), defects of IL-2 synthesis from T-cells correlated with low serum DHEA levels. Addition of exogenous DHEA restored the impaired IL-2 production of T-cells from these patients in vitro. Indeed, our results suggest that immune responses in vivo can be influenced by the opposing effects of DHEA/DHEAS and glucocorticoid. In support, we have shown that the balance between DHEA/DHEAS and glucocorticoid will affect the subsequent inflammatory response, such that increasing the concentration of DHEAS will reverse corticosterone inhibition of the contact sensitization reaction. This is in agreement with the work of several groups showing that DHEA antagonizes the suppressive effect of dexamethasone both in vitro, ex vivo and in vivo, with the effect of DHEA being dominant [22,24,26,32–34].

Conversion of DHEAS to DHEA is necessary to ensure biological activity since DHEAS is itself inactive in the experimental systems studied here. Therefore the converting enzyme responsible (steroid sulfatase) may play an important regulatory role in immune response. Although others and ourselves have shown that biological effects can be obtained by administering exogenous DHEAS to mice, i.e. increasing IL-2 levels [22,24] or increasing contact sensitisation response (Suitters *et al.*, Immunology **91**, in press), only by inhibiting the activity of the steroid sulfatase enzyme itself can a regulatory role be confirmed, possibly by allowing endogenous glucocorticoid effects to predominate. We have shown that an inhibitor of steroid sulfatase, CT2251, has activity in vivo, demonstrating inhibition of liver steroid sulfatase activity for a number of days following administration. In addition, CT2251 will inhibit a contact sensitization response in vivo, when given on its own. Therefore, without the addition of exogenous DHEA/DHEAS, inhibition of steroid sulfatase will attenuate an ongoing immune response, presumably by preventing conversion of endogenous DHEAS to DHEA. Furthermore, steroid sulfatase inhibitor reversed a DHEAS-augmented response, but not a DHEA-augmented response. Since only DHEAS is the substrate for the enzyme, this would suggest that steroid sulfatase plays an important physiological role in regulating immune responses.

It has been shown that much of the effect of DHEA in augmenting a T_H1-type response is mediated in the secondary lymphoid organs, with particularly marked effects in those lymphoid organs which normally give rise to high IL-2 and IFN-γ cytokine profiles, such as spleen and peripheral lymph nodes [24]. The effect of DHEA is less marked in Peyer's patches, which show a predisposition to produce IL-4. Lymphoid organs showed a direct correlation between high expression of steroid sulfatase activity and ability to produce IL-2. More recent evidence by Hennebold and Daynes [35] demonstrated that most steroid sulfatase activity resided within the macrophage population. Therefore it is reasonable to speculate that conversion of DHEAS to DHEA occurs through the activity of steroid sulfatase within macrophages, which itself has a direct effect on T-cells, leading to increased production of IL-2 within the local environment of the spleen or relevant lymph node, thus potentiating the immune response.

A logical extension of these observations is that because DHEA and DHEAS can be metabolized further by a variety of enzymes, DHEA itself may not be the final active molecule. Indeed the activity of 7α-hydroxylase or 17 β-hydroxysteroid dehydrogenase or both can yield steroids such as 7-hydroxy DHEA, AED or androstenetriol (AET) or their sulfated forms. Moreover, in vitro studies by Padgett and Loria [27], suggest that AED and AET can indeed increase the proliferation of mouse splenocytes cultured with mitogen (Con A) and enhance IL-2 production as well as antagonizing the suppressive effects seen with hydrocortisone. Our data extend these studies showing that in vivo, metabolites such as AED or AEDS can enhance a contact sensitization response. Furthermore, the response to AEDS was entirely dependent on the activity of steroid sulfatase since it was abrogated by concomitant treatment with CT2251. These data indicate that although DHEA may not be the final active molecule in this controlling function of the immune response, steroid sulfatase is clearly a key enzyme.

The contact sensitization (DTH-like) response results, upon sensitization of the animals, in an inflammatory end point, namely ear swelling [36]. Because of the profound effects of DHEAS and CT2251, we were interested in their effects on the mechanism of ear swelling. Ear challenge of sensitized animals did not cause much oedema suggesting that fluid influx is not an important component of this response, and none of the treatments used had any effect. This would agree with the evidence demonstrating that a DTH immune response consists primarily of a cellular infiltrate causing a hard induration [36]. In contrast to the effect on oedema, treatment with DHEAS and CT2251 had a marked effect on the cellular infiltrate. Correlating with the increase in ear swelling response, DHEAS increased the number of both immune (T-cells) and inflammatory (monocytes/macrophage) cells. This increase could occur as a direct result of the potentiated production of IL-2. This is a cytokine produced by activated T-cells, which plays a critical role in T-cell biology, stimulating the further proliferation of activated cells and the secretion of other cytokines such as IFN-γ, IL-1 and TNF [37]. In particular IFN-γ is a known activator of macrophage function and other cells, leading to an influx of these cells to the site of an immune reaction. In contrast and again correlating with the ear swelling results, CT2251 reduced cellular infiltration. By inhibiting the conversion of DHEAS to DHEA, CT2251 may be having an indirect effect on the cytokine cascade, thus affecting the stimulus for cellular activation, proliferation and migration.

To study further whether DHEA can influence inflammatory as well as immune responses, animals were treated with the steroid sulfatase inhibitor during the priming phase only or during the inflammatory phase only (i.e. around the time of challenge) in a contact sensitization study. The ear swelling response was decreased with either treatment. Nevertheless, since activated T-cells are required in the effector (inflammatory) phase, it is conceivable that CT2251 was merely interfering with this "secondary" response. However, in a rat carrageenan pleurisy model, in which the only mechanisms contributing to the pathology are inflammatory, CT2251 could partially reverse the response. Again the interpretation could be that in some inflammatory situations as well as in immune responses, endogenous glucocorticoids would normally function to dampen down these effects. By removing the influence of endogenous DHEA or related steroids then these effects begin to become apparent.

It follows therefore that if steroid sulfatase is key in regulating certain immune responses such as contact sensitization, then in a disease state where often there is an inappropriate immune reaction (such as in autoimmune disease), steroid sulfatase activity may be a causative factor. This was examined in a mouse collagen-induced arthritis model in which animals were treated with CT2251 either during the whole of the disease course or only after onset of symptoms. In the former case no disease ensued, and in the latter disease progression was halted with treatment. The pathogenesis of this model involves the activation of T-cells and the development of autoantibodies to mouse collagen. Since steroid sulfatase inhibition could markedly ameliorate disease progression the data indicate that steroid sulfatase, as a regulator of immune function, can indeed contribute to the pathology of this disease.

In conclusion, we have shown that DHEA (or related steroids) is both a proimmune and pro-inflammatory steroid hormone with the capacity to potentiate immune responses in vivo. The conversion of DHEAS to DHEA by the steroid sulfatase enzyme is necessary to allow this effect to occur. These effects of DHEA are consistent with the hypothesis that it is the opposing influences of immunostimulatory and immunosuppressive adrenal steroids that can modify the microenvironment of secondary lymphoid tissues and influence how the immune system, and the T_H cell in particular, responds in the presence of antigen. In this context, the steroid sulfatase enzyme is a key regulator of these effects. Furthermore, although the exact mechanism of action is unproven in vivo, the potential for a therapeutic use of steroid sulfatase inhibitors is clear.

References

1. Abbas AK, Murphy KM, Sher A (1996) Functional diversity of helper T lymphocytes. Nature 383:787–793
2. Snijdewint FGM, Kapsenberg ML, Wauben-Penris PJJ, Bos JD (1995) Corticosteroids class-dependently inhibit in vitro Th1 and Th2-type cytokine production. Immunopharmacology 29:93–101
3. Wilckens T (1995) Glucocortocoids and immune function: physiological relevance and pathogenic potential of hormonal dysfunction. Trends Pharmacol Sci 16:193–197
4. Blalock J E (1994) The syntax of immune-neuroendocrine communication. Immunol Today. 15:504–511
5. Rook GAW, Hernandez-Pando R, Lightman SL (1994) Hormones, peripherally activated prohormones and regulation of the TH1/TH2 balance. Immunol Today 15:301–303
6. Meikle AW, Daynes RA, Araneo BA (1991) Adrenal androgen secretion and biologic effects. Endocrinol Metab Clin North Am 20:381–400
7. Parker LN (1989) Adrenal androgens in clinical medicine. Academic Press New York
8. Liu CH, Laughlin GA, Fischer UG, Yen SSC (1990) Marked attenuation of ultradian and circadian rhythms of dehydroepiandrosterone in post-menopausal women: evidence for a reduced 17,20-desmolase enzymatic activity. J Clin Endocrinol Metab 71:900–906
9. Migeon CJ, Keller AR, Lawrence B, Shepard TH (1957) Dehydroepiandrosterone and androsterone levels in human plasma, effect of age, sex: day-to-day and diurnal variations. J Clin Endocrinol Metab 17:1051–1056

10. Hobkirk R (1985) Steroid sulfotransferases and steroid sulfate sulphatases: characterisation and biological roles. Can J Biochem Cell Biol 63:1127–1144
11. Howarth NM, Purohit A, Reed MJ, Potter BVL (1994) Estrone sulfamates: potent inhibitors of estrone sulphatase with therapeutic potential. J Med Chem 37:219–221
12. Purohit A, Williams GJ, Howarth NM, Potter NVL, Reed MJ (1995) Inactivation of steroid sulphatase by an active site-directed inhibitor, estrone-3-sulfamate. Biochemistry 34:11508–11514
13. Martel C, Melner MH, Gagne D, Simard J, Labrie F (1994) Widespread tissue distribution of steroid sulfatase, 3β-hydroxysteroid dehydrogenase/Δ^5-Δ^4 isomerase (3βHSD), 17β-HSD 5α-reductase and aromatase activities in the rhesus monkey. Mol Cell Endocrinol. 104:103–111
14. Marks PA, Banks JA (1960) Inhibition of mammalian glucose-6-phosphate dehydrogenase by steroids. PNAS USA 46:447–452
15. Regelson W, Loria R, Kalimi M (1988) Hormonal intervention: "Buffer hormones" or "State dependency". The role of dehydroepiandrosterone (DHEA), thyroid hormones, estrogen and hypophysectomy in aging. Ann NY Acad Sci 521:260–273
16. Coleman DL, Leiter EH, Schwizer RW (1982) Therapeutic effects of dehydroepiandrosterone (DHEA) in diabetic mice. Diabetes 31:830–833
17. Coleman DL (1990) Dehydroepiandrosterone (DHEA) and diabetic syndromes in mice. In: Dehydroepiandrosterone (DHEA). Walter de Gruyter Berlin, PP 179–188
18. Cleary MP (1990) The antiobesity effect of dehydroepiandrosterone in rats. Proc Soc Exp Biol Med 196:8–16
19. Punjabi U, Deslypere JP, Verdonck L, Vermeulen A, (1983) Androgen and precursor levels in serum and testes of adult rats under basal conditions and after hCG stimulation. J Steroid Biochem 19:1481–1490
20. Wichmann U, Wichmann G, Krause W, (1984) Serum levels of testosterone precursors, testosterone and estradiol in 10 animals. Exp Clin Endocrinol 83:283–290
21. Orentreich N, Brind JL, Rizer RL, Vogelman JN (1984) Age changes and sex differences in serum dehydroepiandrosterone sulfate concentrations throughout adulthood. J Clin Endocrinol Metab 59:551–555
22. Daynes RA, Dudley DJ, Araneo BA (1990) Regulation of murine lymphokine production in vivo. II. Dehydroepiandrosterone is a natural enhancer of interleukin 2 synthesis by helper T cells. Eur J Immunol 20:793–802
23. Daynes RA, Araneo BA (1989) Contrasting effects of glucocorticoids on the capacity of T-cells to produce the growth factors interleukin 2 and interleukin 4. Eur J. Immunol 19:2319–2325
24. Daynes RA, Araneo BA, Dowell TA, Huang K, Dudley DJ (1990) Regulation of murine lymphokine production in vivo. III. The lymphoid tissue microenvironment exerts regulatory influences over T helper cell function. J Exp Med 171:979–996
25. Suzuki T, Suzuki N, Daynes RA, Engleman EG, (1991) Dehydroepiandrosterone enhances IL2 production and cytotoxic effector function of human T cells. Clin Immunol and Immunopath 61:202–211
26. Blauer KL, Poth M, Rogers WM, Bernton EW (1991) Dehydroepiandrosterone antagonises the suppressive effects of dexamethasone on lymphocyte proliferation. Endocrinology 129:3174–3179
27. Padgett DA, Loria RM, (1994) In vitro potentiation of lymphocyte activation by dehydroepiandrosterone, androstenediol, and androstenetriol. J Immunol 153:1544–1552
28. Morfin R, Courchay G (1994) Pregnenolone and dehydroepiandrosterone as precursors of native 7-hydroxylated metabolites which increase the immune response in mice. J Steroid Biochem Molec Biol 50:91–100
29. Meikle AW, Dorchuck RW, Araneo BA, Stringham JD, Evans TG, Spruance SL, Daynes RA (1992) The presence of a dehydroepiandrosterone-specific receptor binding complex in murine T cells. J Steroid Biochem Mol Biol 42:293–304

30. McLachlan JA, Serkin CD, Bakouche O, (1996) Dehydroepiandrosterone modulation of lipopolysaccharide-stimulated monocyte Cytotoxicity. J Immunol 156:328–335
31. Suzuki T, Suzuki N, Engleman EG, Mizushima Y, Sakane T (1995) Low serum levels of dehydroepiandrosterone may cause deficient IL-2 production by lymphocytes in patients with systemic lupus erythematosus (SLE). Clin Exp Immunol 99:251–255
32. Kalimi M, Shafagoj Y, Loria R, Padgett D, Regelson W, (1994) Anti-glucocorticoid effects of dehydroepiandrosterone (DHEA). Mol Cell Biochem 131:99–104
33. Browne ES, Wright BE, Porter JR, Svec F (1992) Dehydroepiandrosterone: antiglucocorticoid action in mice. Am J Med Sci 303:366–371
34. Araneo B, Daynes R (1995) Dehydroepiandrosterone functions as more than an antiglucocorticoid in preserving immunocompetence after thermal injury. Endocrinology 136:393–401
35. Hennebold JD, Daynes RA (1994) Regulation of macrophage dehydroepiandrosterone sulfate metabolism by inflammatory cytokines. Endocrinology 135:67–75
36. Greene MI, Schatten S, Bromberg JS, (1984) Delayed hypersensitivity. In Paul WE (ed) Fundamental immunology. Raven Press, New York, pp 685–696
37. Austyn JM, Wood KJ, (1993) Principles of cellular and molecular immunology. Oxford University Press, New York

Chapter 7

Corticosterone and the hypothalamic-pituitary-adrenal (HPA) axis in autoimmune diseases

Francisco Ramírez and Don Mason

Susceptibility to experimental autoimmune diseases in Lewis rats

Antigen-specific immune responses, which have evolved to protect the mammalian host from disease-causing organisms, are elicited when lymphocytes recognize pathogen-derived antigens and receive simultaneously the appropriate co-stimuli from the antigen-presenting cell. Following the initial triggering steps the immune system can activate a variety of effector mechanisms to eliminate or control the pathogenic organisms that produce the antigens in question. The activation and the development of effector mechanisms are complex events which require the participation of many components of the immune system. As many of the biological parameters are susceptible to genetic variation, it is not surprising that inbred strains of laboratory animals differ strikingly in the way that they respond to various antigens. The differences concern two aspects of immunity.

First, the ability to recognize a specific antigen as an immunogen – some inbred strains of experimental animals make a weak response to certain antigens [1]. The hereditary nature of this phenomenon led many years ago to the definition of the classical immune response (*Ir*) genes [1]. It was shown that these *Ir* genes are located in the major histocompatibility complex (MHC) [1] and encode the class I and class II MHC molecules which play a central role in the presentation of antigenic peptides to T-cells. Further studies have indicated that the situation is more complicated than the *Ir* gene data show in that genes not related to antigen presentation are also implicated in the differential ability to mount an immune response [2–4].

Second, the quality of the immune response once elicited is a genetically determined variable; the same antigen does not induce in different inbred strains the same effector mechanisms. Here genes outwith the MHC are involved. One of the best-known examples is the different response of mouse strains to the parasite *Leishmania*. Mice are susceptible or resistant to the infection of this micro-organism depending on whether they can induce the appropriate effector mechanisms [5].

As a consequence of the different capabilities to respond to different antigens, animal strains differ in their susceptibility to infections and autoimmune diseases. In some situations the expression of particular MHC alleles determines susceptibility or resistance, but in other circumstances components distinct from the MHC are involved.

Lewis rats are susceptible to the induction of many experimental models of autoimmune and inflammatory diseases such as experimental allergic encephalomyelitis (EAE) following the immunization with antigens from the central nervous system (CNS) [6]; mycoplasmosis after infection with *Mycoplasma pulmonis* [7]; experimental autoimmune myasthenia gravis after immunization with acetylcholine receptor [8]; and arthritis induced after the administration of different substances: complete Freund's adjuvant (CFA) [9], streptococcal peptidoglycan [10], *Yersinia enterocolitica* [11] and collagen [3]. The high susceptibility of Lewis rats to inflammatory and autoimmune diseases has been related to the defective regulation of corticosterone production. In the normal situation the adrenal glands are stimulated to release glucocorticoids during immune or inflammatory responses as a result of activation of the hypothalamic-pituitary-adrenal (HPA) axis by inflammatory and immune mediators such as interleukin-1 (IL-1), IL-6 and tumor necrosis factor (TNF) [12–15]. The glucocorticoid hormones so produced have an inhibitory effect on the synthesis of these cytokines and as a result the amounts produced are regulated to avoid an excessive and dangerous immune and inflammatory response [16]. Lewis rats are not able to regulate appropriately these responses because they manifest a deficiency in this regulatory interaction between the immune and neuroendocrine system [17].

In this chapter, the role of glucocorticoids in the development of EAE will be discussed. First, a brief description of the course of EAE is presented stressing the importance of endogenous corticosterone in the curtailment of the disease. Second, evidence will be presented showing the critical role of hyporesponsiveness of the HPA axis in determining the susceptibility of Lewis rats to EAE. Finally, results from new in vitro experiments designed to understand the role of glucocorticoids in EAE will be shown. In these experiments the long-term effects of glucocorticoids on cytokine production were analysed. The results obtained so far support the hypothesis outlined some years ago about the role of glucocorticoids in the autoregulation of EAE [18]. The model suggests that glucocorticoid hormones not only have an acute inhibitory effect on the synthesis of cytokines but also a long-lasting one, inducing the synthesis of anti-inflammatory or Th-2 cytokines. This hypothesis will be discussed in the last section of this chapter.

Experimental allergic encephalomyelitis

Course of the disease and refractory period

EAE is an experimental paralysing disease that has been studied as a model of multiple sclerosis (MS) in humans [19] and it is a useful system in which to study the regulation of immune and inflammatory responses. The disease can be induced in the appropriate animal strains after immunization with CNS antigens. Lewis rats develop EAE after immunization with guinea-pig myelin basic protein (MBP) emulsified in CFA. This procedure is termed "active EAE" because the animals that develop the disease are themselves the source of the encephalitogenic T-cells. EAE can also be induced by the injection into naive animals of MBP-specific lymphocytes generated from immunized donor animals; this is "passive EAE". EAE in Lewis rats is a monophasic disease characterized by a single episode of paralysis followed by spontaneous recovery. The animals develop a transient paralysis 11–14 days after immunization with MBP in CFA, or 3–6 days after inoculation with MBP-specific cells. The paralysis is caused by the action of $CD4^+$ T lymphocytes that infiltrate the CNS [20]. Four to 6 days after the onset of paralysis the animals recover spontaneously. Animals that have recovered are refractory to attempts to induce further episodes of disease [21,22]. Understanding the mechanisms that promote the spontaneous recovery and the maintenance of the refractory state of EAE may help to interpret some aspects of MS; for example the spontaneous recovery of EAE may be similar to the acute remissions seen in some MS patients. The mechanisms responsible for the spontaneous recovery and refractory state on EAE, although objects of intense research, are not completely understood and different and numerous mechanisms have been proposed: serum suppressor factors [23], suppressor T lymphocytes [24,25], suppressor macrophages [26], production of immunosuppressive factors by glial cells [27], anti-T lymphocyte idiotype responses [28], regulation by IFN-γ [29], by IL-10 [30,31], by TGF-β [32] and neuroendocrine-mediated immunoregulation [33,34].

Several observations have suggested that the level of glucocorticoid hormones are increased in Lewis rats suffering from EAE. These animals showed lymphopaenia and neutrophilia during the time they were paralysed [34] and thymi from these animals were smaller than from controls [33]. It is well known that glucocorticoid hormones can induce these three phenomena: lymphopaenia, neutrophilia and thymocyte death. Determinations of the serum corticosterone concentration in animals developing active or passive EAE showed a marked increase associated with signs of the disease. The increase in corticosteroid concentration occurred 24–48 h before the onset of paralysis, with its peak at the time of maximum clinical score. The animals started to recover from paralysis after the peak in serum corticosterone. When the animals had recovered fully the glucocorticoid levels fell to their normal values. Thymus weight and lymphocyte and neutrophil counts were also reconstituted to normal values after recovery from paralysis [33,34]. The time correlation suggested a cause-effect relationship between the increase in serum corticosterone and the recovery from EAE.

The role of endogenously produced corticosterone was analysed directly by evaluating the effect of adrenalectomy in Lewis rats subjected to EAE. It was observed that, unlike normal rats, adrenalectomized animals did not recover from EAE but instead developed an unremitting and progressive paralysis with a fatal outcome. If these adrenalectomized rats were given corticosterone replacement therapy at the onset of paralysis they recovered within a few days and developed the refractory state like non-adrenalectomized animals [34]. These observations showed that endogenously produced corticosterone plays an essential role in the recovery of rats from EAE. The interpretation of these observations is that the spontaneous recovery occurs because corticosteroids, which as noted, are found at high levels in the sera of paralysed animals, acutely depress the autoimmune response through their immunosuppressive and anti-inflammatory effects. Consistent with this interpretation animals at the peak of the clinical score showed inhibition of the delayed type hypersensitivity (DTH) response to ovalbumin (OVA) but they did not show inhibition of the antibody response to sheep red blood cells (SRBC) [34]. Once animals have recovered from EAE adrenalectomy has no adverse effect and the refractory phase develops as it does in non-adrenalectomized controls. Evidently corticosteroids are not directly responsible for the maintenance of the refractory phase of the disease. This conclusion does not exclude the possibility that the transiently elevated levels of glucocorticoid hormones during paralysis have long-term effects on the immune response to MBP. Relevant to this issue is the suggestion that glucocorticoids may induce some long-lasting effects which influence the balance between cell-mediated immunity and humoral immunity [35–37]. This possibility will be further discussed later in this chapter.

Susceptible and resistant strains

In the previous section results were presented suggesting that the spontaneous recovery of Lewis rats from EAE is dependent on endogenous glucocorticoids. In this section the role of glucocorticoids in the resistance or susceptibility to EAE will be discussed.

Most rat strains are resistant to the induction of active EAE. One reason for this resistance is that these strains display higher corticosterone responses than the susceptible Lewis strain. PVG.RT1^c is a rat strain that is relatively resistant to the induction of EAE. There are several indications which suggest that the corticosterone levels in Lewis rats are lower than in PVG and other rat strains and that this fact is important in the susceptibility to EAE: (a) Lewis rats have smaller adrenal glands than other rat strains (60% smaller than PVG); (b) the basal corticosterone level is 60% lower in Lewis rats than in PVG rats; (c) the corticosterone response of Lewis rats to stress is lower than that of PVG animals. Evidence that these hormonal differences influence susceptibility to EAE is provided by the finding that adrenalectomized PVG rats develop fatal paralysis after administration of MBP indicating that these resistant animals can respond to the encephalitogen but that the adrenal glands normally protect them from EAE [36]. When kinetic studies were performed, changing the time interval between MBP immunization and adrenalectomy in PVG rats, it was observed that the activity of the adrenal glands was only important during the 24–48 hours after the onset of paralysis. Adrenalectomy of animals after the first sign of paralysis did not change the course of disease: after the onset of paralysis the animals recovered spontaneously. This observation confirmed and extended the data with adrenalectomized Lewis rats discussed in the previous section.

Similar observations, regarding the importance of glucocorticoids on the susceptibility of Lewis rats, have been obtained in the experimental model of arthritis induced by injection with streptococcal peptidoglycan. Lewis rats are susceptible to the induction of arthritis and Fisher rats are resistant. When both strains were compared in their ability to regulate corticosterone production it was observed that Lewis rats released relatively lower amounts of corticotropin (ACTH) and corticosterone in response to inflammatory mediators. Lewis adrenal glands are smaller than those in Fisher rats. Contrarily, Lewis thymi are bigger than thymi from Fisher rats. These two indirect measurements of the HPA axis activity suggested again that corticosterone levels in Lewis rats are lower than in other rat strains. More important, and similar to the situation in EAE, Lewis rats are protected from the effect of the streptococcal peptidoglycan after glucocorticoid administration and the resistant Fisher strain become susceptible after administration of the glucocorticoid receptor-antagonist RU-486 [10].

One important observation is that adrenalectomy of PVG.RT1^u rats, unlike the effects with PVG.RT1^c strain animals, does not render them susceptible to the development of active EAE. This observation illustrates the point that MHC genes also control susceptibility to this disease [36]. However MHC-determined resistance to EAE can reflect more subtle differences than a simple failure to respond to the encephalitogen. Lew.RT1^u rats do react to MBP but make a non-pathogenic Th-2 response [38].

Corticosteroid effects in vitro and in vivo

As it has been shown in the preceding sections, the study of some experimental models of autoimmune and inflammatory diseases in the rat shows that the relative inability of Lewis rat to increase glucocorticoid levels during an immune response makes these animals particularly susceptible to the development of these diseases. Additionally, PVG or Fisher rats lose their resistance after the restriction of glucocorticoid action by adrenalectomy or the administration of the glucocorticoid receptor-antagonist RU-486 [10,36]. However, as already described, despite the relative insensitivity of the HPA axis to inflammatory stimuli in the Lewis rat, the elevation in serum corticosterone concentration after the onset of EAE plays an essential role in the spontaneous recovery from the disease. In order to understand the role of corticosterone in all these phenomena the effects have been analysed of corticosteroid hormones on cytokine production and on the expression of surface molecules after CD4$^+$ T cell activation.

Concerning the effects of glucocorticoid hormones on the expression of surface molecules, the induction of the CD8α chain molecule on activated rat CD4 T cells has been described [39]. A brief summary of the results is included to illustrate the diverse effects of glucocorticoid hormones in the immune system. Human and rat CD4 cells express CD8α chain after activation in vitro [40,41]. There are some reports about the in vivo generation of this double-positive population after lymphocyte activation by different ways and in different organs: in the spleen of rats primed with dinitrophenyl-bovine gamma globulin (DNP-BGG) [42], in the blood of cyclosporine-treated heart-allografted rats [43], in the peritoneal exudate from rats immunized with a syngenic tumour cell [44] and in the mesenteric lymph nodes from mice infected with *Salmonella enteriditis* [45]. In humans there are similar findings, the existence of a double-positive population has been described in patients with rheumatoid arthritis [46] and after renal transplantation [47]. During the course of our experiments, evaluating glucocorticoid effects on activated T cells, we found that the level of expression of CD8α and the percentage of rat CD4 cells expressing CD8α were increased by the addition of the synthetic glucocorticoid analogue dexamethasone (DEX) in the culture medium [39]. The significance of these findings remains to be established but it has been shown by others that, in contrast to the double-positive phenotype displayed by some MBP-specific lines obtained from normal donors, MBP-specific lines obtained from MS patients rarely display a $CD4^+CD8^+$ phenotype [48]. Given the beneficial effect of steroids in the therapy of many autoimmune and inflammatory diseases, including MS, we are exploring the possibility that this population of CD4 cells with the ability to express CD8, apparently reduced in MS patients, is important in the regulation of immune responses. The very limited published literature about mature double-positive lymphocytes describes these as regulatory cells [25,49–52].

In vitro effects on cytokine production

Another aspect of our work is the evaluation of the effect of glucocorticoid hormones on cytokine synthesis. Corticosteroids are used in the therapy of many autoimmune and inflammatory diseases and in organ transplantation. Their immunosuppressive and anti-inflammatory effects are believed to depend on their property of depressing the production of many cytokines and inflammatory mediators [53–60]. There are two exceptions: it has been reported that the production of IL-4 [35,61] and TGF-β [62] is increased by glucocorticoid hormones. However, with respect to IL-4 there are contrary data in that other groups have observed the inhibition of IL-4 production in human lymphocytes by corticosteroids [63,64]. IL-4 is a cytokine with anti-inflammatory properties and its induction by glucocorticoid hormones might, in principle, explain their long-term beneficial effects. To examine this possibility and to try to clarify these contradictory results, we analysed IL-4 synthesis by DEX-treated rat $CD4^+$ T-cells activated with the polyclonal mitogen concanavalin A (Con A). We observed a small decrease in IL-4 mRNA expression in these cells after 24 h of activation [37]. As previously described by other groups, we also found inhibition of the production of IL-2 and interferon-γ (IFN-γ) and reduced cellular proliferation thereby confirming the reported negative effect of glucocorticoids during T-cell activation.

To determine whether there were any long-term effects produced by corticosteroids we examined the cytokine production after a second round of in vitro stimulation. This approach is similar to the one employed by several other groups to analyse the effects of the presence of some cytokines on the development of the immune response [65]: T-cells are activated in the presence of the cytokine which is being tested and then the cells are activated a second time in its absence. The effect of the transient presence of the cytokine on subsequent synthesis of various cytokines is then analysed. In our experiments rat CD4+ T lymphocytes were activated with Con A in the presence of a physiological concentration of DEX (10^{-8}M). Seven days later the cells were activated in the absence of the hormone and cytokine production was analysed (Fig. 7.1). This protocol resembles approximately the in vivo situation in EAE, and in immune responses in general, when only a transient increase in blood corticosterone occurs. The results indicate that the exposure of the cells to DEX in the primary stimulation changed the cytokine synthesis induced by the secondary stimulation. We found that the presence of DEX during primary T-cell activation increases IL-4, IL-10 and IL-13 mRNA levels in CD4+ T-cells after secondary activation in the absence of the hormone. Analysis of IL-4 protein synthesis confirmed the data obtained at the mRNA level (Table 7.1). Furthermore, the synthesis of pro-inflammatory or Th-1 cytokines was inhibited in this culture: IFN-γ and TNF synthesis was reduced by 70% compared with control cultures [37]. Glucocorticoid exposure in the primary stage did not significantly affect IL-2 and IL-6 production nor the total numbers of T-cells recovered in the secondary one, although a slight positive effect of the glucocorticoid on cellular proliferation was always observed. These findings, obtained after Con A activation, have been reproduced after a mixed lymphocyte reaction

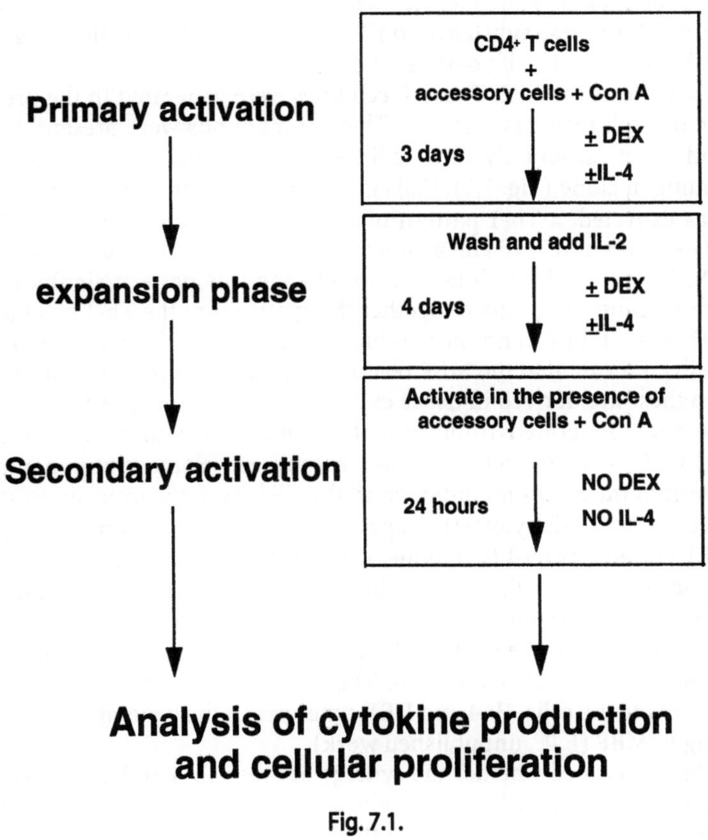

Fig. 7.1.

Table 7.1. Effects of dexamethasone on cytokine gene expression and cellular proliferation[a]

Culture conditions	Cytokines							Proliferation[b]
	IFN-γ	TNF	IL-4	IL-10	IL-13	IL-2	IL-6	
Control	++	++	–	+	+	+	+	+
Dexamethasone	+	+	++	++	++	+	+	++

[a] The production of cytokines is expressed in arbitrary units according to the differences obtained with protein assay except for IL-10 and IL-13 that were analysed by RT-PCR due to the lack of a protein assay.
[b] Proliferation is expressed in arbitrary units.

(MLR) (F.R., unpublished data). These data show that, in addition to their acute immunosuppressive effects, glucocorticoids may display a more subtle and long-lasting influence on the immune response, inducing a Th-2 pattern of cytokine production by $CD4^+$ T lymphocytes on reactivation. Further studies with this in vitro system showed that IL-4 and DEX synergized in their action, in the sense that when both reagents were added together the production of Th-1 cytokines was very strongly inhibited and the synthesis of IL-4 was greatly enhanced.

More recently we have also evaluated the effect of DEX and IL-4 on the cytokine production of MBP-specific lines. The aims of these experiments were to study the effects of these substances on T-cells activated by specific antigen rather than by mitogen and also, if possible, to obtain MBP-specific Th-2 lines to analyse their in vivo effects on EAE. These T-cell lines were generated in the presence of DEX, IL-4 or both reagents together. These compounds were present during the first round of activation only and the lines were activated on subsequent rounds with the antigen alone (Fig. 7.2). Cells grown in the absence of these compounds showed, as expected, a Th-1 pattern of cytokine production with high levels of synthesis of IFN-γ and TNF and a very low amount of IL-4 (Table 7.2). The addition of DEX alone or IL-4 alone did not change this pattern significantly, but when both reagents were added together the synthesis of the Th-1 cytokines IFN-γ and TNF was inhibited and the production of IL-4 was dramatically increased. These results indicate that the MBP reactive cells are predisposed to become Th-1 lines when these are derived in tissue culture medium alone, presumably because they have been generated from animals immunized with MBP emulsified in CFA, and CFA is an inductor of Th-1 responses. (Some components of the *Mycobacterium tuberculosis* contained in the CFA have the capacity to stimulate the production of IL-12 by macrophages and then drive the immune response to Th-1 [66].) The addition of IL-4 alone, which is the classical protocol to obtain a Th-2 response [67], or DEX alone, which shows the capacity to induce a Th-2 response in our experiments [37], did not seem to be enough to change this predisposition induced by the CFA. Only when both reagents were added together was it possible to alter this situation. Then, similarly to the experiments using Con A to activate the cells, IL-4 and DEX synergize in the generation of Th-2 lines responding to MBP (F.R., unpublished work).

It may be questioned whether the synergy between IL-4 and glucocorticoids in vitro also occurs in vivo. However, given the signal transduction events that induce these changes in vitro indicate their existence it seems likely that similar

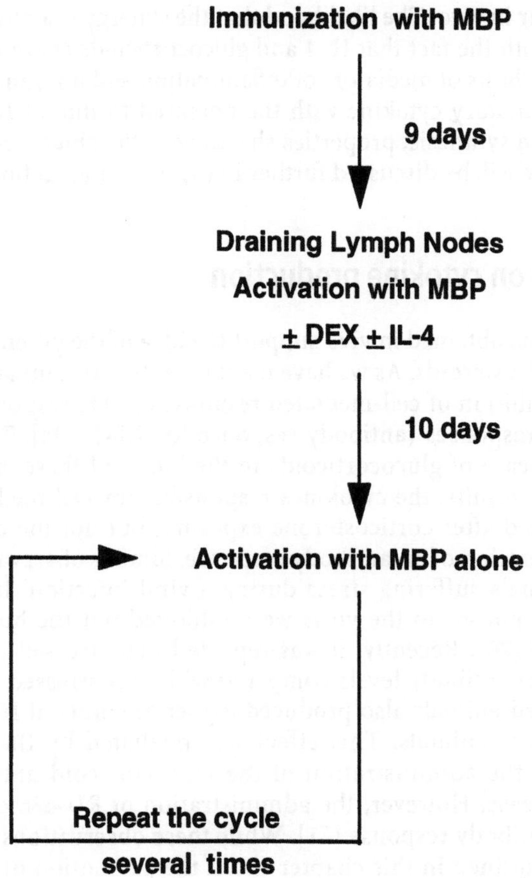

Fig. 7.2

Table 7.2. Effect of IL-4 and dexamethasone (DEX) on cytokine production and the ability to transfer passive experimental allergic encephalomyelitis (EAE) by myelin basic protein (MBP)- specific T-cell lines[a]

MBP lines	Cytokine production		Passive EAE[b]
	IFN-γ	IL-4	
Control lines	+++	+/–	+++
DEX-treated lines	+++	+/–	++
IL-4-treated lines	++	+	+
IL-4- + DEX-treated lines	+	++	–

[a] The production of cytokines is expressed in arbitrary units according to the differences obtained with protein assay.
[b] Animals were injected with MBP-specific T-cell lines and examined for signs of disease. The ability to transfer passive EAE by the different lines is expressed in arbitrary units according to the incidence and severity of the disease. In these experiments both parameters were correlated.

changes can occur in vivo. The likelihood that the synergy is a physiological one is also consistent with the fact that IL-4 and glucocorticoids share many regulatory effects on the synthesis of mediators of inflammation and immune responses: IL-4 is an anti-inflammatory cytokine with the potential to inhibit IFN-γ, TNF, IL-1 and prostaglandin synthesis, properties shared with the glucocorticoid hormones [68,69]. This issue will be discussed further in the following section.

In vivo effects on cytokine production

Some observations obtained in vivo support the idea of the potentiation of a Th-2 response by corticosteroids. As we have mentioned before, animals in the peak of EAE showed inhibition of cell-mediated responses (DTH response to OVA) but not of humoral responses (antibody response to SRBC) [34]. This could be an effect of the increase of glucocorticoids in the blood of these animals because, according to our results, the cytokines responsible for cell mediated responses would be inhibited after corticosterone exposure, but not the cytokines implicated in the generation of an antibody response. Similar observations have been obtained in animals suffering stress during a viral infection. In these animals cell-mediated responses to the virus were inhibited but the humoral response was not affected [70]. Recently, it was reported that stressed mice showed an increase in serum antibody levels compared with non-stressed animals. Spleen cells from stressed animals also produced higher amounts of IL-4 than splenocytes from normal animals. This effect was mediated by the glucocorticoid release, because the administration of the glucocorticoid antagonist RU-486 blocked the increase. However, the administration of RU-486 did not appreciably affect the antibody response [71]. While these observations in vivo support the hypothesis outlined in this chapter about the promotion of Th-2 responses by stress-mediated glucocorticoid release, it is important to point that there are contradictory data with respect to the effect of stress on the humoral immune response [72]. Whether these differences are related to the level of stress, its duration or to some other unidentified factor is unclear at this time. Analysis of the effect of the administration of glucocorticoid hormones to animals on cytokine production has been claimed to show that IL-4 production is increased and IL-2 production inhibited [35] and more work is required to resolve some of the apparent contradictions.

Regulation of EAE by spontaneous corticosterone release

As has been discussed in the previous sections, the spontaneous recovery of Lewis rats from EAE and the resistance to this disease of PVG rats is dependent on the immunosuppressive effects of glucocorticoid hormones released by the adrenal glands [34,36].

EAE in Lewis rats is a disease mediated by $CD4^+$ T cells that produce IL-2 and IFN-γ [73] whilst IL-4 producing MBP-specific cell lines, obtained after treatment with IL-4 and DEX (see Table 7.2), are not able to induce passive EAE (F.R., unpublished work). Similar findings have been described in the mouse EAE model: MBP-specific Th-1 clones can induce disease but Th-2 clones do not and, in some circumstances, can inhibit the development of EAE [74,75]. The analysis of cytokines synthesized by animals with EAE during the different phases of the disease confirms these observations, showing a predominant production of Th-1 cytokines prior to the onset of paralysis and predominance of IL-10 and TGF-β during and after the recovery from paralysis [30,31]. From these observations it seems that EAE is provoked by a Th-1 or cell-mediated response but that after recovery the Th-1 reactivity may become dominated by Th-2 cytokines with the ability to down-regulate cell-mediated responses. If glucocorticoids can tip the balance between Th-1 and Th-2 immune responses, favouring the latter, as suggested by our in vitro experiments [37], then, the refractory phase of EAE can, in principle, be explained if endogenously produced glucocorticoids induce a switch from a Th-1 to a Th-2 response to MBP.

The proposed model to explain the autoregulation of EAE by endogenously produced corticosterone suggests a sequence of events as follows: MBP immunization with CFA induces in Lewis rats a potent cell-mediated immune response and, to a lesser extent, humoral immunity. T-cells that synthesize inflammatory cytokines migrate and attack the CNS inducing the paralysing disease. However, the production of pro-inflammatory cytokines and probably the concomitant stress induced in the animal by the disease activates the HPA axis and corticosterone is released by the adrenal glands. This release has two coordinated effects: it arrests the disease through a direct suppression of inflammation and cell-mediated immunity and, in addition, the long-term effects of glucocorticoids on T-cell differentiation give rise to a Th-2 type response that will prevent the further activation of disease-inducing MBP-reactive Th-1 cells [18].

This hypothesis provides an explanation for the recovery and refractory periods, although a detailed analysis of the cytokines produced during the different phases of EAE, and how endogenous glucocorticoids affect this synthesis, is necessary to verify it.

Acknowledgments

We thank the members of the MRC Cellular Immunology Unit who have contributed to this work and Emma Kenny for reading the manuscript. Work in the laboratory is funded by the Medical Research Council and grants from the Multiple Sclerosis Society of Great Britain and Northern Ireland.

References

1. Benacerraf B, McDevitt HO (1972) Histocompatibility-linked immune response genes. Science 175:273–279

2. Davis BK, Kunz HW, Shonnard JW, Gill TG (1981) Immune response to poly (Glu52 Lys33 Tyr 15) in congenic rats. Transplantation Proc 13:1378–1382
3. Griffiths MM, DeWitt CW (1984) Genetic control of collagen-induced arthritis in rats: the immune response to type II collagen among susceptible and resistant strains and evidence for multiple gene control. J Immunol 132:2830–2836
4. Cole BC, Griffiths MM, Sullivan GJ, Ward JR (1986) Role of non-RTI genes in the response of rat lymphocytes to mycoplasma arthritidis T cell mitogen, concanavalin A and phytohemaglutinin. J Immunol 136:2364–2369
5. Heinzel FP, Sadick MD, Mutha SS, Locksley RM (1991) Production of interferon gamma, interleukin 2, interleukin 4, and interleukin 10 by $CD4^+$ lymphocytes in vivo during healing and progressive murine leishmaniasis. PNAS USA 88:7011–7015
6. Hughes RAC, Stedronska J (1973) The susceptibility of rat strains to experimental allergic encephalomyelitis. Immunology 24:879–884
7. Davis JK, Thorp RB, Maddox PA, Brown MB, Cassell GH (1982) Murine respiratory mycoplasmosis in F344 and LEW rats: evolution of lesions and lung lymphoid cell populations. Infect Immun 36:720–729
8. Biesecker G, Koffler D (1988) Resistance to experimental autoimmune myasthenia gravis in genetically inbred rats. Association with decreased amounts of in situ acetylcholine receptor-antibody complexes. J Immunol 140:3406–3410
9. Hogervorst EJM, Boog CJP, Wagenaar JPA, Wauben MHM, Van der Zee R, Van Eden W (1991) T cell reactivity to an epitope of the mycobacterial 65-kDa heat-shock protein (hsp65) corresponds with arthritis susceptibility in rats and is regulated by hsp 65-specific cellular responses. Eur J Immunol 21:1289–1296
10. Sternberg EM, Hill JM, Chroussos GP, Kamilaris T, Listwak SJ, Gold PW et al. (1989) Inflammatory mediator-induced hypothalamic-pituitary-adrenal axis activation is defective in streptococcal cell wall arthritis-susceptible Lewis rats. PNAS USA 86:2374–2378
11. Hill JL, Yu DTY (1987) Development of an experimental model for reactive arthritis induced by *Yersinia enterocolitica* infection. Infect Immun 55:721–726
12. Besedovsky H, Sorkin E, Keller M, Muller, J (1975) Changes in blood hormone levels during the immune response. Proc Soc Exp Biol Med 150:466–470
13. Besedovsky H, del Rey A, Sorkin E, Dinarello CA (1986) Immunoregulatory feedback between interleukin-1 and glucocorticoid hormones. Science 233:652–654
14. Naitoh Y, Fukata J, Tominaga T, Nakai Y, Tamai S, Mori K et al. (1988) Interleukin-6 stimulates the secretion of adrenocorticotropic hormone in concious, freely moving rats. Biochem Biophys Res Comm 155:1459–1463
15. Warren RS, Starness HF, Alcock N, Calvano S, Brennan MF (1988) Hormonal and metabolic response to recombinant human tumor necrosis factor in rat: in vivo and in vitro. Am J Physiol 255:E206–E212
16. Munck A, Guyre P, Holbrook NJ (1984) Physiological functions of glucocorticoids in stress and their relation to pharmacological actions. Endocrinol Rev 5:25–44
17. Sternberg EM, Young III WS, Bernardini R, Calogero AE, Chrousos GP, Gold PW et al. (1989) A central nervous system defect in biosynthesis of corticotropin-releasing hormone is associated with susceptibility to streptococcal cell wall-induced arthritis in Lewis rats. PNAS USA 86:4771–4775
18. Mason D (1991) Genetic variation in the stress response: susceptibility to experimental allergic encephalomyelitis and implications for human inflammatory disease. Immunol Today 12:57–60
19. Zamvil SS, Steinman L (1990) The T lymphocyte in experimental allergic encephalomyelitis. Annu Rev Immunol 8:579–621
20. Sedgwick J, Brostoff S, Mason D (1987) Experimental allergic encephalomyelitis in the absence of a classical delayed-type hypersensitivity reaction. Severe paralytic disease correlates with the presence of interleukin 2 receptor-positive cells infiltrating the central nervous system. J Exp Med 165:1058–1075

21. MacPhee IAM, Mason DW (1990) Studies on the refractoriness to reinduction of experimental allergic encephalomyelitis in Lewis rats that have recovered from one episode of the disease. J Neuroimmunol 27:9–19
22. Day MJ, Tse A, Mason DW (1991) The refractory phase of experimental allergic encephalomyelitis in the Lewis rat is antigen specific in its induction but not in its effect. J Neuroimmunol 34:197–203
23. Paterson PY, Harwin SM (1963) Suppression of allergic encephalomyelitis in rats by means of antibrain serum. J Exp Med 117:755–774
24. Welch AM, Swanborg RH (1976) Characterisation of suppressor cells involved in regulation of experimental allergic encephalomyelitis. Eur J Immunol 6:910–912
25. Ellerman KE, Powers JM, Brostoff SW (1988) A suppressor T-lymphocyte cell line for autoimmune encephalomyelitis. Nature 331:265–267
26. Welch AM, Swierkosz JE, Swanborg RH (1978) Regulation of self-tolerance in experimental allergic encephalomyelitis. I. Differences between lymph node and spleen suppressor cells. J Immunol 121:1701–1705
27. Fontana A, Kristensen F, Dubs R, Gemsa D, Weber E (1982) Production of Prostaglandin E and an interleukin-1 factor by cultured astrocytes and C6 glioma cells. J Immunol 129:2413–2419
28. Lider O, Reshef T, Beraud E, Ben-Nun A, Cohen IR (1988) Anti-idiotypic network induced by T cell vaccination against experimental allergic encephalomyelitis. Science 239:181–183
29. Billiau A, Heremans H, Vanderkerckhove F, Dijkmans R, Sobis H, Meulepas E et al. (1988) Enhancement of experimental allergic encephalomyelitis in mice by antibodies against IFN-γ. J Immunol 140:1506–1510
30. Kennedy MK, Torrance DS, Picha KS, Mohler KM (1992) Analysis of cytokine mRNA expression in the central nervous system of mice with experimental autoimmune encephalomyelitis reveals that IL-10 mRNA expression correlates with recovery. J Immunol 149:2496–2505
31. Issazadeh S, Ljungdahl A, Hojeberg B, Mustafa M, Olsson T (1995) Cytokine production in the central nervous system of Lewis rats with experimental autoimmune encephalomyelitis: dynamics of mRNA expression for interleukin-10, interleukin-12, cytolysin, tumor necrosis factor alpha and tumor necrosis factor beta. J Neuroimmunol 61:205–212
32. Santambrogio L, Hochwald GM, Saxena B, Leu CH, Martz JE, Carlino JA et al. (1993) Studies on the mechanisms by which transforming growth factor-beta (TGF-beta) protects against allergic encephalomyelitis. Antagonism between TGF-beta and tumor necrosis factor. J Immunol 151:1116–1127
33. Levine S, Sowinski R, Steinetz B (1980) Effects of experimental allergic encephalomyelitis on thymus and adrenal: relation to remission and relapse. Proc Soc Exp Biol Med 165:218–224
34. MacPhee IAM, Antoni FA, Mason DW (1989) Spontaneous recovery of rats from experimental allergic encephalomyelitis is dependent on regulation of the immune system by endogenous adrenal corticosteroids. J Exp Med 169:431–445
35. Daynes RA, Araneo BA (1989) Contrasting effects of glucocorticoids on the capacity of T cells to produce the growth factors interleukin 2 and interleukin 4. Eur J Immunol 19:2319–2325
36. Mason DW, MacPhee I, Antoni F (1990) The role of the neuroendocrine system in determining genetic susceptibility to experimental allergic encephalomyelitis in the rat. Immunology 70:1–5
37. Ramírez F, Fowell D, Puklavec M, Simmonds S, Mason D (1996) Glucocorticoids promote a Th2 cytokine response by $CD4^+$ T cells in vitro. J Immunol 156:2406–2412
38. Mustafa M, Vingsbo C, Olsson T, Issazadeh S, Ljungdahl A, Holmdahl R (1994) Protective influences on experimental allergic encephalomyelitis by MHC class I and class II alleles. J Immunol 153:3337–3344

39. Ramírez F, McKnight AJ, Silva A, Mason D (1992) Glucocorticoids induce the expression of CD8α chains on concanavalin A-activated rat CD4$^+$ T cells: induction is inhibited by interleukin-4. J Exp Med 176:1551–1559
40. Blue M-L, Daley JF, Levine H, Craig KA, Schlossman SF (1986) Biosynthesis and surface expression of T8 by peripheral blood T4$^+$ cells in vitro. J Immunol 137:1202–1207
41. Bevan DJ, Chisholm PM (1986) Co-expression of CD4 and CD8 molecules and de novo expression of MHC class II antigens on activated rat T cells. Immunology 59:621–625
42. Spickett GP, Mason DW (1983) Demonstration of the stability of the membrane phenotype of T helper cells after priming and boosting with a hapten-carrier conjugate. Eur J Immunol 13:785–788
43. Godden U, Herbert J, Stewart RD, Roser B (1985) A novel cell type carrying both T_h and $T_{c/s}$ markers in the blood of cyclosporine-treated, allografted rats. Transplantation 39:624–628
44. Christensen ND, Kreider JW (1988) Helper and suppressor functions of panned populations of immune peritoneal T cells with reactivity to the rat mammary adenocarcinoma 13762A. Cell Immunol 112:200–213
45. Tamauchi H, Sasahara T, Habu S (1993) CD4$^+$CD8$^+$ cells lacking self-Mls reactive T cells are induced in mesenteric lymph nodes of *Salmonella enteriditis*-infected mice. Immunol Lett 37:123–130
46. Lapadula G, Covelli M, Numo R, Tricarico G, Amendoni G, Berlingerio C (1984) Monoclonal antibody investigation in rheumatoid arthritis: presence of a T cell subpopulation bearing a double marker. Clin Rheumatol 3:137–144
47. Burdick JF, Beschorner WE, Smith WJ, McGraw D, Bender WL, Williams GM et al. (1984) Characteristics of early routine renal allograft biopsies. Transplantation 38:679–684
48. Pette M, Fujita K, Kitze B, Whitaker JN, Albert E, Kappos L et al. (1990) Myelin basic protein-specific T lymphocyte lines from MS patients and healthy individuals. Neurology 40:1770–1776
49. Ottenhoff THM, Elfenrik DG, Klatser PR, de Vries RRP (1986) Cloned suppressor T cells from a lepromatous leprosy patient suppress *Mycobacterium leprae* reactive helper T cells. Nature 322:462–464
50. Degwert J, Bettman R, Heuer J, Kolsch E (1987) Isolation of a bovine serum albumine specific T-suppressor cell clone and evaluation of its in vitro functions. Immunology 60:345–352
51. Nanda NK, Thomson E, Mason I (1991) Murine suppressor T cell clones specific for minor histocompatibility antigens express CD4, CD8 and $\alpha\beta$ T cell receptor molecules. Int Immunol 2:1063–1071
52. Shenker BJ, Vitale L, King C (1995) Induction of human T cells that coexpress CD4 and CD8 by an immunomodulatory protein produced by *Actinobacillus actinomycetemcomitans*. Cell Immunol 164:36–46
53. Gillis S, Crabtree GR, Smith KA (1979) Glucocorticoid-induced inhibition of T cell growth factor production. I. The effect on mitogen-induced lymphocyte proliferation. J Immunol 123:1624–1631
54. Snyder DS, Unanue ER (1982) Corticosteroids inhibit murine macrophage Ia expression and interleukin 1 production. J Immunol 129:1803–1805
55. Arya SK, Wong-Staal F, Gallo RC (1984) Dexamethasone-mediated inhibition of human T cell growth factor and γ-interferon messenger RNA. J Immunol 133:273–276
56. Culpepper JA, Lee F (1985) Regulation of IL 3 expression by glucocorticoids in cloned murine T lymphocytes. J Immunol 135:3191–3197
57. Beutler B, Krochin N, Milsark IW, Luedke C, Cerami A (1986) Control of cachectin (tumor necrosis factor) synthesis: mechanisms of endotoxin resistance. Science 232:977–980

58. Waage A, Slupphaug G, Shalaby R (1990) Glucocorticoids inhibit the production of IL6 from monocytes, endothelial cells and fibroblasts. Eur J Immunol 20:2439–2443
59. Flower RJ, Blackwell GJ (1979) Anti-inflammatory steroids induce biosynthesis of a phospholipase A2 inhibitor which prevents prostaglandin generation. Nature 278:456–459
60. Goulding NJ, Guyre PM (1993) Glucocorticoids, lipocortins and the immune response. Curr Opin Immunol 5:108–113
61. Daynes RA, Araneo BA, Dowell TA, Huang K, Dudley D (1990) Regulation of murine lymphokine production in vivo. III. The lymphoid tissue microenviroment exerts regulatory influences over T helper cell function. J Exp Med 171:979–996
62. Ayanlar-Batuman O, Ferrero AP, Díaz A, Jiménez SA (1991) Regulation of transforming growth factor-$\beta 1$ gene expression by glucocorticoids in normal human T lymphocytes. J Clin Invest 88:1574–1580
63. Byron KA, Varigos G, Wootton A (1992) Hydrocortisone inhibition of human interleukin-4. Immunology 77:624–626
64. Wu CY, Fargeas C, Nakajima T, Delespesse G (1991) Glucocorticoids suppress the production of interleukin 4 by human lymphocytes. Eur J Immunol 21:2645–2647
65. Swain SL, McKenzie DT, Weinberg AD, Hancock W (1988) Characterization of T helper 1 and 2 cell subsets in normal mice. Helper T cells responsible for IL-4 and IL-5 production are present as precursors that require priming before they develop into lymphokine-secreting cells. J Immunol 141:3445–3455
66. Scott P (1993) IL-12: initiation cytokine for cell-mediated immunity. Science 260:496–497
67. Swain SL, Weinberg AD, English M, Huston G (1990) IL-4 directs the development of Th2-like helper effectors. J Immunol 145:3796–3806
68. Peleman R, Wu J, Fargeas C, Delespesse G (1989) Recombinant interleukin 4 suppresses the production of interferon γ by human mononuclear cells. J Exp Med 170:1751–1756
69. Hart PH, Vitti GF, Burgess DR, Whitty GA, Piccoli DS, Hamilton JA (1989) Potential anti-inflammatory effects of interleukin 4: suppression of human monocyte tumor necrosis factor alpha, interleukin 1, and prostaglandin E2. PNAS USA 86:3803–3807
70. Sheridan JF, Feng N, Bonneau RH, Allen CM, Huneycutt BS, Glaser R (1991) Restraint stress differentially affects anti-viral cellular and humoral responses in mice. J Neuroimmunol 31:245–255
71. Moynihan JA, Karp JD, Cohen N, Cocke R (1994) Alterations in interleukin-4 and antibody production following pheromone exposure: role of glucocorticoids. J Neuroimmunol 54:51–58
72. Stone AA, Bovbjerg DH (1994) Stress and humoral immunity: a review of the human studies. Adv Neuroimmunol 4:49–56
73. Sedgwick JD, MacPhee IAM, Puklavec M (1989) Isolation of encephalitogenic CD4^{+}T cell clones in the rat. Cloning methodology and interferon-γ secretion. J Immunol Methods 121:185–196
74. Chen Y, Kuchroo VK, Inobe J, Hafler DA, Weiner HL (1994) Regulatory T cell clones induced by oral tolerance: suppression of autoimmune encephalomyelitis. Science 265:1237–1240
75. Kuchroo VK, Das MP, Brown JA, Ranger AM, Zamvil SS, Sobel RA et al. (1995) B7-1 and B7-2 costimulatory molecules activate differentially the Th1/Th2 developmental pathways: application to autoimmune disease therapy. Cell 80:707–718

Chapter 8

Glucocorticoid regulation of *Nramp1* in host resistance to mycobacteria

David H. Brown and Bruce S. Zwilling

Introduction

The ability of glucocorticoid hormones to effectively modulate an immune response has been widely studied. Glucocorticoid production and release from the adrenal cortex is stimulated primarily by adrenocorticotrophin (ACTH), which in turn, is controlled by Corticotrophin-releasing hormone (CRH) derived from the hypothalamus [1]. Circadian rhythms or "episodic" increases and decreases occur during each day. Stress overrides feedback regulation of glucocorticoid levels resulting in elevated levels of glucocorticoids. Alterations in corticosterone levels results in either an enhancement or suppression of defence mechanisms. Glucocorticoids, which freely penetrate the cell, bind to their cytoplasmic receptor [2]. The hormone-receptor complex translocates to the nucleus and binds to regulatory elements associated with certain genes (glucocorticoid response elements), which can activate or inhibit transcription of those genes. Glucocorticoids probably have primary and secondary cell targets. Primary targets are affected directly by glucocorticoids through the binding of the hormone to its receptor, whereas secondary target cells are affected by mediators (cytokines) produced by primary target cells which are regulated, themselves, by glucocorticoids [3]. Mononuclear cells represent one of the best-studied primary target cells of glucocorticoids [3–5]. Mononuclear cells possess high-affinity receptors (type II) for glucocorticoids [3,6]. Many elements of the cellular immune response are altered by glucocorticoids [3,7]. Antigen processing and presentation by macrophages as well as major histocompatibility complex (MHC) class II expression is inhibited by glucocorticoids [8]. The pro-inflammatory cytokines IL-1, IL-6 and TNF-α produced by the activated macrophages, are blocked at the transcriptional and post-transcriptional level [9,10,11] by glucocorticoids. The down-regulation of these potent mediators of inflammation, as a result of stress, underlies the immunosuppressive and anti-inflammatory actions of the glucocorticoids. Glucocorticoids also inhibit interleukin-2 (IL-2) gene expression and the expression of the interleukin-2 (IL-2) receptor by lymphocytes at the transcriptional level [10,11]. Glucocorticoids exert anti-inflammatory effects by increasing synthesis of proteins which inhibit phospholipase A_2 activity which in turn leads to a further inhibition of the arachidonic acid cascade and platelet activating factor (PAF) synthesis [12,13]. Glucocorticoids have also been shown to alter lymphocyte trafficking and inhibit T-cell binding to endothelial cells by modulating expression of adhesion molecules, e.g. intracellular adhesion molecule-1 (ICAM-1), on these cells [14]. The effect of glucocorticoids is not always inhibitory. Treatment of macrophages with corticosterone results in an increase in Fc receptor expression [15] and an increase in the production of migration inhibitory factor [16].

Immunomodulation of microbial immunity

One of the first observations that stressful life events affected pathogenesis of disease was reported by Ishigami, who studied the opsonization of tubercle bacilli among chronically ill tuberculous school children and their teachers [17]. Ishigami found decreased phagocytic cell activity during periods of emotional distress and postulated that the stressful school environment led to an increased susceptibility to tuberculosis. Until recently, little evidence was available that suggested that stress affected the pathogenesis of disease.

Early studies of the in vivo effects of stress on microbial pathogenesis focused primarily on the modulatory effect of glucocorticoids. Many of these early studies were limited by their inability to demonstrate that hypothalamic-pituitary-adrenal (HPA) axis activation was responsible for modulation of certain infectious diseases. Exogenously injected adrenal cortical extracts served as the immunomodulator. These studies, conducted primarily on rodents, led to the conclusion that the injection of exogenous glucocorticoids resulted in an increased susceptibility to the particular pathogen being studied [18]. The availability of synthetic analogues of glucocorticoids, as well as glucocorticoid agonists and antagonists, has led to the identification of the components of the immune system that are affected by products of HPA axis activation.

While exogenous glucocorticoids served as the immunomodulator in many early experiments, various in vivo paradigms were used in order to activate the HPA axis. Friedman et. al. [19] showed that crowding stress altered the resistance of mice to malarial infection. Crowding stress, in another study, resulted in an increase in the susceptibility of mice to infection with *Salmonella typhimurium* [20]. Several researchers showed that cold stress and hypoxia (oxygen deprivation) resulted in the inhibition of clearance of several species of bacteria from the lungs of mice including *Staphylococcus albus* and *Proteus mirabilis* [18]. Gross et al. showed that social stress increased the susceptibility of chickens to aerosol challenge with *Escherichia coli* and to challenge with *Mycobacterium avium* [21,22].

Mycobacterial disease

Prevalence and pathogenesis of tuberculosis

Between 1985 and 1992, active cases of tuberculosis in the United States increased by nearly 20% [23]. This domestic resurgence was due in part to infection of individuals already infected with the Human Immunodeficiency Virus. The immigration of persons from other geographic areas where tuberculosis disease is prevalent, as well other demographic and socioeconomic co-factors also contributed to the increased incidence of disease. Not surprisingly, tuberculosis remains a global health problem. It is estimated that 1.8 billion persons are infected with *Mycobacterium tuberculosis*, that eight million people have active disease, and that seven to nine million new cases develop annually [24]. The World Health Organization has predicted that by the year 2005 four million people will die annually as a result of infection with tuberculosis. It should be noted that recent reports by the U.S. Centers for Disease and Control have determined that of all newly diagnosed cases of tuberculosis, 54% are classified as antibiotic resistant. As the incidence of multidrug-resistant tuberculosis increases, the disease will need to be managed in a way that limits its dissemination. This can only by achieved by preventing reactivation.

The pathogenesis of pulmonary tuberculosis is dependent upon the establishment of a primary lesion upon the inhalation of a droplet nucleus generated by a patient with active tuberculosis [25,26]. Once inhaled, less than 10% of *M. tuberculosis* organisms reach the respiratory bronchioles and alveoli; most will settle in the upper respiratory epithelium, where they are likely to be expelled by the mucociliary escalator [27]. Bacteria that reach the deep lung are phagocytosed by alveolar macrophages. This event alone can result in the killing of the bacilli. In other cases the tubercle bacilli survive to initiate an infection. Over a period of 2 to 3 weeks, the bacilli replicate intracellularly, eventually resulting in the lysis of their host microphage [25,28]. The released mycobacteria are then ingested by newly arrived macrophages. As the cycle is repeated, a primary lesion forms and the bacilli are transported into the draining lymph nodes and eventually into the bloodstream. The asymptomatic bacteraemia coincides with the appearance of enhanced reactivity to the purified protein derivatives of *Mycobacterium tuberculosis* and the onset of cell-mediated immunity [25,29]. As the tubercle-laden macrophages disperse the bacilli into the body, the bacilli are effectively trapped in the tissues and create metastatic foci. These granulomatous focal lesions are composed of macrophage-derived epithelioid giant cells and lymphocytes. T lymphocytes activated by macrophage-derived cytokines migrate to the foci resulting in the further local activation of macrophages and lead to caseous necrosis which results in the gradual sterilization of most metastatic lesions. Although not understood, the sterilization process in the apical-subapical regions of the lung as well as the bone marrow, kidney and meninges does not occur. The caseous necrosis decreases the amount of bacilli to a low steady state level. In most healthy individuals with intact immune function, organisms contained within these foci can remain dormant for decades. The relative strength of the host's cell-mediated immunity ultimately determines if reactivation of disease will occur. In 5% to 10% of infected individuals, a temporary suppression of the cellular arm of the immune system allows reactivation of the disease and resumption in bacterial multiplication. As the immune response re-exerts control, tissue destruction and systemic dissemination of disease occurs.

Innate immunity to mycobacterial infection

Resistance or susceptibility of a host to mycobacterial infection is partly under genetic control [30]. Studies of monozygotic twins found that they were equally susceptible to infection with *M. tuberculosis* [31]. The co-segregation of known chromosomal markers associated with the disease susceptibility as well as epidemiological studies that found that susceptibility to infection corresponded to racial differences further substantiated these findings [32,33]. Animal models of mycobacterial infections also reinforced the concept that the genetic makeup of the host was an important factor in determining resistance or susceptibility to infection. The first studies in animals that demonstrated differential responses to *M. tuberculosis* were carried out in inbred rabbit strains [34]. The genetically resistant rabbits were capable of sequestering tubercle bacilli in the macrophages thus inactivating the pathogens. The results of these early studies served to delineate the two phases of the host response to mycobacterial infections. Inbred strains of mice were segregated into two non-overlapping groups upon intravenous infection with 10^4 CFU of *M. bovis*. Typically, infection of inbred mouse strains with low doses of non-pathogenic *M. Bovis* (BCG) has two distinct phases: an early non-immune phase (0 to 3 weeks) characterized by either rapid proliferation of the bacteria in reticuloendothelial organs (liver, spleen) of susceptible strains (*Bcg*^{s}) or absence of bacterial growth in resistant strains (*Bcg*^{r}), and a late phase (3 to 6 weeks) associated with the development of specific immunity [35,36]. Similar strain variations in innate resistance to other mycobacterial species were also identified by Gros et al. [35]. No intermediate variations were observed in the segregation of the two groups suggesting that resistance to mycobacterial infection was under the control of a single gene. Extensive Mendelian analyses of crosses between resistant and susceptible strains revealed that the trait of innate resistance to *M. bovis* infections was controlled by a single autosomal dominant gene designated as *Bcg* [37,38]. The strain distribution pattern of the *Bcg*^{r} and the *Bcg*^{s} alleles segregated with two isoenzymes, the isocitrate dehydrogenase (*Idh1*) and the dipeptidase 3 (*Pep3*) located on the centromeric portion of murine chromosome 1. Recombination frequencies established the gene order as *Idh1*, *Bcg*, *Pep3* [35,39]. The creation of congenic strains of mice that differ only in the defined portion of chromosome 1 that carries the *Bcg* gene confirmed these results [40] and served to further map and determine the gene order of five cloned genes in the vicinity of the *Bcg* gene [41,42]. The *Bcg* gene was also found to control resistance to *Mycobacterium lepraemurium*, *M. intracellulare* and *M. smegmatis* [30,43,44].

The cellular source expressing the *Bcg/Ity/Lsh* gene during the initial phase of infection was identified as mature tissue macrophages [36,45]. The resistant or susceptible phenotype was expressed in vivo in the absence of functional T lymphocyte population. Depletion of T-cells, B-cells and NK-cells from resistant mice did not alter their ability to control mycobacterial infections indicating that innate resistance is not associated with acquired immunity [36]. The resistance mechanism was thought to be due to the superior antimicrobial ability of macrophages from *Bcg*^{r} mice. Skamene [46] suggested that the superior effector function of macrophages from *Bcg*^{r} mice was due to the fact that these macrophages are at a more advanced level of activation compared with macrophages from *Bcg*^{s} mice. Accordingly, the likely candidate for the *Bcg* gene product would be a protein involved in the regulation of macrophage priming for activation. The expression of this gene was found to be greatly enriched in mature tissue macrophages isolated from spleen and the macrophage cell line J774A.1 [47]. The gene was designated as *Nramp1* (*N*atural *r*esistance *a*ssociated *m*acrophage *p*rotein) and is part of a family of genes with similar sequence homologies [48]. Nucleotide and predicted amino acid sequence analysis revealed that *Nramp1* encodes a novel protein with features associated with integral membrane proteins. Hydropathy plot analysis identified a minimum of ten and a maximum of 12 putative membrane spanning domains. *Nramp1* shares an ancestral relationship with several prokaryotic periplasmic transport proteins through the presence of a consensus transport motif known as the binding protein-dependent inner membrane component transport signature. This motif was also detected in a few eukaryotic membrane proteins, including Crna, a nitrate/nitrite transporter of *Aspergillus nidulans* [47]. Additionally, based upon homology with the gene *SMF1*, which codes for a hydrophobic protein whose deletion renders yeast cells sensitive to low manganese concentration, it was recently proposed that the function of *Nramp1* could be that of a manganese transporter that transports Mn^{2+} from the extracellular milieu into the cytoplasm of a macrophage [49].

The analysis of nucleotide sequence variations within the coding portion of *Nramp1* of 27 inbred mouse strains revealed a non-conservative single glycine to aspartic acid substitution at position 169 within transmembrane region 4 as being associated with susceptibility to mycobacterial infection [47,50]. Vidal et al. [47], using a mutant mouse carrying a null allele at the *Nramp1* locus (*Nramp1*$^-$) on the *Bcg*^{r} genetic background, showed that *Nramp1* and *Bcg* were the same gene. Recent studies indicate that *Nramp1* protein is not produced in macrophages from *Bcg*^{s} mice [51].

The role of the macrophage in mycobacterial infection

The role of the macrophage in controlling mycobacterial infection has been widely studied. For the most part, the killing of mycobacteria occurs within the phagolysosomes of the macrophage. Contained within these structures are toxic constituents including lysosomal hydrolyses, reactive oxygen intermediates, hydrogen peroxide, superoxides, and reactive nitrogen intermediates (RNI). RNI production by murine macrophages is an important effector mechanism that controls the growth of a variety of intracellular pathogens. Many of the biostatic and cytostatic activities of murine macrophages are thought to be dependent upon the formation of RNI. Nitric oxide (NO) is derived from the enzymatic oxidation of the guanidino group of L-arginine controlled by an inducible nitric oxide synthase (iNOS) [52,53,54]. Many of the biological effects of NO are chemically based on direct interactions with iron-containing proteins, such as guanylyl cyclase (haem iron), ribonucleotide reductase (non-haem iron) or aconitase (iron sulfur) [55,56]. Cytokines are powerful modulators of murine macrophage RNI synthesis. rIFNγ and TNFα are potent activators of iNOS while IL-4 and IL-10 serve as effective suppressors of iNOS [57,58]. The production of RNI by murine macrophages has been shown to correlate with the capacity of macrophages to inhibit mycobacterial growth [59,60,61,62]. IFN-γ knock-out mice are unable to restrict the growth of the *M. tuberculosis* organisms. These mice develop granulomas but do not produce RNI. Treatment of mice infected with BCG or *M. tuberculosis* with a neutralizing antibody directed against TNF-α was originally shown to block granuloma formation and enhance bacterial growth [61,63]. Additional studies demonstrated that transgenic mice lacking the 55-kDa TNF-α receptor were unable to clear an *M. tuberculosis* infection. The mice were able to produce granulomas, but in a delayed manner. The granulomas lacked epithelioid cells and contained unusually large numbers of *M. tuberculosis* [64]. Nonetheless, the role of RNI in resistance to mycobacterial infection remains unclear. Other cytokines in addition to rIFN-γ and TNF-α that have been determined to activate murine macrophages to kill mycobacteria include IL-12 and granulocyte macrophage-colony stimulating factor (GM-CSF) both of which fail to induce the production of RNI [57,65,66,67].

Glucocorticoids and mycobacterial resistance

The effect of HPA axis activation on mycobacterial growth

Macrophages from *Bcg*^{r} and *Bcg*^{s} mice express MHC class II glycoproteins differently. Macrophages from *Bcg*^{r} mice can be induced to persistently express I-A while macrophages from *Bcg*^{s} mice will only transiently express I-A [68,69]. This observation and subsequent studies demonstrated that activation of the HPA axis suppressed I-A expression by macrophages from *Bcg*^{s} mice while not affecting I-A expression by macrophages from *Bcg*^{r} mice. This suggested to us that mycobacterial growth, under *Bcg* control, may also be differentially affected by HPA axis activation. These observations became the basis for the studies to determine the role of the HPA axis in regulating the growth of *Mycobacterium avium* in congenic *Bcg*^{r} and *Bcg*^{s} mice. We found that activation of the HPA axis increased the susceptibility of *Bcg*^{s} mice to mycobacterial growth but did not affect the ability of *Bcg*^{r} mice to limit the growth of mycobacteria (Fig. 8.1) [70]. HPA axis activation also resulted in an increased permissiveness of macrophages from *Bcg*^{s} mice to intracellular growth of *M. avium* in vitro while the anti-mycobacterial activity of macrophages from *Bcg*^{r} mice remained unaffected [Fig. 8.2]. The failure of HPA axis activation to increase the susceptibility of *Bcg*^{r} mice was not due to unresponsiveness of the strain to HPA axis activation, equal

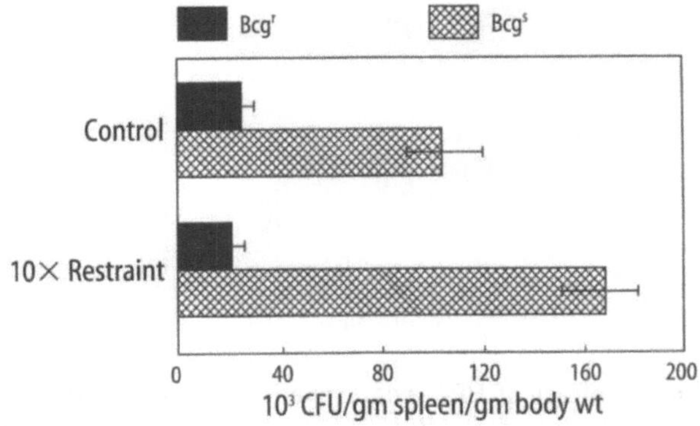

Fig. 8.1. Differential effect of HPA axis activation on mycobacterial resistance of BALB/c.*Bcg*^{r} and BALB/c.*Bcg*^{s} mice. Mice were restrained for 10, 18-h cycles and the CFU of *M. avium* in the spleen determined 12 days after the infection. The number of CFU isolated from the spleens of *Bcg*^{r} and *Bcg*^{s} mice prior to restraint was 19 462 CFU/g spleen per g body weight and 18 083 CFU/g spleen per g body weight, respectively. The data represent the mean ± SD of seven animals per group. The difference between the restraint and control groups for *Bcg*^{s} mice was significant at $P < 0.001$. The differences between the growth of the mycobacteria in the spleens of *Bcg*^{r} and *Bcg*^{s} mice was also significant at $P < 0.001$. Similar observations were made concerning differences in the effect of HPA axis activation for growth of the *Mycobacteria* in the lungs (data not shown). [From Infect Immun (1993) 61: 4793–4800.]

levels of plasma corticosterone and ACTH were produced by the two strains of mice. We also found that HPA axis activation suppressed the induction of TNF-α and NO in macrophages from *Bcg*^{r} and *Bcg*^{s} mice following stimulation with rIFN-γ and lipopolysaccharide (LPS) [70]. This observation was the same as those reported by others who have shown that the simultaneous treatment of macrophages with corticosteroids together with rIFN-γ and LPS suppressed both TNF-α production as well as the production of RNI. Macrophages from both strains of mice produced equal amounts of the cytokine and of reactive nitrogen intermediates in the absence of corticosterone. There have been conflicting reports as to whether or not there are differences in production of these effector molecules between the two strains. Our results could be interpreted as indicating that the production of TNF-α and NO may be independent of *Bcg* control.

The results described above suggested that HPA axis activation could account for the increased susceptibility of individuals to mycobacterial growth. The use of restraint stress served as a tool to produce physiological levels of corticosterone. The role of corticosterone in mediating the effects of HPA axis activation was demonstrated in three ways. First, adrenalectomy abrogated the effect of HPA activation. Second, implantation of time release pellets into adrenalectomized mice, that released levels of corticosterone attained during HPA axis activation, resulted in an increased susceptibility of the *Bcg*^{s} mice to mycobacterial growth. Finally, treatment of the *Bcg*^{s} mice with the glucocorticoid receptor antagonist RU486 abrogated the effects of HPA axis activation [70].

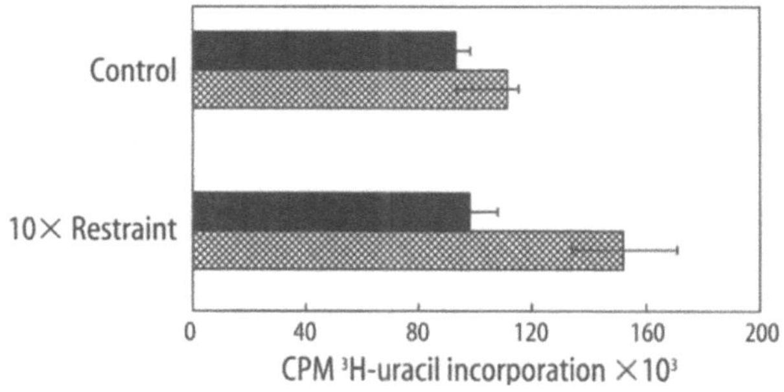

Fig. 8.2. Differential effect of HPA axis activation of anti-mycobacterial activity of macrophages from BALB/c.*Bcg*^{r} and BALB/c.*Bcg*^{s} mice. Splenic macrophages were isolated from mice and infected with *M. avium*. The growth of the *Mycobacteria* was determined by pulsing cultures with ^{3}H-uracil after lysis of the macrophages with saponin. Infected cultures were pulsed immediately following a period of phagocytosis in order to determine the numbers of micro-organisms taken up initially. Replicate cultures were pulsed after 5 days of growth within macrophages. The cpm of ^{3}H-uracil taken up by the *M. avium* following release from the *Bcg*^{s} macrophages immediately after phagocytosis was 574 cpm, while that taken up by *M. avium* following release from *Bcg*^{r} macrophages was 523 cpm and did not differ as a result of HPA axis activation. The *M. avium* released from macrophages immediately after phagocytosis incorporated 45 334 cpm ^{3}H-uracil when pulsed after 5 days of growth and did not differ between those released from macrophages from *Bcg*^{r} and *Bcg*^{s} mice. The data represent the cpm ^{3}H-uracil taken up by the bacteria released from macrophages after 5 days of in vitro culture. The data are from a representative experiment. The effect of HPA axis activation on mycobacterial growth is significant at $P < 0.001$. Solid bars, *Bcg*^{r}; hatched bars, *Bcg*^{s}. [From Infect Immun (1993) 61:4793–4800.]

HPA axis activation and reactivation of *Mycobacterium* tuberculosis

The role of HPA axis activation in the control of the growth of tuberculosis in humans has been the subject of considerable debate [16,25,71]. Several reports have shown that injections of glucocorticoids suppress the antimicrobial activity of macrophages and exacerbate the growth of mycobacteria [72,73,74]. Roach et al. [75] reported that dexamethasone increased the susceptibility of monocytes from some human donors to mycobacterial growth but not monocytes from other donors. These observations were analogous to the observations we made in mice that increased corticosterone levels, which occur as a result of HPA axis activation, increase the susceptibility of *Bcg*^{s} mice, but not *Bcg*^{r} mice, to mycobacterial growth [8,69]. North and Izzo [76] also found that weekly injections of SCID mice with hydrocorticosone acetate increased the susceptibility of the mice to the growth of *Mycobacterium tuberculosis*. In contrast, the resistance of isocongenic immunocompetent mice was unaffected by hydrocortisone, presumably due to the development of specific immunity. Several reports have suggested that the resistance of macrophages to *Listeria*, *Salmonella* or *Toxoplasma* induced by rIFN-γ is also glucocorticoid resistant [73,77]. Thus, it was possible that the insensitivity of the mycobacterial resistance mechanism(s), controlled by *Bcg* (*Nramp1*), to HPA axis activation was due to corticosteroid insensitivity. IFN-γ, may induce anti-mycobacterial mechanisms that are not sensitive to glucocorticoids.

Most studies of the in vivo growth of *M. tuberculosis* in mice utilize relatively large inocula of bacteria, ranging from 10^4 to 10^6 CFU per mouse. These large doses result in progressive growth of the bacilli. In contrast, studies utilizing Bcg^r and Bcg^s mice have clearly demonstrated that *Bcg* (*Nramp1*) can control the growth of several mycobacterial species only when small inoculum doses of bacteria are utilized. Resistance to mycobacteria, as mentioned previously, is defined as the ability to control growth of the bacteria in the spleen following intravenous injection of 10^4 CFU of *M. bovis* BCG. While some reports have indicated that the effect of the resistance gene is limited to spleen [78], Skamene et al. [39] and Goto et al. [44] have shown differences in the growth of mycobacteria in the lungs of Bcg^r and Bcg^s mice as well.

During the course of studies to determine if *Bcg* is involved in the control of early *M. tuberculosis* infection, we found that intravenous infection with small doses (10^2 CFU) of the micro-organisms resulted in the establishment of a latent infection [79]. Growth of the organism was initially detected in the spleens and lungs of both BALB/c (BCG-susceptible) and DBA/2 (BCG-resistant) mice within 15 days of inoculation. The pattern of growth in the spleens of BALB/c.Bcg^s mice was similar to that observed in the spleens of Bcg^r DBA/2 mice except that greater numbers of bacilli were isolated from the spleens of the BALB/c.Bcg^s mice. Growth in the lungs, however, was markedly different. The increased growth of the bacilli in the lungs of BALB/c.Bcg^s mice was related to increasing inoculum size whereas only the largest inoculum dose produced growth in the lungs of the Bcg^r DBA/2 mice. Infection of mice with *M. tuberculosis* resulted in a peak number of bacilli occurring between days 34 and 43 in the lungs of BALB/c.Bcg^s mice (data not shown). The number of micro-organisms detected in both organs returned to low levels by day 57 of infection.

Since the latent disease was reminiscent of the disease in humans, we determined if temporary suppression of immunity, by activation of the HPA axis, would result in reactivation of *M. tuberculosis* growth. The results in Fig. 8.3 show that activation of the HPA axis resulted in the reinitiation of *M. tuberculosis* growth in both the spleens and lungs of BCG-susceptible and BCG-resistant (data not shown) mice. The effect of HPA axis activation was more pronounced in the lungs than in the spleens of both strains of mice with nearly ten times the number of bacilli isolated from the lungs of the mice following HPA axis activation. Our observation is similar to that reported by Cox et al. [72] who found that treatment of mice with pharmacological doses of corticosteroids resulted in reactivation of *M. bovis* BCG growth. The extent of reactivation of *M. tuberculosis* growth was equal in Bcg^s and Bcg^r DBA/2 mice. We interpret these collective data as indicating that reactivation of *M. tuberculosis* growth in Bcg^r and Bcg^s mice was more likely the result of the suppression of specific T-cell-mediated responses since macrophages from resistant mice are refractory to the suppressive effects of glucocorticoids on their ability to control mycobacterial growth (see below). A study by Daynes et al. [80] found that increased levels of corticosterone resulted in a shift from the Th1 to the Th2 type of CD4 cells, resulting in a decrease in

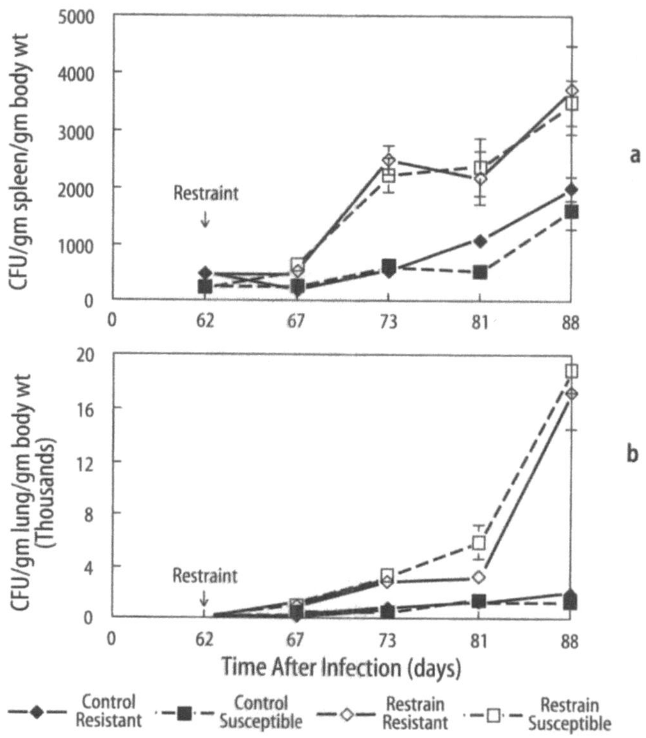

Fig. 8.3. Activation of the HPA axis results in reactivation of *M. tuberculosis* growth in BCG-resistant and -susceptible mice. Mice were infected with 10^2 CFU of *M. tuberculosis*. After 60 days, when a steady-state infection was established, the mice were restrained for 5, 15-h cycles, rested for 2 days, and restrained for 5 additional 15-h cycles. *M. tuberculosis* growth in the spleens (**a**) and lungs (**b**) was monitored by plate counting at the designated times after infection. The data represent the mean ± the standard deviation of five mice for each time point for each treatment. Filled diamonds, control *Bcg*^{r}; filled squares, control *Bcg*^{s}; open diamonds, restraint *Bcg*^{r}; open squares, restraint *Bcg*^{s}; wt, weight. [From Infect Immun (1995) 63: 2243–2247.]

macrophage-activating cytokines. Presumably then, decreases in activating stimuli can also result in the reinitiation of *M. tuberculosis* growth that we observed. Establishment of latent tuberculosis infections provides us with the opportunity to study the immunological events important in the establishment of latency as well as the reactivation of mycobacterial growth in a mouse model. During the course of natural infection in humans, patients present with active disease. Confirming diagnosis of infection in individuals without active disease usually occurs follow skin testing of those who have previously been exposed to others with active disease [81]. Humans have either active or latent tuberculosis disease. It is not feasible to monitor their immunological status during the transition from latent to active disease. A model such as that described above, provides an opportunity to study the role of T-cells and macrophages during the course of disease from the time of initial infection to the establishment of latency and the progression during reactivation to active tuberculosis.

The effect of glucocorticoids on macrophage function

Based upon the observations that HPA axis activation differentially affected resistance to in vivo and in vitro growth of mycobacteria in *Bcg*^{r} and *Bcg*^{s} mice, we extended our studies to evaluate further the effect of HPA activation on anti-mycobacterial mechanisms of rIFN-γ-activated macrophages. The increase in susceptibility to mycobacterial growth, that resulted from HPA axis activation, was determined to be due to the direct effect of corticosterone [70,82]. We also evaluated the effect of corticosterone on macrophage anti-mycobacterial activity. HPA axis activation or treatment of macrophages with corticosterone suppressed the ability of macrophages from susceptible mice to control the growth of *M. avium* [Table 8.1]. In contrast, neither HPA axis activation nor the addition of corticosterone altered the ability of macrophages from BCG-resistant mice to control mycobacterial growth. We found, however, that the induction of TNF-α and of reactive nitrogen intermediates by rIFN-γ and LPS by macrophages from both strains of mice was suppressed by HPA axis activation or the simultaneous addition of corticosterone. The addition of RU486 to macrophages from *Bcg*^{s} mice ameliorated the effect of HPA axis activation on anti-mycobacterial activity. RU486 also partially restored the ability of macrophages to produce TNF-α and NO. Previously we showed that HPA activation suppressed the induction of both TNF-α and NO production by rIFN-γ and LPS stimulated macrophages. This study also showed that HPA axis activation or corticosterone suppressed their induction when the cells were also stimulated by rIFN-γ and *M. avium*. rIFN-γ attenuated the

Table 8.1. Effect of RU486 on corticosterone-mediated suppression of macrophage-mediated anti-mycobacteria activity, TNF-α and NO production in BCG-susceptible mice

Treatment	^{3}H-uracil incorporation[a]	TNF-α (pg/ml)[b]	NO (μM)[b]
Corticosterone	12,321 ± 173	1668 ± 75	18 ± 2
Cort+RU486	10,347 ± 773	2448 ± 114	36 ± 2
Control	9,932 ± 526	2592 ± 52	41 ± 1
Control+RU486	10,105 ± 421	2483 ± 107	40 ± 3

[a]Splenic macrophages from BCG-susceptible mice were infected with 4×10^5 CFU *M. avium* and incubated overnight in the presence of 10^{-8}M corticosterone with or without 62.5 (μg of RU486. After 5 days the macrophages were lysed and the metabolically active bacteria radiolabelled as described. The data represents one of five experiments and is expressed as the mean ± SD of four test wells for each determination. The effect of corticosterone on the growth of *M. avium* was significant by ANOVA ($P \leqslant 0.003$). The effect of RU486 is significant ($P \leqslant 0.003$).

[b]Splenic macrophages were stimulated with 100 U of rIFN-γ and 5×10^5 CFU *M. avium* for 72 h. Corticosterone, with or without RU486, was added to the test wells. The level of TNF-α and NO production in the cell free supernatants was determined by ELISA or by the Griess reagent. Unstimulated macrophages produced 57 ± 6 pg/ml TNF-α and 6 ± 3 μM NO. The data are expressed as the mean ± SD of three separate experiments. The effect of corticosterone and RU486 on TNF-α and NO production is significant by ANOVA ($P \leqslant 0.005$). [From: J Neuroimmunol (1994) 53 : 181–187.]

effect of HPA activation on the induction of both TNF-α and NO. The effect of rIFN-γ was more pronounced on macrophages from BCG-resistant mice.

The differential effects of corticosterone on cytokine-activated macrophages from *Bcg^r^* and *Bcg^s^* mice

When macrophage populations were activated in vitro with rIFN-γ, only the function of macrophages from *Bcg*^{s} mice was suppressed by the addition of corticosterone. In order to determine if the differential effect of glucocorticoids was limited to innate resistance or to resistance induced by rIFN-γ or other macrophage-activating cytokines, we evaluated the anti-mycobacterial activity of macrophages from *Bcg*^{r} and *Bcg*^{s} mice following stimulation of the cells with rIFN-γ or rGM-CSF [83]. In addition, we sought to establish a correlation between differential effects of corticosterone on macrophage anti-microbial activity and expression of the gene, *Nramp1*, by macrophages from both strains of mice. Interestingly, we found that rIFN-γ and rGM-CSF induced similar levels of activation in macrophages from both strains of mice [Fig. 8.4] but that corticosteroids suppressed the ability of cytokine-activated macrophages from *Bcg*^{s} mice to control the growth of mycobacteria, while not affecting the resistance of macrophages from *Bcg*^{r} mice [Fig. 8.5]. The rIFN-γ pathway involves the production of NO and is inhibited by N^G mono-methyl-l arginine (NMMA) while the GM-CSF-induced pathway does not appear to be dependent on NO production and is therefore not affected by NMMA. Both pathways appear to be regulated by *Bcg* because they are differentially affected by corticosterone. The results also appeared to rule out a role for NO as an effector molecule in *Bcg*-mediated resistance for two reasons. First, NMMA did not interfere with innate resistance mechanisms. Second, corticosterone suppressed the anti-mycobacterial activity of macrophages from *Bcg*^{s} mice activated by GM-CSF despite the failure of these macrophages to produce NO [82,83].

Characterization of glucocorticoid effects on *Nramp1* mRNA expression

Nramp1 mRNA expression was up-regulated by both rIFN-γ and GM-CSF [83] [Fig. 8.6]. Corticosterone suppressed *Nramp1* expression only by macrophages from *Bcg*^{s} mice suggesting that the differential effects of corticosterone on macrophages from congenic BCG-resistant and susceptible mice was related to the different effects of the hormone on the gene, *Nramp1*. Corticosterone also suppressed the expression of several rIFN-γ-inducible genes by macrophages from *Bcg*^{s} mice only [84]. Expression of these mRNAs by macrophages from *Bcg*^{r} mice were unaffected by corticosterone. Based upon experiments in which the effects of corticosterone on *Nramp1* expression by macrophages from *Bcg*^{s} mice were abrogated with Actinomycin D and cycloheximide, and upon results of nuclear run-off assays which indicated that the glucocorticoid-mediated alter-

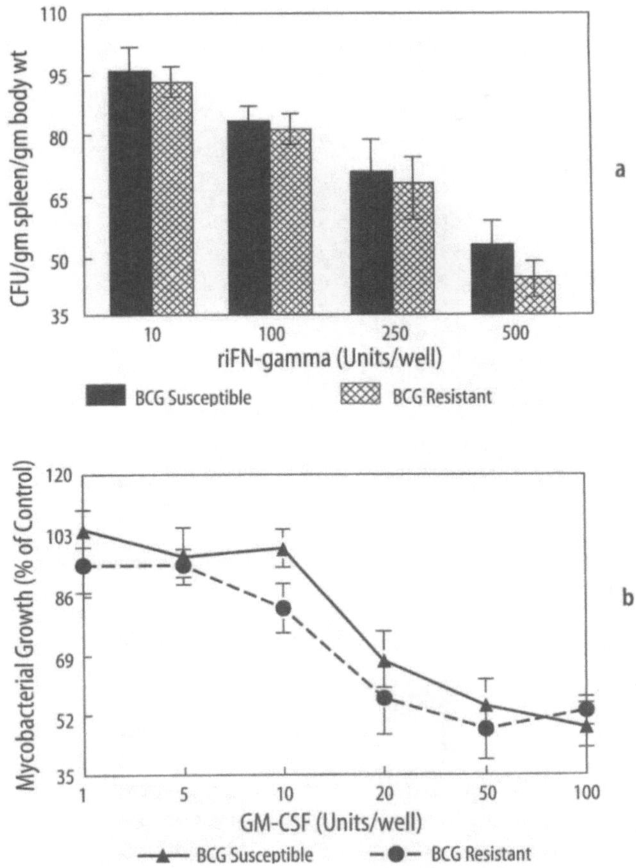

Fig. 8.4. **a** Interferon-gamma induced activation of macrophages from BCG-resistant and susceptible mice. Splenic macrophages were treated with rIFN-γ for 24 h prior to infection with 4×10^5 CFU *M. avium*. After 24 h to allow for phagocytosis and following removal of the unphagocytized bacteria the cultures were incubated in the continued presence of rIFN-γ for 5 days prior to lysis of the macrophages and pulsing of the bacteria with ^{3}H-uracil. The data are expressed as the percentage of control which represents the amount of radioactivity incorporated by the bacteria after infection of cultures of macrophages treated with rIFN-γ divided by that of cultures not treated with rIFN-γ. The effect of rIFN-γ was significant at $P < 0.003$ as determined by ANOVA. There was no statistical difference between macrophages from *Bcg^r* and *Bcg^s* mice. **a** GM-CSF induced activation of macrophages from BCG-resistant and susceptible mice. Macrophages were treated with GM-CSF for 24 h prior to infection as described in **b**. The effect of GM-CSF was significant at $P < 0.02$ by ANOVA. There was no statistical differences between macrophages from *Bcg^r* and *Bcg^s* mice. [From Infect Immun (1995) 63: 2983–2988.]

ation of mRNA was not transcriptionally regulated, we concluded that the loss of *Nramp1* expression was the result of post-transcriptional modifications of *Nramp1* mRNA in *Bcg^s* mice [84].

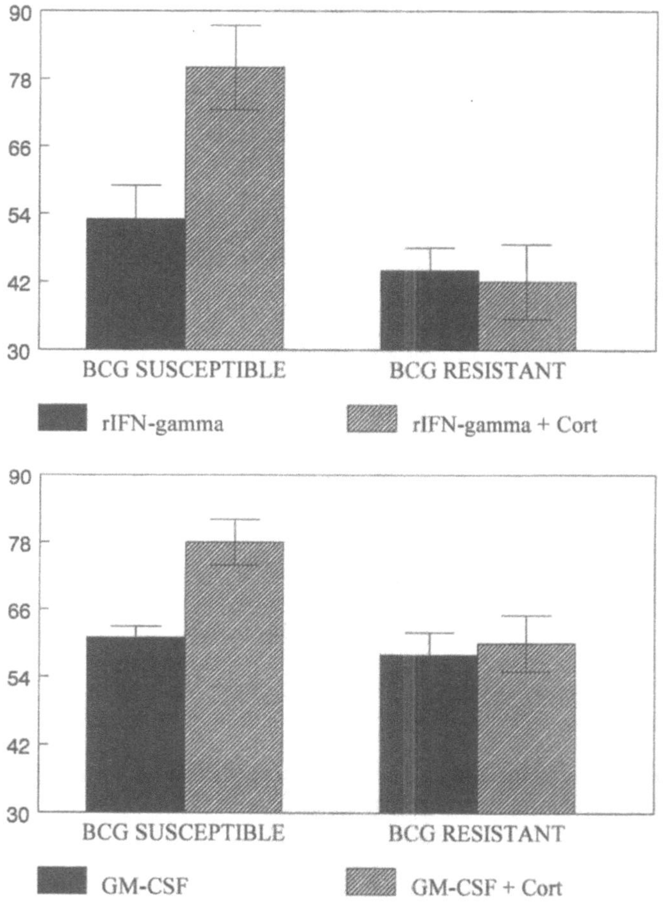

Fig. 8.5. Differential effect of corticosterone on interferon-gamma (**a**) and GM-CSF (**b**) induced anti-mycobacterial activity of macrophages from BCG-resistant and -susceptible mice. Macrophages were treated with 100 U of rIFN-γ or 50 U of GM-CSF and infected as described in the legend for Fig. 8.4. After infection, the cultures were treated with 10^{-6} M corticosterone (Cort) for 5 days prior to the lysis of the macrophages and addition of ^{3}H-uracil. Y-axis represents the percentage of mycobacterial growth compared with untreated (control) macrophages (100%). The effect of corticosterone on the anti-mycobacterial activity of rIFN-γ and GM-CSF activated macrophages from *Bcg*^{s} mice was significant ($P < 0.03$) [From Infect Immun (1995) 63: 2983–2988.]

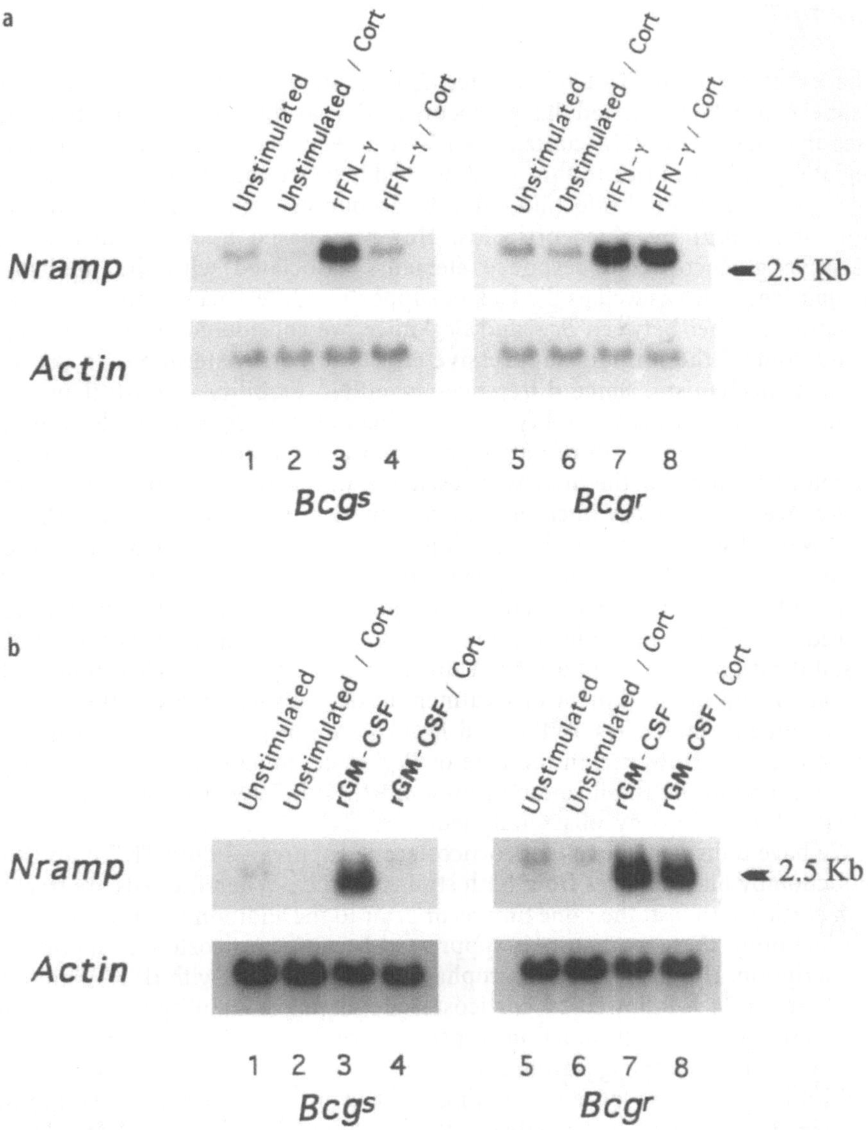

Fig. 8.6. **a** Differential effect of corticosterone on *Nramp* expression by rIFN-γ activated macrophages from Bcg^r and Bcg^s mice. Splenic macrophages (2×10^6) were treated with 2500 U rIFN-γ for 24 h prior to the addition of 10^{-6} M corticosterone (Cort). After 24 h the mRNA was extracted with 8 M guanidine hydrochloride and the expression of *Nramp* mRNA determined by Northern blot analysis. **b** Differential effect of corticosterone on *Nramp* expression by GM-CSF activated macrophages from Bcg^r and Bcg^s mice. Splenic macrophages were treated with 500 U GM-CSF for 24 h prior to the addition of 10^{-6} M corticosterone. The expression of *Nramp* mRNA was determined by Northern blot analysis. [From Infect Immun (1995) 63: 2983–2988.]

Glucocorticoids and mRNA stability: defining the function of *Nramp1*

The exact mechanism of mRNA destabilization is not known. A report by Peppel et al. [85] suggested that glucocorticoids may directly activate ribonucleases that degrade mRNA containing AU-rich sequences in the 3′ untranslated regions of mRNA. Prior to this point, we had believed that differences in resistance to mycobacterial infection could be accounted for by differential regulation of *Nramp1* by glucocorticoids. However, based upon the absence of negative glucocorticoid response elements associated with the published *Nramp1* sequence as well as the lack of supporting differences in glucocorticoid receptor number between *Bcg*^{r} and *Bcg*^{s} mice, we considered whether *Nramp1* could control the immunosuppressive effects of corticosterone by a yet undefined mechanism. Since differences in mRNA stability can affect protein levels, increased mRNA stability in macrophages from *Bcg*^{r} mice could result in the sustained production of anti-mycobacterial effector molecules and could therefore account for the increased resistance of macrophages from *Bcg*^{r} mice. Differences in the decay of several mRNAs in macrophages from *Bcg*^{r} and *Bcg*^{s} mice were observed without the addition of corticosterone [84]. However, accelerated decay of mRNA in macrophages from *Bcg*^{s} mice was induced by treatment of the cells with corticosterone. The glucocorticoid-mediated suppression served to define a possible role for *Nramp1*. Corticosterone induces mRNA destabilization when *Nramp1* is not functional in *Bcg*^{s} mice. When functional *Nramp1* is not induced by prior treatment of the cells with rIFN-γ, the mRNAs of the effector molecules TNF-α and iNOS were not stable in the presence of corticosterone. However, in the case of *Bcg*^{r} mice, functional *Nramp1* expression resulted in the stabilization of these mRNAs thereby leading to prolonged release of these anti-mycobacterial mediators [84].

We have indicated above that corticosterone suppressed both TNF-α and NO induction by macrophages from both strains of mice. When the cells are treated with corticosterone at the same time as or prior to the addition of rIFN-γ then the production of these mediators is suppressed because corticosterone suppresses transcription. However, when macrophages are first treated with rIFN-γ and LPS together for 24 h followed by corticosterone, then the stability of TNF-α and iNOS mRNA and the production of protein are differentially affected, i.e. not suppressed in macrophages from *Bcg*^{r} mice.

With increasing evidence concerning the role of iron in cell-mediated immunity [86–88], we began to evaluate the relationship of iron and mRNA stability in the resistance to mycobacterial growth by macrophages from *Bcg*^{r} and *Bcg*^{s} mice. Our preliminary results suggested that low levels of intracellular iron in the macrophages of *Bcg*^{r} mice resulted in stabilization of mRNA expression. Levels of intracellular iron detected in rIFN-γ stimulated splenic macrophages from *Bcg*^{r} mice were lower when compared with iron levels measured in macrophages from *Bcg*^{s} mice. Additional studies demonstrated that iron chelation with desferroxamine resulted in the abrogation of corticosterone-induced effects on *Nramp1* expression. In contrast, iron loading of macrophages from *Bcg*^{r} mice resulted in unstable *Nramp1* mRNA following corticosterone treatment. From these experiments, we believe that *Nramp1* may function as an iron transport protein that serves to regulate the level of intracellular iron in macrophages.

The role of iron in the microbicidal activity of macrophages is complex [56,89, 90]. Studies have shown that reduction of intracellular iron levels in macrophages can increase the microbicidal activity of those cells, whereas increased iron load can result in decreased microbicidal activity of macrophages. Transferrin remains the key factor in maintaining iron homeostasis in an individual rather than absolute levels of intracellular iron present in host cells [56,89–91]. Iron limitation may result in enhanced immune functions only to the point at which the degree of saturation of transferrin is so low that delivery of critical iron to cells and tissues is diminished. Conversely, iron overload will only affect immune function when transferrin becomes fully saturated and non-transferrin-bound iron is present in the circulation. Excess free iron could affect immune function by the production of damaging levels of oxygen radicals such as superoxide and hydrogen peroxide or by other mRNA destabilizing mechanisms [92]. Interestingly, of all the nutrients required by or available to *M. tuberculosis*, only iron poses serious acquisition problems [93–95]. If iron is suddenly made available to pathogens such as *M. tuberculosis*, the organism would need to acquire, and hold the iron obtained from major iron-containing molecules of the host (including transferrin and ferritin) until it could initiate synthesis of necessary haem groups and iron-containing proteins.

The role of glucocorticoids in mycobacterial infection: a proposed model of resistance

Our model [Fig. 8.7] suggests that BCG-susceptible macrophages, expressing a non-functional *Nramp1* protein, that is associated with high levels of intracellular iron, allow access to iron by the bacilli. The iron is effectively mobilized into the bacterium by the exochelin-mycobactin-mediated mechanisms. Sufficient levels of iron enable growth of the bacteria to occur. In BCG-resistant macrophages, functional *Nramp1* protein lowers levels of intracellular iron and withholds iron from the mycobacteria. The resultant low levels of intracellular iron increase the affinity of iron-regulated proteins for sequence elements in the 3′UTR of mRNAs. The presence of these bound proteins inhibits mRNA degradation by glucocorticoid-induced nucleases. Non-functional *Nramp1* is unable to control intracellular iron levels in the macrophage, allowing free, non-ferritin-bound, iron to decrease the interaction of mRNA-stabilizing proteins with sequence elements within the 3′UTR making these mRNA susceptible to the degradative effects of corticosterone. The stabilization of mRNA and subsequent prolonged production of cytokine effector molecules results in resistance to mycobacterial growth.

Acknowledgments

This work is supported by MIMH grants MH45679 and MH54966 from the National Institutes of Mental Health.

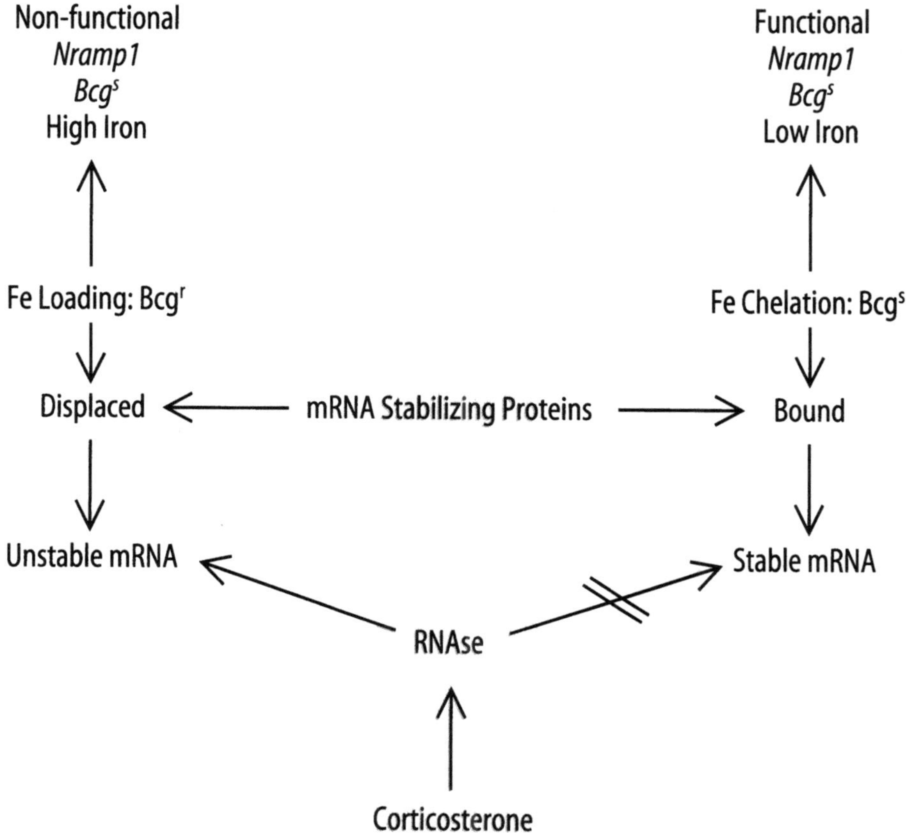

Fig. 8.7. A proposed model for the role of iron and *Nramp1* in mycobacterial resistance. Under conditions of low intracellular iron (functional *Nramp1* expressed by *Bcg^r* macrophages), iron-regulated proteins bind to sequence elements along the mRNA message. Bound proteins stabilize mRNA and prevent degradation by corticosterone-induced RNAses. Under conditions of high intracellular iron (non-functional *Nramp1* expressed by *Bcg^s* macrophages), iron displaces bound proteins resulting in unstable mRNA susceptible to degradation by corticosterone-induced RNAses.

References

1. Johnson EO, Kamilaris TC, Chrousos GP, Gold PW (1992) Mechanisms of stress: a dynamic overview of hormonal and behavioral homeostasis. Neurosci Biobehav Rev 16:115–123
2. Payvar F, Wrange O, Carlstedt-Duke J, Okret S, Gustafsson JA, Yamamoto KR (1981) Purified glucocorticoid receptors bind selectively in vitro to a cloned DNA fragment whose transcription is regulated by glucocorticoids in vivo, PNAS USA 78:6628–6633
3. Munck A, Mendel DB, Smith LI, Orti E (1990) Glucocorticoid receptors and actions, Am Rev Respir Dis 141:S2–S7

4. Crabtree GR, Munck A, Smith KA (1988) Glucocorticoids and lymphocytes: increased glucocorticoid receptor levels in antigen stimulated lymphocytes, J Immunol 124:2430–2437
5. Werb Z, Foley R, Munck A (1978) Interaction of glucocorticoids with macrophages: identification of glucocorticoid receptors in monocytes and macrophages, J Exp Med 147:1684–1690
6. Miesfeld RL (1990) Molecular genetics of corticosteroid action. Am Rev Respir Dis 141:S11–S17
7. Beato M (1989) Gene regulation by steroid hormones. Cell 56:335–344
8. Zwilling BS, Brown DH, Pearl D (1992) Induction of major histocompatibility complex class II glycoproteins by interferon-γ: attenuation of the effects of restraint stress. J Neuroimmunol 37:115–122
9. Kern JA, Lamb RJ, Reed JC, Daniele RP, Nowell PC (1988) Dexamethasone inhibition of interleukin-1 β production by human monocytes, J Clin Invest 81:237–242
10. Lee SW, Tso AP, Chan H, Thomas J, Petrie K, Eugui EM et al. (1988) Glucocorticoids selectively inhibit the transcription of the interleukin 1-β gene and decrease the stability of interleukin 1-β mRNA. PNAS USA 85:1204–1209
11. Northop JP, Crabtree GR, Mattila PS (1992) Negative regulation of interleukin 2 transcription by the glucocorticoid receptor. J Exp Med 175:1235–1241
12. Fraser CM, Venter JC (1990) Beta-adrenergic receptors: relationship of primary structure, receptor function and regulation. Am Rev Respir Dis 141:S22–S29
13. Svedmyr N (1990) Action of corticosteroids on beta-adrenergic receptors, Am Rev Respir Dis 141:S31–S35
14. Eguch KA, Kawakami A, Nakashima M, Ida H, Sakito S, Matsuoka N et al. (1992) Interferon-alpha and dexamethasone inhibit adhesion of T cells to endothelial cells and synovial cells. Clin Exp Immunol 88:448–453
15. Warren MK, Vogel SN (1985) Opposing effects of glucocorticoids on rIFN-g induced murine macrophage Fc receptors and Ia antigen expression. J Immunol 134:2462–2469
16. Calandra T, Bernhagen J, Metx CN, Spiegel La, Bacher M, Donnelly T et al. (1995) MIF as a glucocorticoid-induced modulator of cytokine production. Nature 377:68–71
17. Ishigami T (1919) The influence of psychic acts on the progress of pulmonary tuberculosis. Am Rev Tuberculosis 2:470–484
18. Brown DH, Zwilling BS (1996) Neuroimmunology of host-microbial interactions. In: Psychoneuroimmunology, stress and infection. CRC Press, Boca Raton, pp. 153–171
19. Friedman SB, Ader R, Grota LJ (1973) Protective effect of noxious stimulation in mice infected with rodent malaria. Psychosom Med 35:535–540
20. Edwards EA, Dean LM (1977) Effects of crowding of mice on humoral antibody formation and protection to lethal antigenic challenge. Psychosom Med 39:19–24
21. Gross WB (1984) Effect of a range of social stress severity on *Escherichia coli* challenge infection. Am J Vet Res 45:2074–2079
22. Gross WB, Falkinham JD, Payeur JB (1989) Effect of environmental-genetic interactions on *Mycobacterium avium* challenge infection. Avian Dis 33:411–417
23. Bloom BR (1992) Tuberculosis, back to the frightening future. Science 358:538–539
24. Daniel TM, Bates JH, Downes KA (1994) History of tuberculosis. In: Bloom BR (ed) Tuberculosis:pathogenesis, protection, and control. ASM Press, Washington, DC, pp 13–24
25. Wiegeshaus E, Balasubramanian V, Smith DW (1989) Immunity to tuberculosis from the perspective of pathogenesis, Infect Immun 57:3671–3676
26. Dannenberg AM (1993). Immunopathogenesis of pulmonary tuberculosis. Hosp Pract 1:51–58
27. Nardell E (1993) Pathogenesis of tuberculosis. In: Reichman LB, Hirschfield E (eds) Lung biology in health and disease. Marcel Dekker, New York, pp 103–123

28. Kaplan G, Laal S, Sheffel G, Nusrat A, Nath I, Mathur NK et al. (1988) The nature and kinetics of delay-immune response to purified protein derivative of tuberculosis in the skin of lepromatous leprosy patients. J Exp Med 168:893–899
29. Taylor R (1990) TB or not TB. The answer may be written in black or white. J NIH Res 2:51–57
30. Buschman E, Schurr E, Gros P, Skamene E (1990). In: Ayoub EM et al. (eds) Microbial determinants of virulence and host resistance. ASM Press, Washington, DC, pp 93–111
31. Comstock GW (1978) Tuberculosis in twins: a reanalysis of the prophit survey. Am Rev Res Dis 117:624–629
32. Stead WW, Lofgren JD, Sinner JW, Riddick WT (1990) Racial differences in susceptibility to infection with *M. tuberculosis*. N Engl J Med 322:422–427
33. Crowle AJ, Elkins N (1990) Relative permissiveness of macrophages from black and white people for virulent tubercle bacilli. Infect Immun 58:632–638
34. Lurie MB, Dannenberg AM (1965) Macrophage function in infectious disease with inbred rabbits, Bacteriol Rev 29:466–471
35. Gros P, Skamene E, Forget A (1981) Genetic control of natural resistance to *Mycobacterium bovis* (BCG) in mice, J Immunol 127:2417–2422
36. Gros P, Skamene E, Forget A (1983) Cellular mechanisms of genetically controlled host resistance to *Mycobacterium bovis* (BCG). J Immunol 131:1966–1972
37. Lynch CJ, Pierce-Chase CH, Dubois R (1965) A genetic study of susceptibility to experimental tuberculosis in mice infected with mammalian tubercle bacilli. J Exp Med 120:105–111
38. Forget A, Skamene E, Gros P, Miailke AC, Turcotte R (1981) Differences in response among inbred mouse strains to infection with small doses of *Mycobacterium bovis*. Infect Immun 32:42–48
39. Skamene E, Gros P, Forget A, Kongshavn PAL, St Charles C, Taylor BA (1982) Genetic regulation of resistance to intracellular pathogens. Nature 297:506–510
40. Potter MA, O'Brian AD, Skamene E, Gros P, Forget A, Kongshavn P et al. (1983) A BALB/c congenic strain of mice that carries a genetic locus (*Ity*) controlling resistance to intracellular parasites. Infect Immun 40:1234–1235
41. Schurr E, Skamene E, Forge A, Gros P (1989) Linkage analysis of the *Bcg* gene on mouse chromosome. I. Identification of a closely linked marker. J Immunol 142:4507–4513
42. Schurr E, Buschman E, Gros P, Skamene E (1989) Genetic aspects of mycobacterial infections in mouse and man. Prog Immunol 7:994–999
43. Skamene E, Gros P, Forget A, Patel PT, Nesbitt M (1984) Regulation of resistance to leprosy by chromosome 1 locus in the mouse. Immunogenetics 19:17–22
44. Goto Y, Buschman E, Skamene E (1989) Regulation of host resistance to *Mycobacterium intracellulare* in vivo and in vitro by the *Bcg* gene. Immunogenetics 30:218–223
45. Lissner CR, Weinstein DL, O'Brien AD (1985) Mouse chromosome I *Ity* locus regulates microbicidal activity of isolated peritoneal macrophages against a diverse group of intracellular and extracellular bacteria. J Immunol 135:544–550
46. Skamene E (1989) Genetic control of susceptibility to mycobacterial infections. Rev Infect Dis 11:S394–399
47. Vidal SM, Malo D, Vogan K, Skamene E, Gros P (1993) Natural resistance to infection with intracellular parasites: isolation of a candidate for *Bcg*. Cell 73:469–476
48. Gruenheid S, Cellier M, Vidal S, Gros P (1995) Identification and characterization of a second mouse *Nramp* gene. Genomics 25:514–525
49. Supek F, Supekova L, Nelson H, Nelson N (1996) A yeast manganese transporter related to the macrophage protein involved in conferring resistance to mycobacteria. PNAS USA 93:5105–5110

50. Malo D, Vogan K, Vidal S, Hu J, Cellier M, Schurr E et al. (1994) Haplotype mapping and sequence analysis of the mouse *Nramp* gene predict susceptibility to infection with intracellular parasites. Genomics 23:51–61
51. Govoni G, Vidal S, Gauthier S, Skamene E, Malo D, Gros P (1996) The *Bcg/Ity/Lsh* locus: genetic transfer of resistance to infections in C57BL/6J mice transgenic for the *Nramp1*^{Gly169} allele. Infect Immun 64:2923–2929
52. Drapier JC, Hirling H, Wietzerbin J. Kaldy P, Kuhn LC (1993) Biosynthesis of nitric oxide activates iron regulatory factor in macrophages. EMBO J 12:3643–3650
53. Green LC, Wagner DA, Glogowski J, Skipper PL, Wishnok JS, Tennebaum SR (1982) Analysis of nitrate, nitrate and ^{13}N nitrate in biological fluids. Anal Biochem 126:134–138
54. Nathan CF, Hibbs JB (1991) Role of nitric oxide synthesis in macrophage antimicrobial activity. Curr Opin Immunol 3:65–70
55. Drapier JC, Hibbs JB (1988) Differentiation of murine macrophages to express non-specific cytotoxicity from tumor cells results in l-arginine dependent inhibition of mitochhondrial iron-sulfur enzymes in the macrophages effector cells. J Immunol 140:2829–2838
56. Weinberg ED (1992) Iron depletion: a defense against intracellular infection and neoplasia. Life Sci 50:1289–1292
57. Gazzinelli R, Oswald I, James S, Sher A (1992) IL-10 inhibits parasite killing and nitrogen oxide production by IFN-gamma activated macrophages. J Immunol 148:1792–1796
58. Ozwald I, Gazzinelli R, Sher A, James S (1992) IL-10 synergizes with IL-4 and transforming growth factor beta to inhibit macrophage cytotoxic activity. J Immunol 148:3578–3582
59. Chan J, Xing Y, Magliozzo RS, Bloom BR (1992) Killing of virulent *Mycobacterium tuberculosis* by reactive nitrogen intermediates produced by activated murine macrophages. J Exp Med 175:1111–1122
60. Chan J, Tanaka K, Carroll D, Flynn J, Bloom BR (1995) Effects of nitric oxide synthase inhibitors on murine infection with *Mycobacterium tuberculosis*. Infect Immun 63:737–740
61. Denis M (1991) Tumor necrosis factor and granulocyte macrophage-colony stimulating factor stimulate human macrophages to restrict growth of virulent *Mycobacterium avium* and to kill avirulent *M. avium*: killing effector mechanism depends on the generation of reactive nitrogen intermediates. J Leuk Biol 49:380–387
62. Flesch IEA, Kaufman SHE (1991) Mechanisms involved in mycobacterial growth inhibition by gamma interferon-activated bone marrow macrophages : role of reactive nitrogen intermediates. Infect Immun 59:3213–3218
63. Kindler V, Sappino AP, Grau CJ, Piquet PE, Vassalli P (1989) The inducing role of tumor necrosis factor in the development of bactericidal granulomas during BCG infection. Cell 56:731–740
64. Flynn JL, Goldstein MM, Chan J, Triebold KJ, Pfeffer K, Lowenstein CL, Schreiber RD, Mak TW, Bloom BR (1995) Tumor necrosis factor alpha is required in the protective immune response against *Mycobacterium tuberculosis* in mice. Immunity 2:561–572
65. Blanchard DK, Michelini-Norris MB, Pearson CA, McMillen S, Djeu JY (1991) Production of granulocyte-macrophage colony stimulating factor (GM-CSF) by monocytes and large granular lymphocytes stimulated with *Mycobacterium-M.intracellulare*: activation of bactericidal activity by GM-CSF. Infect Immun 59:2396–2402
66. Denis M, Ghadirian E (1991) Transforming growth factor beta (TGF-β1) plays a detrimental role in the progression of experimental *Mycobacterium avium* infection; in vivo and in vitro evidence. Microb Pathog 11:367–372

67. Zhang Y, Doerfler M, Lee TC, Guillemin B, Rom WN (1993) Mechanisms of stimulation of interleukin-1-beta and tumor necrosis factor alpha by *Mycobacterium tuberculosis* components. J Clin Invest 91:2076–2083
68. Johnson SC, Zwilling BS (1985) Continuous expression of I-A antigen by peritoneal macrophages from mice resistant to *Mycobacterium bovis* (strain BCG). J Leuk Biol 38:635–645
69. Zwilling BS, Brown DH, Christner R, Faris M, Hilburger M, McPeek M et al. (1990) Differential effect of restraint stress on MHC class II expression by murine peritoneal macrophages. Brain Behav Immun 4:330–338
70. Brown DH, Sheridan J, Pearl D, Zwilling BS (1993) Regulation of mycobacterial growth by the hypothalamic-pituitary-adrenal axis: Differential responses of *Mycobacterium bovis* BCG resistant and susceptible mice. Infect Immun 61:4793–4800
71. Collins FM (1989) Mycobacterial disease: immunosupression and acquired immunodeficiency syndrome. Clin Microbiol Rev 2:360–365
72. Cox J, Knight BC, Ivanyi J (1989) Mechanisms of recrudescence of *Mycobacterium bovis* BCG infection in mice. Infect Immun 57:1719–1724
73. Schaffner A (1985) Therapeutic concentrations of glucocorticoids suppress the antimicrobial activity of human macrophages without impairing their responsiveness to gamma interferon. J Clin Invest 76:1755–1761
74. Stokvis H, Langermans JAM, DeBaker-Vledder E, Van Der Hurst MEB, Van-Furth R (1992) Hydrocortisone treatment of BCG infected mice impairs the activation and enhancement of antimicrobial activity of peritoneal macrophages. Scand J Immunol 36:299–305
75. Roach TIA, Kiderlen AF, Blackwell JM (1991) Role of inorganic nitrogen oxides and tumor necrosis factor alpha in killing *Leishmania donovani* amastigotes in gamma interferon-lipopolysaccharide activated macrophages from Lsh^s and Lsh^r congenic mouse strains. Infect Immun 59:3935–3944
76. North RJ, Izzo AA (1993) Mycobacterial virulence. Virulent strains of *Mycobacterium tuberculosis* have faster in vivo doubling times and are better equipped to resist growth-inhibiting functions of macrophages in the presence and absence of specific immunity. J Exp Med 177:1723–1733
77. Masur H, Murry HW, Jones TC (1982) Effect of hydrocortisone on macrophage response to lymphokine. Infect Immun 35:709–714
78. Orme IM, Collins FM (1984) Demonstration of acquired resistance in Bcg^r inbred mouse strains infected with a low dose of BCG Montreal. Clin Exp Immunol 54:56–62
79. Brown DH, Miles BA, Zwilling BS (1995) Growth of *Mycobacterium tuberculosis* in BCG-resistant and susceptible mice: establishment of latency and reactivation. Infect Immun 63:2243–2247
80. Daynes RA, Meikle AW, Araneo BA (1991) Locally active steroid hormones may facilitate compartmentalization of immunity by regulating the types of lymphokines produced by helper T cells. Res Immunol 142:40–45
81. Hopewell PC (1994) Overview of clinical tuberculosis. In: Bloom BR (ed) Tuberculosis: pathogenesis, protection, and control. ASM Press, Washington, DC, pp 25–46
82. Brown DH, Zwilling BS (1994) Activation of the hypothalmic-pituitary-adrenal axis differentially affects the anti-mycobacterial activity of macrophages from BCG-resistant and susceptible mice. J Neuroimmunol 53:181–187
83. Brown DH, Lafuse WP, Zwilling BS (1995) Cytokine-mediated activation of macrophages from *Mycobacterium bovis* BCG-resistant and -susceptible mice: differential effects of corticosterone on antimycobacterial activity and expression of the *Bcg* gene (candidate *Nramp*), Infect Immun 63:2983–2988
84. Brown DH, Lafuse WP, Zwilling BS (1997) The stabilized expression of mRNA is associated with mycobacterial resistance controlled by *Nramp1*. Infect Immun 65:597–603

85. Peppel K, Vinci JM, Baglioni C (1991) The AU-rich sequences in the 3′ untranslated region mediate the increased turnover of interferon mRNA induced by glucocorticoids. J Exp Med 173:349–355
86. Weiss G, Goosen B, Doppler W. Fuchs D, Pantopoulos K, Werner-Felmayer G et al. (1993) Translational regulation via iron responsive elements by the nitric oxide/NO-synthase pathway. EMBOJ 12:3651–3657
87. Weiss G, Wachter H, Fuchs D (1995) Linkage of cell-mediated immunity to iron metabolism. Immunol Today 16:495–500
88. Pantopoulos K, Hentze MW (1995) Nitric oxide signalling to iron-regulatory protein: direct control of ferritin mRNA translation and transferrin receptor mRNA stability in transfected fibroblasts. PNAS USA 92:1267–1271
89. Alford CE, King TE, Campbell PA (1991) Role of transferrin, transferrin receptors and iron in macrophage listericidal activity. J Exp Med 174:459–466
90. Lane TE, Wu-Hsieh BA, Howard DH (1991) Iron limitation and the gamma interferon mediated antihistoplasma state of murine macrophages. Infect Immun 59:2274–2278
91. Hamilton TA, Weiel JE, Adams DO (1984) Expression of the transferrin receptor in murine peritoneal macrophages is modulated in the different stages of activation. J Immunol 132:2285–2291
92. Lima MF, Kierszenbaum F (1987) Lactoferrin effects on phagocytic function. The presence of iron is required for the lactoferrin molecule to stimulate intracellular killing by macrophages but not to enhance the uptake of particles and microorganisms. J Immunol 139:1647–1652
93. Wheeler PR, Rateledge C (1994) Metabolism of *Mycobacterium tuberculosis*. In: Bloom BR (ed) Tuberculosis: pathogenesis, protection and control. ASM Press, Washington, DC, pp 353–386
94. Koch AL (1971) The adaptive responses of *Escherichia coli* to a feast and famine existence. Adv Microb Physiol 6:147–217
95. Gobin J, Horwitz MA (1996) Exochelins of *Mycobacterium tuberculosis* remove iron from human iron-binding proteins and donate iron to mycobactins in the *M. tuberculosis* cell wall. J Exp Med 183:1527–1532

Chapter 9

Human and murine tuberculosis as models for immuno-endocrine interactions

G.A.W. Rook, R. Hernandez-Pando, R. Baker, H. Orozco, K. Arriaga, L. Pavon and M. Streber

Introduction

Recent studies of human and murine tuberculosis have revealed striking changes in adrenal steroid output and metabolism. In this review we use tuberculosis as a model on which to base discussion of the ways in which these changes, even when not disease-specific, may impact upon the function of T lymphocytes during chronic inflammation. Other chapters in this volume amplify further several of the topics highlighted in this one.

Numerous clinicians have suspected abnormal adrenal function in tuberculosis. [1,2] Deficient cortisol production could account for susceptibility to tuberculin shock following initiation of treatment in severely ill patients, and for sudden unexpected deaths [1,3]. Similarly, changes in the balance of regulatory glucocorticoid and "anti-glucocorticoid" steroids produced by the adrenal could contribute to the perturbation of T lymphocyte numbers and function, and to the susceptibility to cytokine-mediated tissue damage that are seen in tuberculosis [4–9].

Previous investigations of adrenal function in human tuberculosis were inconclusive. Serum cortisol levels were variable, with a mean value that was close to normal (reviewed by Post and colleagues) [2]. But serum cortisol levels are a poor indication of subtle changes of adrenal function unless a complete 24-h study of the diurnal rhythm is performed, and a "mesor" calculated. In some studies the response of the adrenal to vastly supraphysiological [10,11] doses of adrenocorticotrophin (ACTH) was also documented but the results were equivocal, and can be regarded as normal or subnormal according to the criteria used [2].

Further confusion has been caused by the fact that the adrenals are sometimes directly infected by *Mycobacterium tuberculosis*, leading to destruction and to Addison's disease. Awareness of this point has obscured the evidence that the adrenals are large in early TB, but can be small in later disease, even in the absence of radiographic evidence for direct infection (calcification) [12]. In a rabbit model, Lurie observed striking changes in adrenal weight that were related to the innate susceptibility to tuberculosis of the rabbit strain used [13], and we found that in tuberculous mice the adrenals first increase in weight and then atrophy to a low plateau value of 50% of normal by day 50. The adrenals in these mice were not infected [9], and the size changes correlate precisely with changes in T lymphocyte function and cytokine profile.

We therefore regarded the data supporting normal adrenal steroid metabolism in tuberculosis as unsatisfactory, and in direct contradiction to clinical and experimental evidence. As a first step towards reassessing this topic, gas chromatography and mass spectrometry were used to quantify and identify adrenal steroid metabolites in 24-h urine collections.

Preliminary findings from analysis of adrenal steroid metabolites in the urine of tuberculosis patients

The results of analysis by gas chromatography and mass spectrometry of the adrenal steroid metabolites in the urine of untreated tuberculosis patients are shown in Table 9.1. The findings from this study, and from Sarma and colleagues [14] that will be discussed below can be summarized in the following statements:

- Loss of the diurnal rhythm of plasma cortisol (from Sarma et al. [14])
- Reduced total output of adrenal androgens [dehydroepiandrosterone (DHEA) metabolites]
- Increased ratio of metabolites of cortisol relative to metabolites of cortisone

Table 9.1. Steroid metabolites in 24-h urine samples from seven male tuberculosis patients before treatment, compared with nine normal male donors

	TB ($n=7$)			Controls ($n=9$)	
	Mean mg/24 hrs	SD	U-test	mg/24 hrs	SD
DHEA	80.0	100.8	0.029	320.0	343.1
Androstenetriol	322.9	465.7	0.15	413.3	116.5
16 α (OH) DHEA	287.1	338.7	0.46	255.6	85.9
Androsterone	411.4	309.9	0.00087	1216.7	190.9
Aetiochlanolone	432.8	179.1	0.0013	1037.7	369.5
Pregnanediol	132.9	175.8	0.63	94.4	56.4
Pregnanetriol	157.1	157.7	0.4	233.3	86.5
Tetrahydrocortisol, THF	1080.0	928.6	0.46	1023.0	261.5
Tetrahydrocortisone, THE	474.3	377.8	0.00086	2188.8	595.5
α-cortolone	245.7	212.8	0.013	615.6	277.1
β-cortol and β-cortolone	235.7	115.0	.00005	854.4	231.7
allo THF	518.6	665.2	0.017	727.8	148.1
11β (OH) androsterone	305.7	167.5	0.0018	641.1	95.7
11β (OH) aetiocholanolone	155.7	93.8	0.79	176.7	99.9
Total androgen derivatives	1534.3	1094.2	0.007	3243.4	525.5
Total cortisol derivatives	3015.7	1939.9	0.007	6227.8	1301.5
Androgen/cortisol ratio	0.62	0.48	0.7	0.53	0.09

Adapted from [53] with permission.

- Reduced ratio of metabolites of DHEA relative to metabolites of cortisol
- Increased ratio of 16α-hydroxylated to reduced metabolites of DHEA.

These findings have been confirmed in a subsequent study (Baker et al., in preparation), and are discussed below in relation to recent observations on the physiology of glucocorticoid function.

The mode of action of glucocorticoids

Full accounts of this topic are to be found in chapter 4 and 5. Briefly, glucocorticoid hormones (GC) bind to intracellular receptors that exist complexed to the 90 kDa heat shock protein (hsp90) and to other proteins of lower molecular weight. Dissociation of the receptor-GC complex from hsp90 then occurs, and receptor dimers are formed [15]. In the nucleus these then interact with "glucocorticoid response elements" (GRE) that are involved in the regulation of expression of numerous genes including some that encode cytokines and other genes involved in the regulation of inflammation and immunity.

However, this relatively simple picture is complicated by numerous factors that will be considered below. Those factors operating at the receptor level include:

- Interaction of glucocorticoid-receptor complexes with other transcription factors such as AP-1, NFkB and JAK STAT [15,16]
- Modulation of GC action via alternative receptors for glucocorticoids. These include the mineralocorticoid receptors (discussed below in several immunological contexts [17,18], and an alternative splicing of the glucocorticoid receptor yielding the β-isoform that may act as a physiological inhibitor [19]. Little is known about the role of the latter, which is not considered further.

The role of GC in T-cell repertoire selection in the thymus

The fundamental importance of GC as regulators of T lymphocyte function is highlighted by the discovery that radio-resistant thymic epithelial cells contain steroidogenic enzymes [20]. These cells actively secrete steroids that may play a crucial role in thymic T lymphocyte repertoire selection [20]. Exposure to GC, or engagement of the T cell receptor both provide signals for apoptosis in immature thymocytes. However when both signals are present together they can antagonize each other, and if the balance of the two signals is within a critical range, apoptosis does not occur. These observations have led Ashwell's group to propose the following hypothesis. T-cells of which the receptors fail to bind major histocompatibility complex (MHC) and peptide with significant avidity (i.e. useless cells) die as a result of apoptosis triggered by GC alone. T-cells that engage MHC/peptide with moderate avidity are preserved by the antagonism of the two signals, while T-cells that engage MHC/peptide with very high avidity (i.e. potentially autoreactive T-cells) are eliminated because the GC signal is unable to wholly antagonize the signal for apoptosis mediated via the antigen receptor [20,21].

When apoptosis of mouse thymocytes is induced in vitro with GC, without engagement of the T-cell receptor, programmed cell death can be inhibited by cytokines, particularly by IL-4 which in this system was more effective that IL-2 or IL-1 [22].

These GC-mediated effects must also be modulated by the thymic 11β-hydroxysteroid dehydrogenase (11β-HSD) discussed later [23]. Moreover it is important to remember that GC are not the only steroid hormones to influence thymic function. Nevertheless, in summary two conclusions can be drawn. First, GC influence T-cell repertoire selection. Second, GC can either cause or inhibit apoptosis of T-cells, depending on the presence of other signals, and on the nature of the T-cells concerned.

The effects of GC on T-cell function in vitro

The data from in vitro work using co-culture of mixed T-cell populations with GC are conflicting and difficult to interpret. This is often due to the use of vastly supraphysiological levels of GC. However, the major problem can be deduced from the previous section on the role of GC in the thymus. The precise role of GC clearly depends on the nature of the T-cells being considered, and on the other stimuli present.

T-cell subpopulations and susceptibility to the effects of GC in vitro

In a particularly illuminating study, T-cells were sorted by a fluorescence-activated cell sorter into immunologically “naive” and committed “memory” subsets, before culture with GC. This was done so that the effects of GC could be tested separately in a system that mimicked the recruitment of new T-cells to an ongoing response, and in a system that mimicked short-term influences of GC on established inflammation. Human peripheral blood lymphocytes were sorted on the basis of expression of the cell membrane glycoprotein variants, CD4+CD45RO– (naive) and CD4+CD45RO+ (memory). They were then primed with anti-CD3 (solid phase) in the presence of IL-2 for 9 days, resulting in considerable clonal expansion. (It may be significant that transcription of the IL-2 gene is down-regulated by the GC receptor [24]. Therefore the added IL-2 makes these experiments somewhat contrived, but nevertheless they lead to some important observations.) The cells were then washed, put into fresh medium, and restimulated for 72 h with anti-CD3 and IL-2. GC was added to these cultures, either in the priming phase, or during the restimulation. Then supernatants from the restimulation phase were assayed for cytokines [25]. The results, summarized in Table 9.2, were quite different with naive and primed cells, and also depended critically on whether the GC were present during priming or restimulation. Briefly:

(i) If cultured without GC, both T-cell types showed a Th0-like pattern, secreting IL-5, IL-10, IFNγ, and in the case of the CD45RO+ memory cells, IL-4 as well.
(ii) Naive ($CD45RO^-$) cells cultured with GC always tended to switch towards Th2. They eventually secreted IL-10 if the GC was present during the priming phase, and both IL-4 and IL-10 if GC was present during the restimulation phase.
(iii) Memory ($CD45RO^+$) cells also switched to exclusive IL-10 production if the GC was present during the priming phase, but if present during *restimulation*, the only cytokine produced was IFNγ [25].

This careful analysis resolves many of the previous controversies that arose when mixed cell populations were studied. For instance it explains why when T-cell clones were grown from bronchoalveolar lavage (BAL) samples using phytohaemaggluscinin (PHA) and IL-2 as the stimulus, dexamethasone (at supraphysiologic doses) inhibited anti-CD3-induced production of IL-4 and IL-5 more than it inhibited production of IFNγ [26]. BAL T-cells are known to be activated memory cells, so this is compatible with the results of Brinkmann et al. [25] and does not alter the fact that if an immune response in naive cells is allowed to develop in the presence of GC, a Th2 line will develop. This has been rather clearly shown with spleen cells from "clean" laboratory rodents [27], which have few memory cells under normal circumstances, and is described in detail in Chapter 7.

GC and the balance of Th1 to Th2 cytokine profiles

Overall the "bottom line" may be that GC favour the *development* of a Th2 cytokine profile from naive cells, although cytokine secretion by established Th2

Table 9.2. Differing effects of glucorticoids (GC) on the cytokine profiles of naive and memory T-cells

	Proliferation	IL-4[a]	IL-5	IL-10	IFN-γ
CD4 + $CD45RO^-$ T-cells (naive)					
No GC	++	–	++	++	++
+ GC during priming phase (anti-CD3 + IL-2, 9 days)	–	–	–	+++	–
+ GC during restimulation (anti-CD3 + IL-2, 72 h)	++	+++	–	++	–
CD4 + $CD45RO^+$ T-cells (memory)					
No GC	++	++	++	++	++
+ GC during priming phase (anti-CD3 + IL-2, 9 days)	+	–	–	+++	–
+ GC during restimulation (anti-CD3 + IL-2, 72 h)	++	–	–	–	+[b]

[a] Cytokines were measured in the supernatants of the restimulation cultures.
[b] A reduction related to the concentration of GC, but always >50% of the levels seen in the absence of GC. Adapted from [25].

cells is readily inhibited. The ability of some Th1 cells to continue to secrete Th1 cytokines after exposure to GC does not alter the fact that Th1 *function* is blocked by GC. This is partly due to other GC effects such as blockage of microphage function [28, and see Chapter 8], and enhanced expression of TGFβ [29], and partly due to the fact that in a dynamic in vivo situation what is important is the diversion of subsequently recruited T-cells towards Th2. Thus conventional treatments for Th-2-mediated diseases such as eczema, asthma and hay fever may work via anti-inflammatory effects, and by reducing cytokine production by Th2 cells [30], and yet at the same time encourage perpetuation of the underlying problem by driving newly recruited T cells towards Th2.

Factors that modulate glucocorticoid function in vivo. I. Diurnal rhythm

As can be deduced from the previous section, the overall physiological effect of GC (like the effect of stress which is considered later) [28,31,32] seems to be to drive the immune response towards Th2 [33–36]. However, the eventual extent and nature of glucocorticoid-mediated effects in vivo depend not only on the conventionally measured circulating GC levels, but also on the diurnal rhythms, and on a series of metabolic events that regulate GC action in individual tissues.

The diurnal rhythm of glucocorticoid production; disturbances of the cortisol diurnal rhythm in disease

In normal individuals there is a diurnal rhythm of serum cortisol. The level peaks in the early morning and then declines throughout the day. The first endocrine abnormality in tuberculosis patients listed in the introduction is a striking loss of this diurnal rhythm [14, Baker et al., in preparation] (Fig. 9.1). This effect is not disease-specific. The rhythm can be lost during chronic or subacute stress or chronic infection. For instance exposure of military personnel to physical and psychological stress caused by strenuous physical exertion accompanied by sleep and energy deprivation resulted in loss of the rhythm [37], and it can also be lost during HIV infection [38]. This lack of disease specificity does not detract from the probability that it has important immunological consequences.

The immunological significance of the loss of diurnal rhythm

If the normal rhythm of plasma cortisol is a physiological clock that affects T-cell and macrophage function, then changes in that rhythm must have important effects on the immune system. Unfortunately the endocrinological "fashion" has

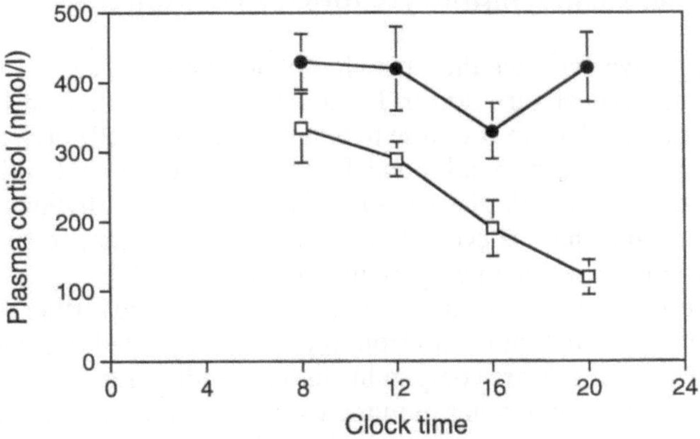

Fig. 9.1. Diurnal rhythm of plasma cortisol concentrations (mean ± SE) in healthy subjects (open squares) and newly diagnosed cases of pulmonary tuberculosis (filled circles). (From [14] with permission.)

been to measure peak levels of cortisol in the early morning, which alter very little during disease and stress, and can be immunologically uninformative. A sample at 8.00 p.m., nearer the late night nadir of cortisol secretion, is much more likely to reveal stress or disease-mediated abnormalities and can be much more revealing to the immunologist. Despite the paucity of appropriate data there is enough to support the view that the diurnal cortisol rhythm is of fundamental importance to the immune system.

MR receptors and the evening trough in cortisol levels

Since cortisol has a ten-fold higher affinity for the mineralocorticoid receptor (MR) than it has for the glucocorticoid receptor (GR), the cortisol diurnal rhythm tends to result in occupation of the MR *and* GR in the morning, but only the MR later in the day when cortisol levels are low. T-cells in the thymus and peripheral blood express mostly or exclusively GR. These tend to mediate a lymphocyte suppressive effect, so these T-cells alternate between steroid-suppressed and non-steroid-suppressed states in each 24-h cycle.

In other organs such as the spleen the situation can be more complex. The rat spleen expresses MR as well as GR, so the trough in the diurnal cycle of GC levels can result not just in diminished GC effects, but even in their replacement with contrary effects mediated via the MR. In this model, effects mediated via the MR can stimulate T-cell function [17,18]. Similarly human monocytes are said to express MR as well as GR, and aldosterone is able to inhibit lipopolysaccharide (LPS)-induced IL-1ra secretion by monocytes in vitro [39]. The balance of MR- and GR-mediated effects on the immune system must vary naturally during the 24-h cycle.

Cyclical variation in lymphocyte function in humans

There is suggestive evidence that this physiological cycle is sufficient to cause a corresponding cyclical variation in lymphocyte function in man [40–43]. For instance, release of IFNγ in response to tetanus toxoid or PPD [41], and the proliferative response of peripheral blood T-cells to tetanus toxoid [42], both show an inverse correlation with plasma cortisol levels. Diurnal rhythms can also be shown in vivo, and there is cyclical variation in the response to challenge with oxazolone, a contact sensitizing agent, in sensitized rats [44].

Clearly these correlations could be coincidental but some further support for a role for the cortisol rhythm comes from the observation that diurnal changes in circulating T-cell subsets are comparable to those induced by prednisolone [43]. However, there are several other immunologically relevant diurnal rhythms, that may not be secondary to the cortisol rhythms. There is a diurnal rhythm of IL-6, with the highest levels throughout the night (peaking at 1.00 a.m.) with a nadir at 10.00 a.m. [45]. The complexity of the diurnal rhythms found with TNFα, IL-10 and granulocyte macrophage-colony stimulating factor (GM-CSF) was considerable, with unique biphasic patterns for each one, while IL-2 showed a single peak at noon [46]. Soluble IL-2 receptors show a peak at 12.29 p.m. and a nadir at 4.14 a.m., so the peak at least correlates with IL-2 itself [47].

A speculation about the importance of the cortisol diurnal rhythm in HIV and TB

Loss of the cortisol rhythm is not confined to tuberculosis, and may be a feature of many chronic infections. In view of the evidence that it is immunologically significant, it is worth thinking about it in relation to HIV infection. The diurnal rhythms of ACTH and cortisol are blunted in some HIV positive subjects [38]. This may relate to the observation that the diurnal rhythm in CD4+ cell numbers (rise between 8.00 a.m. and 10.00 p.m.) is also blunted in HIV [24] or according to one report, lost altogether [48]. It is therefore interesting that one author has claimed that treating HIV seropositive individuals with a single daily dose of prednisolone at 8.00 a.m. or 9.00 a.m. has a beneficial effect [49]. Once this regime was established, plasma cortisols measured in the morning, immediately before each dose of prednisolone, were found to be reduced. Unfortunately the 8.00 p.m. cortisols and prednisolones were not measured. Nevertheless, it is possible that this regimen was in fact establishing a diurnal rhythm of prednisolone, peaking in the morning soon after the daily tablet, and then diminishing throughout the day, in accordance with the half-life of this compound. The low early morning cortisols imply suppression of endogenous adrenal activity. Therefore this artifically imposed prednisolone rhythm may have partly mimicked a normal cortisol rhythm (particularly since unlike dexamethasone, prednisolone tends to bind both to MR and GR), replacing the faulty endogenous cortisol rhythm associated with the disease [38]. This may be beneficial because it stops the exposure of the patients' T-cells to a 24-h unchanging GC influence. We [7] and subsequently others [50] have emphasized the possibility that glucocorticoids play a role in the loss of T lymphocytes in HIV infection.

Factors that modulate glucocorticoid function in vivo. II. Metabolism in the periphery

The regulatory role of cortisol metabolism in the periphery

The local concentration of cortisol in any cell or tissue is not dependent only on the concentration reaching that tissue from the circulation. Most tissues need to regulate local concentrations to suit their own particular needs and mechanisms exist to do this that may be disturbed in tuberculosis.

Glucocorticoid catabolism by target cells

Clearly differences in receptor numbers, or affinity, or in the signalling pathways can alter the GC sensitivity of individual cells. The ability to catabolize the cortisol may be another factor. It was noted many years ago that strains of fibroblasts that were resistant to cortisol-mediated growth inhibition were able to catabolize the cortisol via a variety of oxidations and reductions [23,51]. There may be a similar variation in the sensitivity of individual lymphocytes to GC. Catabolism of cortisol by lymphocytes from patients with systemic lupus erythematosis (SLE) was found to be significantly increased relative to normal. Lymphocytes from rheumatoid arthritis (RA) patients were not different from normal [52].

Reversible glucocorticoid inactivation in target organs

A better-studied type of inactivation of cortisol that can occur locally within a given tissue, is reversible conversion to inactive cortisone (Fig. 9.2). We pointed out above that in tuberculosis, analysis of GC metabolites in urine reveals a massive switch in the cortisol/cortisone balance towards the active compound, suggesting an inhibition of cortisol inactivation, or enhanced reactivation of cortisone [53,54]. This is crucial because inhibition of the enzyme that causes this inactivation of GC in the tissues increases susceptibility of mice to *Listeria monocytogenes* [55].

The anatomical site of the metabolic abnormality leading to this change in cortisol/cortisone balance in tuberculosis is not known, but a cytokine-mediated effect on cortisol/cortisone interconversion in the lung is an obvious candidate [56]. This concept is expanded below.

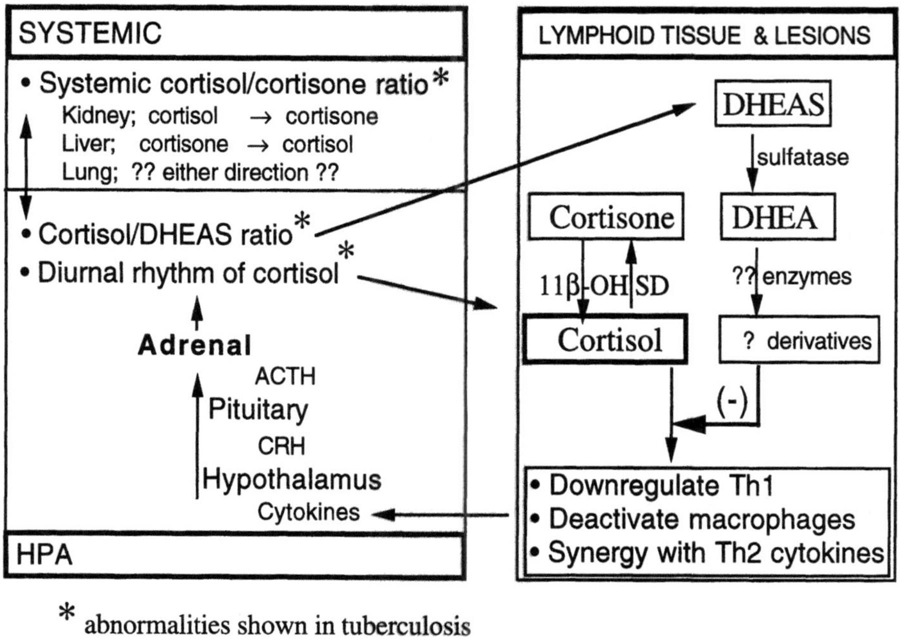

Fig. 9.2. Factors that determine the effective cortisol function in lymphoid tissue and lesions. Cytokines from the immune response (bottom right) contribute additional drive to the hypothalamo-pituitary-adrenal axis. The eventual output of the adrenal (left) can vary in diurnal rhythm and in cortisol/dehydroepiandrosterone sulfate (DHEAS) ratio. The circulating cortisol level can then be modulated by changes in the cortisol/cortisone ratio, due to the "cortisol-cortisone shuttle" enzymes in the kidney, liver, and other tissues. Within lesions and lymphoid tissues (box on right), both DHEAS and cortisol are subject to regulated local metabolism. The pathway of conversion of DHEA to the putative active anti-glucocorticoid (indicated as (–), is unknown, because we do not know which are the active metabolites.

The cortisol-cortisone shuttle and the tuberculous lung

The enzymes involved in the cortisol-cortisone shuttle are described in detail in Chapters 4 and 5. In general the importance of the enzymes regulating this shuttle is as a local tissue regulatory mechanism, rather than a mechanism to regulate plasma levels. The highest levels of enzyme occur in two sites, the kidney and the liver (Fig. 9.2). In the kidney the enzyme 11β-hydroxysteroid dehydrogenase type 2 (11β-HSD2) converts cortisol into inactive cortisone, and so stops the MR from binding cortisol [57]. This allows mineralocorticoid functions to be mediated almost exclusively by aldosterone in spite of the fact that this hormone is present at much lower levels than GC which have a similar affinity for MR. The liver enzyme 11β-HSD1, on the other hand, predominantly converts cortisone to cortisol, though it has the capacity to function in either direction.

In fact most tissues possess enzymes of this class, and can therefore adjust the local GC concentration. In the current context, the crucial organs are the lymph nodes and the lungs, both of which contain 11β-HSD activities [51,56,58]. The enzyme found in lymphoid tissue is discussed in detail in Chapter 5.

The enzyme in the lung may be particularly relevant to tuberculosis. It resembles the one in the liver, though in the lungs of normal donors it is running in reverse, converting cortisol into inactive cortisone [56]. It has been observed that in ovarian granulosa cell preparations, removal of contaminating T-cells alters the activity of the enzyme, suggesting that it may be regulated by cytokines [59]. We postulate therefore that in the cytokine-rich environment of the tuberculous lung, the 11β-HSD1 may convert cortisone back to cortisol. This hypothesis is accessible to investigation.

Factors that modulate glucocorticoid function in vivo. III. Modulation by other agonists

Modulation of the effect of GC on T-cells in the presence of other agonists

In vivo T-cells are exposed to GC in the presence of multiple agonists, and a few examples will serve to illustrate the complexity of the resulting interactions. For instance, low concentrations of dexamethasone or prostaglandin E2 that did not inhibit the responsiveness of human T-cells to stimulation by anti-CD3, acted synergistically to inhibit both IL-2 secretion and subsequent proliferation [60]. Anti-CD28 overcame much of this suppression [61].

The interaction of GC with other mediators is particularly apparent when GC are considered as inducers of T-cell apoptosis. Thus GC-induced apoptosis of T-cells can be inhibited by ligands for CD44 [62]. It is sometimes stated that peripheral T-cells are relatively insensitive to GC-induced apoptosis. However peripheral blood T-cells showed sensitivity to GC-induced apoptosis soon after the proliferative response to mitogen stimulation. They could then be protected from apoptosis by several cytokine (IL-2>IL-4>IL-10), while puromycin and cycloheximide increased susceptibility [63].

Modulation of the effects of GC on macrophages by MIF

Migration inhibition factor (MIF) is likely to play a crucial role in tuberculosis, because it modulates release of pro-inflammatory cytokines from macrophages by reducing the down-regulatory effects of GC. Thus LPS-triggered macrophages pretreated with low doses of MIF became refractory to GC-mediated inhibition of release of IL-1β, IL-6, IL-8 and TNFα [64,65]. This may account for the observation that in a murine model of delayed type hypersensitivity (DTH) to tuberculin, where MIF mRNA and protein were shown to be abundant (mostly in macrophages), a neutralizing antibody to MIF inhibited the DTH response [66]. We have argued elsewhere that pro-inflammatory cytokines contribute significantly to tissue damage in this infection [67], and this is demonstrated in the murine model later.

However, MIF is not only released locally by inflammatory cells. It is an abundant pre-formed constituent of the anterior pituitary where it is found in granules with ACTH or thyroid stimulating hormone (TSH) [68]. Moreover corticotrophin-releasing hormone (CRH) is a potent secretagogue for MIF. Thus while ACTH is being released in response to stimulation of the cytokine-hypothalamo-pituitary-adrenal axis, MIF can also be released and can then oppose the effect of GC on macrophages in the periphery. This effect appears to be physiologically relevant. For instance MIF potentiates LPS-mediated shock, as can be demonstrated directly by the administration of recombinant MIF. Similarly, neutralizing antibody to MIF is protective in the same model [65]. It is therefore probable that in tuberculosis MIF is playing a role in the regulation of GC-mediated suppression of pro-inflammatory cytokine release, and this is central to the pathogenesis of the disease. Unfortunately, we currently know nothing about its abundance in tuberculosis, or how its release is affected by the changes in diurnal rhythm already outlined.

Antiglucocorticoid functions of dehydroepiandrosterone

There is a further category of regulatory pathways that modulate the functions of GC in the periphery (Fig. 9.2). These are mediated by "anti-glucocorticoid" effects of dehydroepiandrosterone sulfate (DHEAS), or perhaps of unknown metabolites of this steroid [4,6,58]. The mechanisms are not understood, but the effects are striking and as will be demonstrated below, DHEA is clearly central to the theme of this review. We have outlined above the fall in output of DHEA, and the fall in the ratio of DHEA to cortisol that occur in tuberculosis (Table 9.1) [53].

The pattern of metabolism of DHEA also changes in tuberculosis, so that less than normal is reduced to androsterone or aetiocholanolone, while more of it is converted to 16α-hydroxylated derivatives (even in the absence of anti-tuberculosis therapy; this is not an effect induced by rifampicin, though rifampicin can activate the relevant enzyme, CYP 3A7) [53]. Since we do not know which metabolites exert the anti-glucocorticoid effects, interpretation of this change is difficult, but it may be significant that these 16α-hydroxylated derivatives are not active in vivo as anti-glucocorticoids (Al-Nakhli et al. in preparation). Moreover, the same 16α-hydroxylated derivatives are formed during pregnancy and by premature neonates [69]. Our hypothesis is that this represents one of several mechanisms that tend to drive the immune response towards Th2 in pregnancy [70], since Th1 responses to placental antigens are associated with abortion [71]. Perhaps tuberculosis is "attempting" to do the same thing.

The physiology of DHEA: differences between rodents and man

DHEA is the most abundant product of the human adrenal after adrenarche. The adrenal secretes 10–15 mg of DHEAS per day in healthy young adults, but serum levels then fall steadily with increasing age. It is strongly bound to albumin and undergoes renal tubular reabsorption. In young adults it is present at concentrations close to 4 μg/ml. Eventually most of it is converted to DHEA. Thus in man the major source of DHEA is DHEAS. Most DHEA circulates free but some is weakly bound to albumin.

One problem that must be faced when assessing the literature on DHEA is the fact that its physiology in rodents is clearly quite different. Thus whereas there are about 4 μg/ml of DHEAS in adult human plasma, rat plasma contains <1 ng/ml of this hormone [72]. Unexpectedly, in rodents the brain is the site where concentrations of DHEAS (and at a lower level, DHEA) are highest. Thus rat brain contained an average of 3.5 ng/g of tissue over a 24-h period [72]. DHEAS in brain has clear neurological functions that are outside the scope of this review. The DHEAS in brain is thought to be locally synthesized, and it is possible that it is not synthesized at all in rodent adrenals since this organ lacks 17 α-hydroxylase. Nevertheless DHEAS and fatty acid esters of DHEA were found in the adrenals of rats at 5.1 ± 2.6 and 16.1 ± 6.4 ng/g of tissue respectively [72]. If the rodent adrenal does make DHEA it must use an unusual pathway. Alternatively in rodents the immunological functions of DHEA may be performed by some other derivative that can be reached via a different route.

Effects of DHEA on immune responses in rodents

In view of the points outlined in the previous paragraph, it is clear that much of the literature on the effects of DHEA or DHEAS in rodents should be interpreted with caution, or discounted altogether. Some authors have used doses of 1 g/kg [73], in a species with physiological levels <1 ng/ml of plasma or gram of tissue! Doses of this magnitude result in rapid formation of androgens and oestrogens, with permanent priapism in male mice, and unpleasantly aggressive behaviour in the females. Sex hormones are known to have immunological effects of their own (reviewed in [74]) so such protocols clearly cause confusion. Other authors using almost equally excessive doses (10 mg DHEA/mouse, i.e. about 500 mg/kg) showed that DHEA enhances influenza immunization in aged mice [75]. In rats immunosuppressed with dexamethasone, 120 mg/kg per day of DHEA was able to partially restore spleen cell mitogenic responses to endotoxin and concanavalin A, and in vitro IgG production. DHEA also reduced the colonization of the gut by *Cryptosporidium parvum* in these animals [76]. Continuing down towards more appropriate dose levels, three daily injections of 1.2 mg DHEA (i.e. 50–60 mg/kg per day) were found to stop dexamethasone from rendering peripheral lymphocytes of mice unresponsive to mitogens, or causing involution of the thymus [4]. In fact even this dose is neither required nor optimal, and doses in the microgram range are more effective. Larger doses actually diminish the effect in the thymus protection assay used by Blauer and colleagues [4], probably because of conversion to testosterone. Gonadectomy is known to increase thymic bulk, while estrogen and testosterone decrease it. (This may be because testosterone increases expression of TGFβ in the thymus [77].)

As an example of the efficacy of lower doses, optimal inhibition of GC-induced thymic involution in mice requires only 10–20 μg/mouse of DHEA or of its derivative 3,17-androstenediol (i.e. ~0.5 mg/kg) [Al-Nakhli et al., in preparation]; this is a hundredth of the dose used by Blauer and colleagues [4]. Similarly immunization with pneumococcal polysaccharide yielded higher antibody titres and increased numbers of plaque-forming cells, in old (but not young) mice if they were given a single s.c. injection of 100 μg DHEAS in propylene glycol, or 10 μg DHEA dissolved in ethanol, 3 h before the immunization [78]. This is a T-cell-independent response, so the mode of action is unclear. In general DHEA, using acceptable doses in the low microgram range, appears to oppose GC, and to promote a Th1 cytokine pattern. It has been shown to restore immune functions in aged mice, and to correct the dysregulated spontaneous cytokine release seen in old animals [78,79]. It has been tested for similar properties in aged humans [80]. It also enhances production of Th1 cytokines such as IL-2 and IFNγ [5,58,81,82]. This is the reverse of the effect of glucocorticoids which often enhance Th2 activity and synergize with Th2 cytokines [33–36].

DHEA also enhances IL-2 secretion from human peripheral blood T-cells [83], and we have found that DHEA or 3β, 17β-androstenediol (AED) will enhance mitogen-stimulated production of IFNγ from murine or human T-cells in the range 10^{-7} to 10^{-8} M [Al-Nakhli et al., in preparation].

Mechanism of the "anti-glucocorticoid" actions of DHEA

We do not know how DHEA exerts its anti-glucocorticoid effects. It is not a competitive antagonist of the GC receptor. There is a report of a specific DHEA-binding protein (? receptor) in T lymphocytes [84] but this remains unconfirmed.

In its role as a "neurosteroid" DHEAS clearly binds to several cell-membrane-associated receptors in the brain, and it is an antagonist of γ-aminobutyric acid $(GABA)_A$ receptors [72]. The only hint at the possibility of cell-membrane-associated receptors for DHEAS in the periphery has emerged from the observation that DHEAS very rapidly inhibits archidonic acid-induced platelet aggregation [85]. The effect was seen at concentrations that are within the physiological range in man but not in rodents.

When large doses of DHEAS are given to rodents it acts as a peroxisome proliferator. To do this DHEAS requires the presence of peroxisome proliferator-activated receptor alpha (PPARα) [86]. There is no direct evidence that it is a ligand for this receptor, but it is interesting that there is a protein that binds DHEAS in liver, and PPARα is expressed at particularly high levels in this organ and in the immune system (reviewed in [87,88]). The PPARs are a family of transcription factors analogous to the steroid receptors. They form heterodimers with RXR, the receptor for 9-*cis*-retinoic acid, and bind to DNA motifs known as PPAR-response elements. This results in up-regulation of expression of enzymes responsible for lipid homoeostasis, fatty acid degradation, and destruction of leukotriene B4 (LTB4). It has been suggested that this effect could explain the restoration of immunological competence by DHEAS in old animals, via changes in cell membrane fluidity, phospholipid-dependent cell signalling pathways, and arachidonate-dependent mediators [87]. This is an interesting possibility, though PPARα up-regulates the enzymes that degrade LTB4 and this in turn limits inflammation [88] whereas DHEA actually increases non-specific inflammation (see chapter 6). A hypothesis that provides a potential solution to this paradox is described in detail in chapter 5

A further obvious explanation is that the active "anti-glucocorticoid" is an undiscovered metabolite of DHEA, acting via, perhaps, one of the many "orphan" steroid receptors. It would be biologically "logical" for there to be anti-glucocorticoid metabolites that are beyond the point in the metabolic pathways that can give rise to sex steroids, so that regulation of inflammation and immunity could occur independently of sexual and reproductive functions. It has been suggested that 7-hydroxylated derivatives perform this function, but the supporting in vivo data are unacceptable, and the in vitro data are not convincing [73,89].

Stress and T-cell function: changes in GC : DHEA ratios and rhythms

Stress provides a pathway that can lead to changes in the GC/DHEA ratio, and in the diurnal rhythm (see Chapter 2). Psychological and physical stress can activate the hypothalamo-pituitary-adrenal axis, and so lead to an increase the production of cortisol. However, the immunological effects of stress are complicated by the activation of other pathways, such as the adrenal medulla, so not everything observed is due to cortisol. Moreover, most studies of the immunological effects of stress have tended not to include the types of immunological investigation that cast any light on the Th1/Th2 balance. Nevertheless it is clear that stress increases Th2 and decreases Th1 activity. Excessive exercise and deprivation of food and sleep resulted in raised cortisol, but no fall in DHEA, (though testosterone fell to castrate levels). The DHEA/GC ratio therefore fell and correlated with a fall in delayed hypersensitivity (DTH) responsiveness [32]. The same study documented a stress-induced rise in serum IgE levels.

In chronic illness or after burn injury, DHEA levels fall dramatically, so DHEA/GC ratios can be more severely affected than in stressed but otherwise healthy army peronnel.

A classical example of the Th1->Th2 switching effect of stress is the increase in antibody to Epstein–Barr virus in students reacting in a stressed manner to their exams. This virus is usually controlled by a type 1 response and cytotoxic T-cells, and loss of control results in virus replication and increased antibody [31]. Similarly, peripheral blood leukocytes from medical students during exam periods showed lower mRNA for IFNγ and for the glucocorticoid receptor [90].

Stress and the reactivation of tuberculosis

It has been known for many years that the incidence of tuberculosis increases in war zones and in areas of poverty. Similarly, moving cattle from one farm to another in trucks is sufficient to cause reactivation of latent bovine tuberculosis. Stress is likely to play a role in these phenomenona, presumably by driving a Th1->Th2 shift, and down-regulating macrophage function. This point can be demonstrated in a more controlled manner in laboratory animals. Stress due to crowding or restraint can increase mycobacterial growth in tuberculosis in mice [28,91]. Some of the mechanisms involved are discussed in Chapter 8.

Murine tuberculosis provides a model which is acutely sensitive to the presence of even a small Th2 component [67]. Recent observations in this model will be used in the following sections to illustrate the interconnections between adrenal function and immunology in tuberculosis.

The adrenals in murine tuberculosis

When mice were infected with virulent *Mycobacterium tuberculosis* H37Rv by the intra-tracheal route, there was an early phase of adrenal hyperplasia, histologically resembling the ACTH-driven changes seen in Cushing's disease. This was followed at 3 weeks by progressive atrophy until the weight of the adrenals was ~50% of that seen in control uninfected mice, in spite of the fact that the adrenals were not infected (Fig. 9.3a). All layers of the adrenal cortex were affected, but the medulla was normal. Electron microscope studies revealed apoptosis of adrenal cells [9,67].

Adrenals in murine tuberculosis; correlations with Th1/Th2 balance

The switch from adrenal hyperplasia to adrenal atrophy corresponded to onset of an IgG1 response recognizing a wide range of mycobacterial components in Western blots [9]. It also correlated with other signs of a switch from a

Fig. 9.3. Events in murine pulmonary tuberculosis that correlate with changes in adrenal weight. Following intratracheal infection with *Mycobacterium tuberculosis*, the adrenals increase in weight for 21 days (**a**) while the T-cells in the interstitial inflammation (**b**) are dominated by Th1 cells (IL-2 +ve) (**b**, open squares). During this phase delayed hypersensitivity (DTH) to tuberculin is increasing (**c**, open squares) and DTH sites are not sensitive to TNFα (**c**, filled circles). After 21 days, the adrenals start to atrophy (**a**), and the inflammatory zones become populated with Th2 (IL-4 + ve) cells (**b**, filled circles), while Th1 cells decrease (**b**, open squares). Simultaneously DTH decreases (**c**, open squares), and then DTH sites, become sensitive to TNFα (**c**, filled circles).

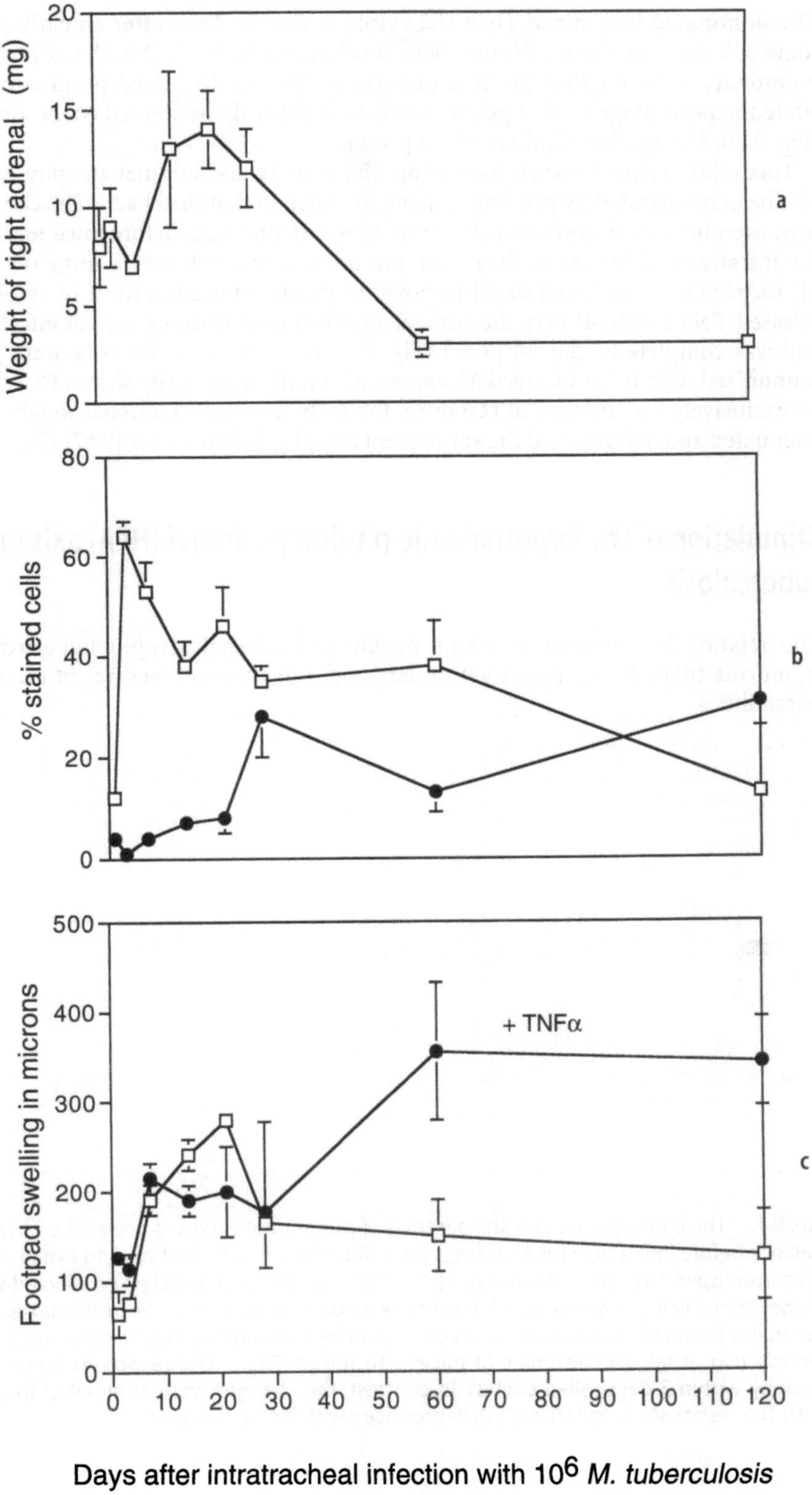
Weight of right adrenal (mg)
20
15
10
5
0
a
% stained cells
80
60
40
20
0
b
Footpad swelling in microns
500
400
300
200
100
0
+ TNFα
c
0 10 20 30 40 50 60 70 80 90 100 110 120
Days after intratracheal infection with 10^6 M. tuberculosis

Th1-dominated to a mixed Th1+Th2 cytokine profile. Thus after an early peak (days 3–7 after infection), the numbers of cells positive for IL-2 in the interstitial pulmonary inflammation (by immunocytochemistry) decreased progressively, while the percentage of IL-4 positive cells rose abruptly between days 21 and 28 (Fig. 9.3b; Hernandez-Pando *et al.*, in press).

This relationship between increasing Th2 activity and adrenal atrophy could also be demonstrated by pre-immunizing the mice so that they had a Th2 component even before infection with *M. tuberculosis*. If 2 months before mice received the intratracheal infection, they were pre-immunized with a very large dose of *M. vaccae* (10^9 autoclaved bacilli) known to prime a mixed pattern of cytokine release (IFNγ and IL-4) [92], the adrenal atrophy began within 4 days of infection, and was complete by day 14 [9,67] (Fig. 9.4). In contrast, if the mice were pre-immunized with 10^7 autoclaved *M. vaccae*, a stimulus previously shown to induce an exclusively Th1 pattern of response, the early increase in adrenal weight was attenuated and delayed, and the subsequent atrophy did not occur [9,67] (Fig. 9.4).

Stimulation of the hypothalamic-pituitary-adrenal (HPA) axis in tuberculosis

The relationship between cytokine profile and adrenal atrophy/hypertrophy in murine tuberculosis is not yet understood, but there are several interesting possibilities.

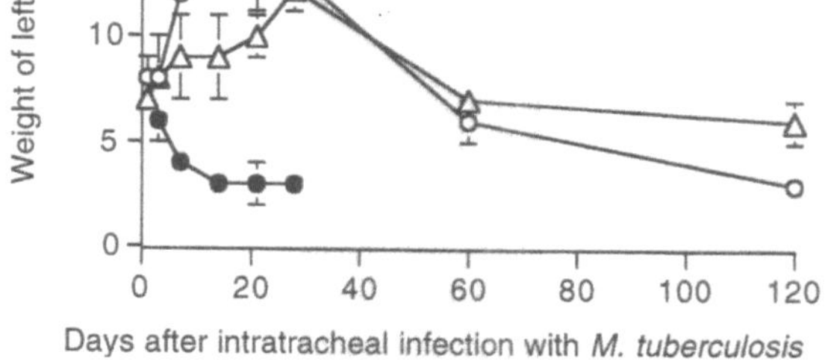

Fig. 9.4. The consequences for the adrenals of pre-immunization to preset the Th1/Th2 balance before infection with *M. tuberculosis*. Mice were immunized so as to evoke a Th1 response (open triangles) or a mixed Th1 + Th2 response (filled circles) or injected with saline (open circles). Two months later they were given intratracheal *M. tuberculosis*. The adrenals of control (saline) mice (open circles) showed changes similar to those seen previously (Fig. 9.3a). The adrenals of mice with mixed Th1 + Th2 responses started to atrophy within 3 days (filled circles). In contrast, the changes were attenuated in mice with Th1 responses (open triangles). Reproduced with kind permission.

A "trivial" explanation is that when there is a Th2 component, tuberculosis makes the mice so ill that the adrenals are damaged by a pathway similar to that operating in the Waterhouse–Friderichsen syndrome. However, this is unlikely. The adrenals do not show haemorrhage or cortical necrosis. There is a progressive atrophy over many days, with apoptosis, not necrosis [9].

Second, it is possible that the mixed Th1+Th2 cytokine pattern is inherently toxic to the adrenal. Numerous cytokines have been shown to have direct effects on adrenal function, but we cannot draw conclusions about this possibility from the available data, and information on the Th2 cytokines is lacking.

Third, it is possible that the presence of a Th2 component compromises activation of the HPA axis, and so reduces ACTH-dependent drive to the adrenal. IL-1 is a major activator of the HPA and is released from inflammatory sites. However there is also a physiological inhibitor of IL-1, known as IL-1 receptor antagonist or IL-1ra. Production of IL-1ra is enhanced by the classical Th2 cytokines (IL-10. [93], soluble CD23 [94], IL-13 [95], and IL-4 [96,97]). This is potentially an explanation for adrenal atrophy in tuberculous mice in which Th2 cytokine production is prominent [98]. The relevance of this hypothesis is emphasized by the observation that induction of glucocorticoid production by endotoxin is effectively blocked by administration of IL-1ra in rodents [99–101]. Therefore, activation of inappropriate Th2 cells will tend to block the feedback from the cytokine network via the hypothalamus.

Finally, it is now clear that some of the feedback to the HPA axis is mediated by nerves rather than cytokines. For instance LPS injected into the peritoneal cavity may induce expression of IL-1 locally within the hypothalamus. The signal to the hypothalamus passes via the vagus nerve, bypassing the need for IL-1 from the blood to enter the brain [102] (see Chapter 1). We know nothing about neural signals from the lungs to the HPA axis, but they may exist. When there is a mixed Th1+Th2 response, the lungs of tuberculous mice are undergoing rapid destructive immunopathology, and such signals could be important.

Adrenals in murine tuberculosis and the toxicity of cytokines

A further important correlate of the late phase of murine tuberculosis, (when the adrenals are atrophic and the immune response characterized by a mixed Th1+Th2 cytokine profile), is increased toxicity of TNFα. DTH responses to tuberculin were seen throughout the infection, but differed in their sensitivity to TNFα in a manner that correlated closely with adrenal size. Thus if TNFα was injected at 24 h into DTH sites elicited during the phase of adrenal hyperplasia, there was no increment in swelling at 48 h. However, similar injections of TNFα resulted in a doubling of the swelling in DTH sites elicited during the phase of adrenal atrophy (Fig. 9.3c). It may be relevant that the toxicity of cytokines such as TNFα is closely regulated by rapid feedback from the cytokine-HPA axis. Failure of this feedback enhances toxicity [103,104]. Studies in this [105] and other laboratories [106] have led us to argue that although TNFα has essential protective macrophage-activating roles in certain Th1 responses to mycobacteria in mice, [107] it nevertheless has important toxic roles in the *disease* where immunity is failing, and when an inappropriate Th2 component is present [92]. We do not know whether the adrenal in tuberculosis is capable of responding appropriately to a sudden "acute-on-chronic" cytokine-mediated stimulus, though this is testable.

The critical importance of the ratio GC to DHEA in tuberculosis

At first sight the adrenal changes in murine tuberculosis seem rather different from those seen in the human disease. However, when we consider the relevance of the ratio of GC to DHEA some striking parallels emerge. A deficit in DHEA relative to cortisol can be striking in human tuberculosis (Table 9.1 and [53]), and a fall in DHEA levels is a good marker of progression to AIDS in HIV-infected individuals [108]. In both diseases this correlates with a defect in Th1-mediated immunity, as seen in late tuberculosis in mice. Studies in mouse and man lead to the possibility that the GC/DHEA ratio is fundamental in these diseases.

The ratio of GC to DHEA in murine tuberculosis

The importance of the ratio of glucocorticoid to DHEA has been studied in a murine model of pulmonary tuberculosis (Hernandez-Pando et al., in preparation) and is expressed diagramatically in Fig. 9.5. If there is very little DHEA (as in normal mice) the T lymphocyte response shifts progressively towards a Th2 cytokine profile, and the animals die from pulmonary consolidation and pneumonia.

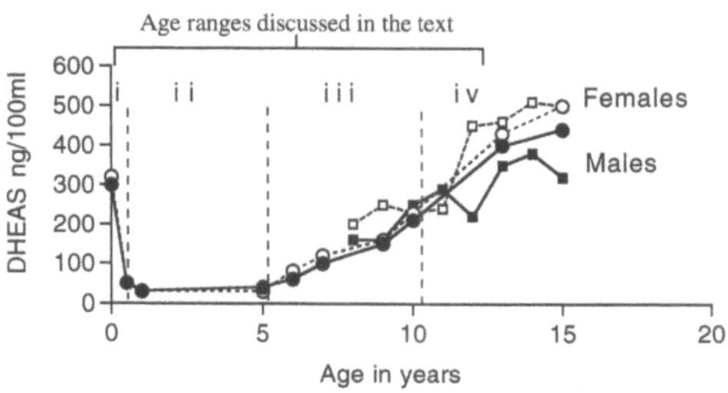

Fig. 9.5. Changes in DHEAS levels at different ages in humans. The age ranges indicated differ in their susceptibility to tuberculosis, or in the nature of the disease that develops. This is discussed in full in the text.

However, if there is too much DHEA relative to corticosterone, the animals also die, but from a totally different type of disease. There is tissue-destructive pathology that is not characteristic of the disease in mice, but rather resembles adult human disease (see later). This disease pattern can be seen if DHEA or AED are administered from day 60, when the adrenals are atrophic and presumably releasing reduced quantities of corticosterone.

There is, however, an optimal ratio of the two steroids that is protective. Thus if the animals are treated with DHEA or AED early in the disease while adrenal hypertrophy is present, they are partially protected. Thus DHEA can be protective during the phase of adrenal hypertrophy, although the same dose is rapidly fatal during the phase of adrenal atrophy in late disease.

Still more striking is the observation that DHEA or AED can be protective if given during late disease (i.e. during the phase of adrenal atrophy) but only if accompanied by corticosterone supplements (Hernandez-Pando et al., in preparation). It should be noted that the corticosterone supplements are themselves detrimental if given without DHEA. Thus either hormone kills mice with late tuberculosis, but given together there is an "emergent" property that is protective. The crucial factor seems to be the ratio of the two hormones.

The ratio of GC to DHEA in human tuberculosis

The possibility that a critical appropriate ratio of glucocorticoid to "anti-glucocorticoid" can protect against TB can also be deduced from the age-related changes in susceptibility to tuberculosis, and in the histopathology of tuberculosis seen in human children of different ages (Fig. 9.5). Although GC levels are roughly constant throughout life, DHEA levels are strikingly age-related, and appear to correlate well with these changes in pathogenesis:

(i) Tuberculosis that develops *in utero* or within days of birth can occasionally be cavitatory and necrotizing. DHEA levels are high during pregnancy, and since the foetal adrenal is the major source, they remain high (though rapidly falling) during the first weeks of life.

(ii) Tuberculosis is common in small infants less than 5 years old, but it is characterized by consolidation and pneumonia and is quite unlike the tissue-destructive cavitatory adult type of disease [109]. In fact it resembles the disease seen in mice not given DHEA supplements. In infants of this age group DHEA levels are, by human standards, very low [110], as they are in mice.

(iii) In contrast, TB is rare in the 5–10-year age group when DHEA levels, which start rising at adrenarche, are about 50% of adult values. This resistance to tuberculosis occurs in spite of evidence from skin-test surveys of continuing exposure to the infection [109].

(iv) Finally the disease is common, and of adult type (i.e. with cavitation and necrosis), in adolescents at puberty when adult DHEA levels are achieved [109].

These relationships and parallels between GC/DHEA ratio that occur naturally in humans, and can be induced experimentally in mice, are illustrated in Fig. 9.5.

Conclusions

The overall effect of increases in GC activity above the normal is to deactivate the Th1 system, and drive the immune response towards a Th2 cytokine profile. In tuberculosis even a minor Th2 component is sufficient to undermine the efficacy of immunity. Preimmunization to establish a Th2 component before infection, results in animals that are more susceptible to the disease than are non-immunized controls [67]. In progressive tuberculosis there are a number of subtle changes in diurnal rhythms, steroid output and steroid metabolism, all of which will tend to increase GC-mediated effects in the periphery. It is probable that these endocrine changes contribute to the loss of immunological control of the infection, and that new therapies can be devised that exploit our increasing knowledge of the importance of diurnal rhythms and of "anti-glucocorticoid" hormones.

Acknowledgements

We are grateful to the Wellcome Trust for supporting R.B.'s studies of the endocrinology of tuberculosis, the National University of Mexico and the CONACYT for supporting R.H.-P.'s visits to the UK, and laboratory work in Mexico.

References

1. Scott GM, Murphy PG, Gemidjioglu ME (1990) Predicting deterioration of treated tuberculosis by corticosteroid reserve and C-reactive protein. J Infect 21:61–69
2. Post FA, Soule SG, Willcox PA, Levitt NS (1994) The spectrum of endocrine dysfunction in active pulmonary tuberculosis. Clin Endocrinol 40:367–371
3. Onwubalili JK, Scott GM, Smith H (1986) Acute respiratory distress related to chemotherapy of advanced pulmonary tuberculosis: a study of two cases and review of the literature. Q J Med 59:599–610
4. Blauer KL, Poth M, Rogers WM, Bernton EW (1991) Dehydroepiandrosterone antagonises the suppressive effects of dexamethasone on lymphocyte proliferation. Endocrinology 129:3174–3179
5. Daynes RA, Araneo BA, Dowell TA, Huang K, Dudley D (1990) Regulation of murine lymphokine production in vivo. III. The lymphoid tissue microenvironment exerts regulatory influences over T helper cell function. J Exp Med 171:979–996
6. Wright BE, Porter JR, Browne ES, Svec F (1992) Antiglucocorticoid action of dehydroepiandrosterone in young obese Zucker rats. Int J Obes 16:579–583
7. Rook GAW, Onyebujoh P, Stanford JL (1993) TH1->TH2 switch and loss of CD4 cells in chronic infections; an immuno-endocrinological hypothesis not exclusive to HIV. Immunol Today 14:568–569
8. Rook GAW, Hernandez-Pando R, Lightman S (1994) Hormones, peripherally activated prohormones, and regulation of the TH1/TH2 balance. Immunol Today 15:301–303

9. Hernandez-Pando R, Orozco H, Honour JP, Silva J, Leyva R, Rook GAW (1995) Adrenal changes in murine pulmonary tuberculosis; a clue to pathogenesis? FEMS Immunol Med Microbiol 12:63–72
10. Crowley S, Holownia P, Hindmarsh P, Brook CGD, Honour JW (1991) The acute adrenal response to physiological ACTH. J Endocrinol 130:475–479
11. Roberts NA, Barton RN, Horan MA (1990) Ageing and sensitivity of the adrenal gland to physiological doses of ACTH in man. J Endocrinol 126:507–513
12. Reznek RH, Armstrong P (1994) The adrenal gland. Clin Endocrinol 40:561–576
13. Lurie MB (1964) Resistance to tuberculosis; experimental studies in native and acquired defensive mechanisms Harvard University Press, Cambridge, Mass.
14. Sarma GR, Chandra I, Ramachandran G, Krishnamurthy PV, Kumaraswami V, Prabhakar R (1990) Adrenocortical function in patients with pulmonary tuberculosis. Tubercle 71:277–282
15. Wilckens T (1995) Glucocorticoids and immune function: physiological relevance and pathogenic potential of hormonal dysfunction. Trends Pharmacol Sci 16:193–197
16. Stöcklin E, Vissler M, Gouilleux F, Groner B (1996) Functional interactions between Stat5 and the glucocorticoid receptor. Nature 383:726–728
17. Wiegers GJ, Croiset G, Reul JM, Holsboer F, de Kloet ER (1993) Differential effects of corticosteroids on rat peripheral blood T-lymphocyte mitogenesis in vivo and in vitro. Am J Physiol 265:E825–E830
18. Wiegers GJ, Reul JM, Holsboer F, de Kloet ER (1994) Enhancement of rat splenic lymphocyte mitogenesis after short-term preexposure to corticosteroids in vitro. Endocrinology 135:2351–2357
19. Oakley RH, Sar M, Cidlowski JA (1996) The human glucocorticoid receptor beta isoform. Expression, biochemical properties, and putative function. J Biol Chem 271:9550–9559
20. Vacchio MS, Papdopoulos V, Ashwell JD (1994) Steroid production in the thymus: implications for thymocyte selection. J Exp Med 179:1835–1846
21. King LB, Vacchio MS, Dixon K, Hunziker R, Margulies DH, Ashwell JD (1995) A targeted glucocorticoid receptor antisense transgene increases thymocyte apoptosis and alters thymocyte development. Immunity 3:647–656
22. Migliorati G, Pagliacci C, Moraca R, Crocicchio F, Nicoletti I, Riccardi C (1992) Interleukins modulate glucocorticoid-induced thymocyte apoptosis. Int J Clin Lab Res 21:300–303
23. Dougherty TF, Berliner ML, Berliner DL (1960) 11β-hydroxy dehydrogenase system activity in thymi of mice following prolonged cortisol treatment. Endocrinology 66:550–558
24. Malone JL, Simms TE, Gray GC, Wagner KF, Burge JR, Burke DS (1990) Sources of variability in repeated T-helper lymphocyte counts from human immunodeficiency virus type 1-infected patients: total lymphocyte count fluctuations and diurnal cycle are important. J Acquir Immune Defic Syndr 3:144–151
25. Brinkmann V, Kristofic C (1995) Regulation by corticosteroids of Th1 and Th2 cytokine production in human CD4+ effector T cells generated from CD45RO- and CD45RO+ subsets. J Immunol 155:3322–3328
26. Krouwels FH, van der Heijden JF, Lutter R, van Neerwen RJJ, Jansen HM, Out TA (1996) Glucocorticoids affect functions of airway- and blood-derived human T cell clones, favoring the Th1 profile through two mechanisms. Am J Respir Cell Mol Biol 14:388–397
27. Ramirez F, Fowell DJ, Puklavec M, Simmonds S, Mason D (1996) Glucocorticoids promote a Th2 cytokine response by CD4+ T cells in vitro. J Immunol 156:2406–2412
28. Brown DH, Sheridan J, Pearl D, Zwilling BS (1993) Regulation of mycobacterial growth by the hypothalamus-pituitary-adrenal axis: differential responses of *Mycobacterium bovis* BCG-resistant and -susceptible mice. Infect Immun 61:4793–4800

29. Batuman OA, Ferrero AP, Diaz A, Jimenez SA (1991) Regulation of transforming growth factor-beta 1 gene expression by glucocorticoids in normal human T lymphocytes. J Clin Invest 88:1574–1580
30. Corrigan CJ, Hamid Q, North J, Barkans J, Moqbel R, Durham S et al. (1995) Peripheral blood CD4 but not CD8 T-lymphocytes in patients with exacerbation of asthma transcribe and translate messenger RNA encoding cytokines which prolong eosinophil survival in the context of a Th2-type pattern: effect of glucocorticoid therapy. Am J Respir Cell Mol Biol 12:567–578
31. Zwilling BS (1992) Stress affects disease outcomes. ASM News 58:23–25
32. Bernton E, Hoover D, Galloway R, Popp K (1995) Adaptation to chronic stress in military trainees. Adrenal androgens, testosterone, glucocorticoids, IGF-1 and immune function. Ann NY Acad Sci 774:217–231
33. Fischer A, Konig W (1991) Influence of cytokines and cellular interactions on the glucocorticoid-induced Ig (E, G, A, M) synthesis of peripheral blood mononuclear cells. Immunology 74:228–233
34. Wu CY, Sarfati M, Heusser C, Fournier S, Rubio-Trujillo M, Peleman R et al. (1991) Glucocorticoids increase the synthesis of immunoglobulin E by interleukin 4-stimulated human lymphocytes. J Clin Invest 87:870–877
35. Guida L, O'Hehir RE, Hawrylowicz CM (1994) Synergy between dexamethasone and interleukin-5 for the induction of major histocompatibility complex class II expression by human peripheral blood eosinophils. Blood 84:2733–2740
36. Padgett DA, Sheridan JF, Loria R (1995) Steroid hormone regulation of a polyclonal Th2 immune response. Ann NY Acad Sci 774:323–325
37. Opstad PK (1994) Circadian rhythm of hormones is extinguished during prolonged physical stress, sleep and energy deficiency in young men. Eur J Endocrinol 131:56–66
38. Lortholary O, Christeff N, Casassus P, Thobie N, Veyssier P, Trogoff B et al. (1996) Hypothalamo-pituitary-adrenal function in human immunodeficiency virus-infected men. J Clin Endocrinol Metab 81:791–796
39. Sauer J, Castren M, Hopfner U, Holsboer F, Stalla GK, Arzt E (1996) Inhibition of lipopolysaccharide-induced monocyte interleukin-1 receptor antagonist synthesis by cortisol: involvement of the mineralocorticoid receptor. J Clin Endocrinol Metab 81:73–79
40. Petrovsky N, Harrison LC (1995) Thl and Th2: swinging to a hormonal rhythm. Immunol Today 16:605
41. Petrovsky N, McNair P, Harrison LC (1994) Circadian rhythmicity of interferon-gamma production in antigen-stimulated whole blood. Chronobiologia 21:293–300
42. Hiemke C, Brunner R, Hammes E, Muller H, Meyer zBK, Lohse AW (1995) Circadian variations in antigen-specific proliferation of human T lymphocytes and correlation to cortisol production. Psychoneuroendocrinology 20:335–342
43. Fukuda R, Ichikawa Y, Takaya M, Ogawa Y, Masumoto A (1994) Circadian variations and prednisolone-induced alterations of circulating lymphocyte subsets in man. Intern Med 33:733–738
44. Pownall R, Kabler PA, Knapp MS (1979) The time of day of antigen encounter influences the magnitude of the immune response. Clin Exp Immunol 36:347–354
45. Sothern RB, Roitman-Johnson B, Kanabrocki EL, Yager JG, Roodell MM, Weatherbee JA et al. (1995) Circadian characteristics of circulating interleukin-6 in men. J Allergy Clin Immunol 95:1029–1035
46. Young MR, Matthews JP, Kanabrocki EL, Sothern RB, Roitman JB, Scheving LE (1995) Circadian rhythmometry of serum interleukin-2, interleukin-10, tumor necrosis factor-alpha, and granulocyte-macrophage colony-stimulating factor in men. Chronobiol Int 12:19–27
47. Lemmer B, Schwulera U, Thrun A, Lissner R (1992) Circadian rhythm of soluble interleukin-2 receptor in healthy individuals. Eur Cytokine Netw 3:335–336

48. Martini E, Muller JY, Doinel C, Gastal C, Roquin H, Douay L et al. (1988) Disappearance of CD4-lymphocyte circadian cycles in HIV-infected patients: early event during asymptomatic infection. Aids 2:133–134
49. Andrieu JM, Lu W, Levy R (1995) Sustained increases in CD4 cell counts in asymptomatic human immunodeficiency virus type 1-seropositive patients treated with prednisolone for 1 year. J Infect Dis 171:523–530
50. Clerici M, Bevilacqua M, Vago T, Shearer GM, Norbiato G (1994) An immunoendocrinological hypothesis of HIV infection. Lancet 343:1552–1553
51. Berliner DL, Dougherty TF (1961) Hepatic and extrahepatic regulation of corticosteroids. Pharmacol Rev 13:329–359
52. Klein A, Buskila D, Gladman D, Bruser B, Malkin A (1990) Cortisol catabolism by lymphocytes of patients with systemic lupus erythematosus and rheumatoid arthritis. J Rheumatol 17:30–33
53. Rook GAW, Honour J, Kon OM, Wilkinson RJ, Davidson R, Shaw RJ (1996) Urinary steroid metabolites in tuberculosis; a new clue to pathogenesis. Q J Med 89:333–341
54. Baker R, Zumla A, Rook GAW (1996) Tuberculosis, steroid metabolism and immunity. QJ Med 89:387–394
55. Hennebold JD, Mu H-H, Poynter ME, Chen X-P, Daynes RA (1997) Active catabolism of glucocorticoids by 11β-hydroxytseroid dehydrogenase in vivo is a necessary requirement for natural resistance to infection with *Listeria monocytogenes*. Int Immunol 9:105–115
56. Hubbard WC, Bickel C, Schleimer RP. Simultaneous quantitation of endogenous levels of cortisone and cortisol in human nasal and bronchoalveolar lavage fluids and plasma via gas chromatography-negative ion chemical ionization mass spectrometry. Anal Biochem 221:109–117
57. Walker BR (1994) Organ-specific actions of 11-beta-hydroxysteroid dehydrogenase in humans: implications for the pathophysiology of hypertension. Steroids 59:84–89
58. Daynes RA, Araneo BA, Hennebold J, Enioutina J, Mu HH (1995) Steroids as regulators of the mammalian immune response. J Invest Dermatol 105:14S–19S
59. Evangelatou M, Antoniw J, Cooke BA (1996) The effect of leukocytes on 11β-HSD activity in human granulosa cell cultures. J Endocrinol 148 (Suppl):abstract P55
60. Elliott LH, Levay AK, Sparks B, Miller M, Roszman TL (1996) Dexamethasone and prostaglandin E2 modulate T-cell receptor signaling through a cAMP-independent mechanism. Cell Immunol 169:117–24
61. Elliott L, Brooks W, Roszman T (1992) Inhibition of anti-CD3 monoclonal antibody-induced T-cell proliferation by dexamethasone, isoproterenol, or prostaglandin E2 either alone or in combination. Cell Mol Neurobiol 12:411–427
62. Ayroldi E, Cannarile L, Migliorati G, Bartoli A, Nicoletti I, Riccardi C (1995) CD44 (Pgp-1) inhibits CD3 and dexamethasone-induced apoptosis. Blood 86:2672–2678
63. Brunetti M, Martelli N, Colasante A, Piantelli M, Musiani P, Aiello FB (1995) Spontaneous and glucocorticoid-induced apoptosis in human mature T lymphocytes. Blood 86:4199–4205
64. Calandra T, Bernhagen J, Mitchell RA, Bucala R (1994) The macrophage is an important and previously unrecognized source of macrophage migration inhibitory factor. J Exp Med 179:1895–1902
65. Calandra T, Bernhagen J, Metz CN, Spiegel LA, Bacher M, Donnelly T et al. (1995) MIF as a glucocorticoid-induced modulator of cytokine production. Nature 377:68–71
66. Bernhagen J, Bacher M, Calandra T, Metz CN, Doty SB, Donnelly T et al. (1996) An essential role for macrophage migration inhibitory factor in the tuberculin delayed-type hypersensitivity reaction. J Exp Med 183:277–282
67. Rook GAW, Hernandez-Pando R (1996) The pathogenesis of tuberculosis. Annu Rev Microbiol 50:259–284

68. Nishino T, Bernhagen J, Shiiki H, Calandra T, Dohi K, Bucala R (1995) Localization of macrophage migration inhibitory factor (MIF) to secretory granules within the corticotrophic and thyrotrophic cells of the pituitary gland. Mol Med 1:781–788
69. Kitada M, Kamataki T, Itahashi K, Rikihisa T, Kanakubo Y (1987) P-450 HFLa, a form of cytochrome P-450 purified from human foetal livers, is the 16α-hydroxylase of dehydroepiandrosterone 3-sulphate. J Biol Chem 262:13534–13537
70. Wegmann TG, Lin H, Guilbert L, Mosmann TR (1993) Bidirectional cytokine interactions in the maternal-fetal relationship: is successful pregnancy a Th2 phenomenon? Immunol Today 14:353–356
71. Hill JA, Polgar K, Anderson DJ (1995) T-helper 1-type immunity to trophoblast in women with recurrent spontaneous abortion. JAMA 273:1933–1936
72. Robel P, Baulieu EE (1995) Dehydroepiandrosterone (DHEA) is a neuroactive neurosteroid. Ann NY Acad Sci 774:82–110
73. Morfin R, Courchay G (1994) Pregnenolone and dehydroepiandrosterone as precursors of native 7-hydroxylated metabolites which increase the immune response in mice. J Steroid Biochem Molec Biol 50:91–100
74. Wilder RL (1995) Neuroendocrine-immune system interactions and autoimmunity. Annu Rev Immunol 13:307–338
75. Danenberg HD, Ben-Yehuda A, Zakay-Rones Z, Friedman G (1995) Dehydroepiandrosterone enhances influenza vaccination in aged mice. Ann NY Acad Sci 774:297–299
76. Rasmussen KR, Martin EG, Healey MC (1993) Effects of dehydroepiandrosterone in immunosuppressed rats infected with *Cryptosporidium parvum*. J Parasitol 79:364–370
77. Olsen NJ, Zhou P, Ong H (1993) Testosterone induces expression of transforming growth factor-β in the murine thumus. Steroid Biochem Mol Biol 45:327–332
78. Garg M, Bondada S (1993) Reversal of age-associated decline in immune response to Pnu-immune vaccine by supplementation with the steroid hormone dehydroepiandrosterone. Infect Immun 61:2238–2241
79. Daynes RA, Araneo BA, Ershler WB, Maloney C, Li G-Z, Ryu S-Y (1993) Altered regulation of IL-6 production with normal aging; possible linkage to the age-associated decline in dehydroepiandrosterone and its sulphated derivative. J Immunol 150:5219–5230
80. Morales AJ, Nolan J, Nelson J, Yen AS (1994) Effects of replacement dose of DHEA in men and women of advancing age. J Clin Endocrinol Metab 78:1360–1367
81. Daynes RA, Araneo BA (1989) Contrasting effects of glucocorticoids on the capacity of T cells to produce the growth factors IL-2 and IL-4. Eur J Immunol 19:2319–2325
82. Daynes RA, Meikle AW, Araneo BA (1991) Locally active steroid hormones may facilitate compartmentalization of immunity by regulating the types of lymphokines produced by helper T cells. Res Immunol 142:40–45
83. Suzuki T, Suzuki N, Daynes RA, Engleman EG (1991) Dehydroepiandrosterone enhances IL2 production and cytotoxic effector function of human T cells. Clin Immunol Immunopathol 61:202–211
84. Meikle AW, Dorchuck RW, Araneo BA, Stringham JD, Evans TG, Spruance SL et al. (1992) The presence of a dehydroepiandrosterone-specific receptor-binding complex in murine T cells. J Steroid Biochem Molec Biol 42:293–304
85. Jesse RL, Loesser K, Eich DM, Qian YZ, Hess ML, Nestler JE (1995) Dehydroepiandrosterone inhibits human platelet aggregation in vitro and in vivo. Ann NY Acad Sci 774:281–290
86. Peters JM, Yuan-Chun Z, Ram PA, Lee SST, Gonzalez FJ, Waxman DJ (1996) Peroxisome proliferator-activated receptor α required for gene induction by dehydroepiandrosterone-3β-sulphate. Mol Pharmacol 50:67–74

87. Spencer NFL, Poynter ME, Henebold JD, Mu HH, Daynes RA (1995) Does DHEAS restore immune competence in aged animals through its capacity to function as a natural modulator of peroxisome activities? Ann NY Acad Sci 774:201–216
88. Devchand PR, Keller H, Peters JM, Vasquez M, Gonzalez FJ, Wahli W (1996) The PPARα-leukotriene B4 pathway to inflammation control. Nature 384:39–43
89. Padgett DA, Loria RM (1994) In vitro potentiation of lymphocyte activation by dehydroepiandrosterone, androstenediol and androstenetriol. J Immunol 153:1544–1552
90. Glaser R, Lafuse WP, Bonneau RH, Atkinson C, Kiecolt GJ (1993) Stress-associated modulation of proto-oncogene expression in human peripheral blood leukocytes. Behav Neurosci 107:525–529
91. Tobach E, Bloch H. Effect of stress by crowding prior to and following tuberculous infection. Am J Physiol 187:399–402
92. Hernandez-Pando R, Rook GAW (1994) The role of TNFα in T cell-mediated inflammation depends on the Th1/Th2 cytokine balance. Immunology 82:591–595
93. Jenkins JK, Malyak M, Arend WP. The effects of interleukin-10 on interleukin-1 receptor antagonist and interleukin-1 beta production in human monocytes and neutrophils. Lymphokine Cytokine Res 13:47–54
94. Herbelin A, Elhadad S, Ouaaz F, de-Groote D, Descamps-Latscha B (1994) Soluble CD23 potentiates interleukin-1-induced secretion of interleukin-6 and interleukin-1 receptor antagonist by human monocytes. Eur J Immunol 24:1869–1873
95. Muzio M, Re F, Sironi M, Polentarutti N, Minty A, Caput D et al. (1994) Interleukin-13 induces the production of interleukin-1 receptor antagonist (IL-1ra) and the expression of the mRNA for the intracellular (Keratinocyte) form of IL-1ra in human myelomonocytic cells. Blood 83:1738–1743
96. Sone S, Orino E, Mizuno K, Yano S, Nishioka Y, Haku T et al. (1994) Production of IL-1 and its receptor antagonist is regulated differently by IFN-gamma and IL-4 in human monocytes and alveolar macrophages. Eur Respir J 7:657–663
97. Jenkins JK, Arend WP (1993) Interleukin 1 receptor antagonist production in human monocytes is induced by IL-1 alpha, IL-3, IL-4 and GM-CSF. Cytokine 5:407–415
98. Hernandez-Pando R, Orozco H, Sampieri A, Pavón L, Velasquillo C, Larriva-Sahd J et al. (1996) Correlation between the kinetics of Th1/Th2 cells and pathology in a murine model of experimental pulmonary tuberculosis. Immunology 89:26–33
99. Schotanus K, Tilders FJ, Berkenbosch F. Human recombinant interleukin-1 receptor antagonist prevents adrenocorticotropin, but not interleukin-6 responses to bacterial endotoxin in rats. Endocrinology 133:2461–2468
100. Ebisui O, Fukata J, Murakami N, Kobayashi H, Segawa H, Muro S et al. (1994) Effect of IL-1 receptor antagonist and antiserum to TNF-alpha on LPS-induced plasma ACTH and corticosterone rise in rats. Am J Physiol 266:E986–E992
101. Kakucska I, Qi Y, Clark BD, Lechan RM (1993) Endotoxin-induced corticotropin-releasing hormone gene expression in the hypothalamic paraventricular nucleus is mediated centrally by interleukin-1. Endocrinology 133:815–821
102. Laye S, Bluthe RM, Kent S, Combe C, Medina C, Parnet P et al. (1995) Subdiaphragmatic vagotomy blocks induction of IL-1 beta mRNA in mice brain in response to peripheral LPS. Am J Physiol 268:R1327–1331
103. Zuckerman SH, Shellhaas J, Butler JD (1989) Differential regulation of lipopolysaccharide-induced interleukin 1 and tumour necrosis factor synthesis; effects of endogenous and exogenous glucocorticoids and the role of the pituitary-adrenal axis. Eur J Immunol 19:301–305
104. Bertini R, Bianchi M, Ghezzi P (1988) Adrenalectomy sensitizes mice to the lethal effects of interleukin 1 and tumor necrosis factor. J Exp Med 167:1708–1712
105. Rook GAW, Bloom BR (1994) Mechanisms of pathogenesis in tuberculosis. In: Bloom BR (ed) Tuberculosis; pathogenesis, protection and control. ASM Press, Washington DC, pp 485–501

106. Kaplan G. Cytokine regulation of disease progression in leprosy and tuberculosis. Immunobiology 191:564–568
107. Chan J, Xing Y, Magliozzo RS, Bloom BR (1992) Killing of virulent *Mycobacterium tuberculosis* by reactive nitrogen intermediates produced by activated murine macrophages. J Exp Med 175:1111–1122
108. Wisniewski TL, Hilton CW, Morse EV, Svec F (1993) The relationship of serum DHEA-S and cortisol levels to measures of immune function in human immunodeficiency virus-related illness. Am J Med Sci 305:79–83
109. Donald PR, Beyers N, Rook GAW (1996) Adolescent tuberculosis. S Afr Med J 86:231–233
110. de-Peretti E, Forest MG (1978) Pattern of plasma dehydroepiandrosterone sulfate levels in humans from birth to adulthood: evidence for testicular production. J Clin Endocrinol Metab 47:572–577

Subject Index

MIX
Papier aus verantwortungsvollen Quellen
Paper from responsible sources
FSC® C105338
FSC
www.fsc.org

If you have any concerns about our products,
you can contact us on
ProductSafety@springernature.com

In case Publisher is established outside the EU,
the EU authorized representative is:
Springer Nature Customer Service Center GmbH
Europaplatz 3, 69115 Heidelberg, Germany

Printed by Libri Plureos GmbH
in Hamburg, Germany